Grundlehren der mathematischen Wissenschaften 252

A Series of Comprehensive Studies in Mathematics

Thierry Aubin

Nonlinear Analysis on Manifolds. Monge–Ampère Equations

Springer-Verlag
New York Heidelberg Berlin

Thierry Aubin
Universite de Paris VI
Mathematiques
4, place Jussieu
75230 Paris, Cedex 05
France

AMS Subject Classification (1980): 53 CXX

Library of Congress Cataloging in Publication Data
Aubin, Thierry.
Nonlinear analysis on manifolds, Monge-Ampère equations.
(Grundlehren der mathematischen Wissenschaften; 252)
Bibliography: p.
Includes index.
1. Riemannian manifolds. 2. Monge-Ampère equations.
I. Title. II. Series.
QA649.A83 1982 516.92 82-19165

Printed in the United States of America

9 8 7 6 5 4 3 2 1

ISBN 0-387-90704-1 Springer-Verlag New York Heidelberg Berlin
ISBN 3-540-90704-1 Springer-Verlag Berlin Heidelberg New York

Preface

This volume is intended to allow mathematicians and physicists, especially analysts, to learn about nonlinear problems which arise in Riemannian Geometry.

Analysis on Riemannian manifolds is a field currently undergoing great development. More and more, analysis proves to be a very powerful means for solving geometrical problems. Conversely, geometry may help us to solve certain problems in analysis.

There are several reasons why the topic is difficult and interesting. It is very large and almost unexplored. On the other hand, geometric problems often lead to limiting cases of known problems in analysis, sometimes there is even more than one approach, and the already existing theoretical studies are inadequate to solve them. Each problem has its own particular difficulties.

Nevertheless there exist some standard methods which are useful and which we must know to apply them. One should not forget that our problems are motivated by geometry, and that a geometrical argument may simplify the problem under investigation. Examples of this kind are still too rare.

This work is neither a systematic study of a mathematical field nor the presentation of a lot of theoretical knowledge. On the contrary, I do my best to limit the text to the essential knowledge. I define as few concepts as possible and give only basic theorems which are useful for our topic. But I hope that the reader will find this sufficient to solve other geometrical problems by analysis.

The book is intended to be used as a reference and as an introduction to research. It can be divided into two parts, with each part containing four chapters. Part I is concerned with essential background knowledge. Part II develops methods which are applied in a concrete way to resolve specific problems.

Chapter 1 is devoted to Riemannian geometry. The specialists in analysis who do not know differential geometry will find, in the beginning of the chapter, the definitions and the results which are indispensable. Since it is useful to know how to compute both globally and in local coordinate charts, the proofs which we will present will be a good initiation. In particular, it is important to know Theorem 1.53, estimates on the components of the metric tensor in polar geodesic coordinates in terms of the curvature.

Chapter 2 studies Sobolev spaces on Riemannian manifolds. Successively, we will treat density problems, the Sobolev imbedding theorem, the Kondrakov theorem, and the study of the limiting case of the Sobolev imbedding theorem. These theorems will be used constantly. Considering the importance of Sobolev's theorem and also the interest of the proofs, three proofs of the theorem are given, the original proof of Sobolev, that of Gagliardo and Nirenberg, and my own proof, which enables us to know the value of the norm of the imbedding, an introduction to the notion of best constants in Sobolev's inequalities. This new concept is crucial for solving limiting cases.

In Chapter 3 we will find, usually without proof, a substantial amount of analysis. The reader is assumed to know this background material. It is stated here as a reference and summary of the versions of results we will be using. There are as few results as possible. I choose only the most useful and applicable ones so that the reader does not drown in a host of results and lose the main point. For instance, it is possible to write a whole book on the regularity of weak solution for elliptic equations without discussing the existence of solutions. Here there are six theorems on this topic. Of course, sometimes other will be needed; in those cases there are precise references.

It is obvious that most of the more elementary topics in this Chapter 3 have already been needed in the earlier chapters. Although we do assume prior knowledge of these basic topics, we have included precise statements of the most important concepts and facts for reference. Of course, the elementary material in this chapter could have been collected as a separate "Chapter 0" but this would have been artificial, and probably less useful to the reader. And since we do not assume that the reader knows the material on elliptic equations in Sobolev spaces, the corresponding sections should follow the two first chapters.

Chapter 4 is concerned with the Green's function of the Laplacian on compact manifolds. This will be used to obtain both some regularity results and some inequalities that are not immediate consequences of the facts in Chapter 3.

Chapter 5 provides some of the most useful methods for nonlinear analysis. As an exercise we use the variational method to solve an equation studied by Yamabe. The sketch of the proof is typical of the method. Then we solve Berger's problem and a problem posed by Nirenberg, for which we also use the results from Chapter 2 concerning the limiting case of the Sobolev imbedding theorem.

Chapter 6 is devoted to the Yamabe problem concerning the scalar curvature. Here the concept of best constants in Sobolev's inequalities plays an essential rôle. We close the chapter with a summary of the status of related problems concerning scalar curvature.

Chapter 7 is concerned with the complex Monge–Ampère equation on compact Kählerian manifolds. The existence of Einstein–Kähler metrics and the Calabi conjecture are problems which are equivalent to solving such equations.

Lastly, Chapter 8 studies the real Monge–Ampère equation on a bounded convex set of $\mathbb{R}^n$. There is also a short discussion of the complex Monge–Ampère equation on a bounded pseudoconvex set of $\mathbb{C}^n$.

Throughout the book I have restricted my attention to those problems whose solution involves typical application of the methods. Of course, there are many other very interesting problems. For example, we should at least mention that, curiously, the Yamabe equation appears in the study of Yang–Mills fields, while a corresponding complex version is very close to the existence of complex Kähler–Einstein metrics discussed in Chapter 7.

It is my pleasure and privilege to express my deep thanks to my friend Jerry Kazdan who agreed to read the manuscript from the beginning to end. He suggested many mathematical improvements, and, needless to say, corrected many blunders of mine in this English version. I also have to state in this place my appreciation for the efficient and friendly help of Jürgen Moser and Melvyn Berger for the publication of the manuscript. Pascal Cherrier and Philippe Delanoë deserve special mention for helping in the completion of the text.

May 1982 Thierry Aubin

Contents

Chapter 4
Green's Function for Riemannian Manifolds

Chapter 5
The Methods

Chapter 6
The Scalar Curvature

Chapter 7

Complex Monge–Ampère Equation on Compact Kähler Manifolds

Chapter 8

Monge–Ampère Equations

Chapter 1

Riemannian Geometry

§1. Introduction to Differential Geometry

1.1 A *manifold* M_n, of dimension n, is a Haussdorff topological space such that each point of M_n has a neighborhood homeomorphic to $\mathbb{R}^n$. Thus a manifold is locally compact and locally connected. Hence a connected manifold is pathwise connected.

1.2 A *local chart* on M_n is a pair (Ω, φ), where Ω is an open set of M_n and φ a homeomorphism of Ω onto an open set of $\mathbb{R}^n$.

A collection $(\Omega_i, \varphi_i)_{i \in I}$ of local charts such that $\bigcup_{i \in I} \Omega_i = M_n$ is called an *atlas*. The *coordinates* of $P \in \Omega$, related to φ, are the coordinates of the point $\varphi(P)$ of $\mathbb{R}^n$.

1.3 An atlas of class C^k (respectively, C^∞, C^ω) on M_n is an atlas for which all changes of coordinates are C^k (respectively, C^∞, C^ω). That is to say, if $(\Omega_\alpha, \varphi_\alpha)$ and $(\Omega_\beta, \varphi_\beta)$ are two local charts with $\Omega_\alpha \cap \Omega_\beta \neq \emptyset$, then the map $\varphi_\alpha \circ \varphi_\beta^{-1}$ of $\varphi_\beta(\Omega_\alpha \cap \Omega_\beta)$ onto $\varphi_\alpha(\Omega_\alpha \cap \Omega_\beta)$ is a diffeomorphism of class C^k (respectively, C^∞, C^ω).

1.4 Two atlases of class C^k on M_n $(U_i, \varphi_i)_{i \in I}$ and $(W_\alpha, \psi_\alpha)_{\alpha \in \Lambda}$ are said to be equivalent if their union is an atlas of class C^k.
By definition, a *differentiable manifold* of class C^k (respectively, C^∞, C^ω) is a manifold together with an equivalence class of C^k atlases, (respectively, C^∞, C^ω).

1.5 A mapping f of a differentiable manifold C^k: W_p into another M_n, is called differentiable C^r ($r \leq k$) at $P \in U \subset W_p$ if $\psi \circ f \circ \varphi^{-1}$ is differentiable C^r at $\varphi(P)$, and we define the *rank* of f at P to be the rank of $\psi \circ f \circ \varphi^{-1}$ at $\varphi(P)$. Here (U, φ) is a local chart of W_p and (Ω, ψ) a local chart of M_n with $f(P) \in \Omega$.

A C^r differentiable mapping f is an *immersion* if the rank of f is equal to p for every point P of W_p. It is an *imbedding* if f is an injective immersion such that f is a homeomorphism of W_p onto $f(W_p)$ with the topology induced from M_n.

1.1. Tangent Space

1.6 Let (Ω, φ) be a local chart and f a differentiable real-valued function defined on a neighborhood of $P \in \Omega$. We say that f is flat at P if $d(f \circ \varphi^{-1})$ is zero at $\varphi(P)$.

A *tangent vector* at $P \in M_n$ is a map $X: f \to X(f) \in \mathbb{R}$ defined on the set of functions differentiable in a neighborhood of P, where X satisfies:

(a) If $\lambda, \mu \in \mathbb{R}$, $X(\lambda f + \mu g) = \lambda X(f) + \mu X(g)$.
(b) $X(f) = 0$, if f is flat.
(c) $X(fg) = f(P)X(g) + g(P)X(f)$; this follows from (a) and (b).

1.7 The *tangent space* $T_P(M)$ at $P \in M_n$ is the set of tangent vectors at P. It has a natural vector space structure. In a coordinate system $\{x^i\}$ at P, the vectors $(\partial/\partial x^i)_P$ defined by $(\partial/\partial x^i)_P(f) = [\partial(f \circ \varphi^{-1})/\partial x^i]_{\varphi(P)}$, form a basis.

The tangent space $T(M)$ is $\bigcup_{P \in M} T_p(M)$. It has a natural vector fiber bundle structure. If $T_P^*(M)$ denotes the dual space of $T_P(M)$, the cotangent space is $T^*(M) = \bigcup_{P \in M} T_P^*(M)$. Likewise, the fiber bundle $T_s^r(M)$ of the tensor of type (r, s) is $\bigcup_{P \in M} \overset{r}{\otimes} T_P(M) \overset{s}{\otimes} T_P^*(M)$.

1.8 Let $P \in M_n$ and Φ be a differentiable map of M_n into W_p. Set $Q = \Phi(P)$. The map Φ induces a linear map Φ_* of the tangent space $T_P(M)$ into $T_Q(W)$ defined by $(\Phi_* X)(f) = X(f \circ \Phi)$, where $X \in T_p(M)$ and f is a differentiable function in a neighborhood of Q. We call Φ_* the *linear tangent mapping* of Φ.

By duality, we define the *linear cotangent mapping* Φ^* of $T^*(W)$ into $T^*(M)$ as follows: $T_Q^*(W) \ni \omega \to \Phi^*(\omega) \in T_P^*(M)$, which satisfies

$$\langle \Phi^*(\omega), X \rangle = \langle \omega, \Phi_*(X) \rangle, \text{ for all } X \in T_P(M).$$

One verifies easily that $\Phi^*(df) = d(f \circ \Phi)$.

1.9 A differentiable vector field is a section of $T(M)$. A section of vector fiber bundle (E, π, M) is a differentiable map ξ of M into E, such that $\pi \circ \xi =$ identity. If $E = T(M)$, π is the mapping of E onto M: $T_P(M) \ni X \to P$.

The *bracket* $[X, Y]$ of two vector fields X and Y is the vector field defined by

$$[X, Y](f) = X[Y(f)] - Y[X(f)].$$

A *differentiable tensor field* of type (r, s) is a section of $T_s^r(M)$.

1.10 An exterior differential *p-form* η is a section of $\Lambda^p T^*(M)$. In a local chart

$$\eta = \sum_{j_1 < j_2 < \cdots < j_p} a_{j_1 \cdots j_p} \, dx^{j_1} \wedge dx^{j_2} \wedge \cdots \wedge dx^{j_p},$$

and the exterior differentiation $d\eta$ of η is

$$d\eta = \sum_{j_1 < j_2 < \cdots < j_p} da_{j_1 \cdots j_p} \wedge dx^{j_1} \wedge \cdots \wedge dx^{j_p}.$$

Clearly $dd\eta = 0$.

Denote by $\Gamma_1(M)$ the space of differentiable vector fields and by $\Lambda^p(M)$ the space of exterior differential p-forms. For $\alpha \in \Lambda^p(M)$ and $\beta \in \Lambda^q(M)$, $d(\alpha \wedge \beta) = d\alpha \wedge \beta + (-1)^p \alpha \wedge d\beta$, as it is easy to verify.

1.2. Connection

1.11 A *connection* is a bilinear map D (called the covariant derivative) of $T(M) \times \Gamma_1(M)$ into $T(M)$ such that:

(a) $D(X_p, Y) = D_{X_P}(Y) \in T_P(M)$ when $X_P \in T_P(M)$.
(b) If f is a differentiable function

$$D_{X_P}(fY) = X_P(f)Y + f(P)D_{X_P}(Y).$$

(c) If X and Y belong to $\Gamma_1(M)$, X of class C^r and Y of class C^{r+1}, then $D_X Y$ is of class C^r.

In a local chart (Ω, φ), denote $\nabla_i Y = D_{\partial/\partial x^i} Y$. Conversely, if we are given, for all pairs (i, j),

$$\nabla_i\left(\frac{\partial}{\partial x_j}\right) = \Gamma_{ij}^k \frac{\partial}{\partial x^k},$$

then a unique connection D is defined.

The functions Γ_{ij}^k are called the *Christoffel symbols* of the connection D with respect to the local coordinate system $x^1, \ldots, x^n$.

1.12 The *torsion* of the connection is the map of $\Gamma_1 \times \Gamma_1$ into Γ_1 defined by

$$T(X, Y) = D_X Y - D_Y X - [X, Y].$$

$T^k(\partial/\partial x^i, \partial/\partial x^j) = \Gamma_{ij}^k - \Gamma_{ji}^k$ are the components of a tensor.

1.3. Curvature

1.13 The *curvature* of the connection is the 2-form with values in $\mathrm{Hom}(\Gamma_1, \Gamma_1)$ defined by:

$$R(X, Y) = D_X D_Y - D_Y D_X - D_{[X, Y]}.$$

One verifies that $R(X, Y)Z$ at P depends only upon the values of X, Y, and Z at P.

In a local chart, denote by R^l_{kij} the lth component of $R(\partial/\partial x^i, \partial/\partial x^j)\partial/\partial x^k$. R^l_{kij} are the components of a tensor, called the curvature tensor, and

(1) $$R^l_{kij} Z^k = \nabla_i \nabla_j Z^l - \nabla_j \nabla_i Z^l.$$

It follows that

$$R^l_{kij} = \partial_i \Gamma^l_{jk} - \partial_j \Gamma^l_{ik} + \Gamma^l_{im} \Gamma^m_{jk} - \Gamma^l_{jm} \Gamma^m_{ik}.$$

1.14 The definition of *covariant derivative* extends to differentiable tensor fields as follows:

(a) For functions, $D_X f = X(f)$.
(b) D_X preserves the type of the tensor.
(c) D_X commutes with the contraction.
(d) $D_X(u \otimes v) = (D_X u) \otimes v + u \otimes (D_X v)$, where u and v are tensor fields.

For simplicity, we set $\nabla_{\alpha_1 \alpha_2 \cdots \alpha_l} u = \nabla_{\alpha_1} \nabla_{\alpha_2} \cdots \nabla_{\alpha_l} u$.

§2. Riemannian Manifold

1.15 A C^∞ *Riemannian manifold* is a pair (M_n, g), where M_n is a C^∞ differentiable manifold and g a C^∞ Riemannian metric. A Riemannian metric is a twice-covariant tensor field g (that is to say, a section of $T^*(M) \otimes T^*(M)$), such that at each point $P \in M$, g_P is a positive definite bilinear symmetric form:

$$g_P(X, Y) = g_P(Y, X) \quad \text{and} \quad g_P(X, X) > 0 \quad \text{if } X \neq 0.$$

Hereafter, unless otherwise stated, a Riemannian manifold M_n is a connected C^∞ Riemannian manifold of dimension n.

1.16 Theorem. *On a paracompact C^∞ differentiable manifold, there exists a C^∞ Riemannian metric g.*

Proof. Let $(\Omega_i, \varphi_i)_{i \in I}$ be an atlas and $\{\alpha_i\}$ a C^∞ partition of unity subordinate to the covering $\{\Omega_i\}$. Such $\{\alpha_i\}$ exists since the manifold M_n is paracompact. Set $\mathscr{E} = (\mathscr{E}_{jk})$ be the Euclidean metric on $\mathbb{R}^n$ (in an orthonormal basis $\mathscr{E}_{jk} = \delta^k_j$, Kronecker's symbol). Then $g = \sum_{i \in I} \alpha_i \varphi^*_i(\mathscr{E})$ is a Riemannian metric on M_n, as one can easily verify. ∎

For an alternate proof of Theorem (1.16) one can also use *Whitney's theorem* and give M_n the imbedded metric. Whitney's theorem asserts that every differentiable manifold M_n has an immersion in $\mathbb{R}^{2n}$ and an imbedding

in $\mathbb{R}^{2n+1}$. So let Φ be an imbedding of M_n in $\mathbb{R}^{2n+1}$. On M_n define the Riemannian metric g by $g(X, Y) = \mathscr{E}(\Phi_* X, \Phi_* Y)$. This is the metric induced by the imbedding.
Let $\{x_i\}$, $(i = 1, 2, \ldots, n)$, be a local coordinate system at a point $P \in M_n$ and $\{y^\alpha\}$ $(\alpha = 1, 2, \ldots, 2n + 1)$ the coordinates of $\Phi(P) \in \mathbb{R}^{2n+1}$. The components of g can be expressed as follows:

$$g_{ij} = \sum_{\alpha=1}^{2n+1} \frac{\partial y^\alpha}{\partial x^i} \frac{\partial y^\alpha}{\partial x^j}.$$

By definition, g^{ij} are the components of the inverse matrix of the metric matrix $((g_{ij}))$: $g_{ij} g^{ik} = \delta^k_j$.

2.1. Metric Space

1.17 Definition. Let C be a differentiable curve in a Riemannian manifold (M_n, g): $\mathbb{R} \supset [a, b] \ni t \to C(t) \in M_n$ with C differentiable (namely the restriction of a differentiable mapping of a neighborhood of $[a, b]$ into M_n).

Define the arc length of C by:

$$(2) \qquad L(C) = \int_a^b \sqrt{g_{C(t)}\left(\frac{dC}{dt}, \frac{dC}{dt}\right)}\, dt = \int_a^b \sqrt{g_{ij}[C(t)] \frac{dC^i}{dt} \frac{dC^j}{dt}}\, dt,$$

where $C^i(t)$ are the coordinates of $C(t)$ in a local chart, and dC^i/dt the components of the tangent vector at C: $dC/dt = C_*(\partial/\partial t)$, $\partial/\partial t$ being the unit vector of $\mathbb{R}$.

One verifies easily that the definition of $L(C)$ makes sense; the integral depends neither on the local chart, nor on a change of parametrization $s = s(t)$ with $ds/dt \neq 0$. Henceforth we suppose that the manifold is connected. This implies that it is pathwise connected. Two points P and Q of M_n are the endpoints of a differentiable curve. Indeed, a continuous curve from P to Q is covered by a finite number of open sets Ω_i homeomorphic to $\mathbb{R}^n$, and in each Ω_i one replaces the continuous curve by a differentiable one.

Set $d(P, Q) = \inf L(C)$ for all differentiable curves from P to Q.

1.18 Theorem. *$d(P, Q)$ defines a distance on M_n, and the topology determined by d is equivalent to the topology of M_n as a manifold.*

Proof. Clearly $d(P, Q) = d(Q, P)$ and $d(P, Q) \leq d(P, R) + d(R, Q)$. Since $d(P, P) = 0$, the only point remaining to be proved is that $d(P, Q) = 0 \Rightarrow P = Q$. Assume that $P \neq Q$ and let (Ω, φ) be a local chart with $\varphi(P) = 0$, $P \in \Omega$. There exists a ball of radius r, $B_r \subset \mathbb{R}^n$, with center 0, such that $\bar{B}_r \subset \varphi(\Omega)$ and $Q \notin \varphi^{-1}(\bar{B}_r)$. At a point M define $\lambda(M) = \inf_{\|\xi\|=1} g_M(\xi, \xi)$ and $\mu(M) = \sup_{\|\xi\|=1} g_M(\xi, \xi)$, where $\|\xi\|$ is the Euclidean norm of $\xi \in R^n$ and

$\tilde{\xi} = \varphi_*^{-1}\xi$. $\lambda(M)$ and $\mu(M)$ are strictly positive real numbers, because the sphere $\mathbb{S}_{n-1}(1)$ is compact. Clearly $\lambda = \inf \lambda(M)$ and $\mu = \sup \mu(M)$ for $M \in \varphi^{-1}(\bar{B}_r)$, satisfy $0 < \lambda \leq \mu < \infty$, because $\lambda(M)$ and $\mu(M)$ are strictly positive continuous functions on the compact set $\varphi^{-1}(\bar{B}_r)$.
Let Γ be the connected component in $\varphi^{-1}(\bar{B}_r)$ of P on the curve C from P to Q. Γ has P and R as extremities, $C(a) = P$, $C(b) = R$. We have

$$L(C) \geq L(\Gamma) = \int_a^b \sqrt{g_{ij}\frac{dC^i}{dt}\frac{dC^j}{dt}}\, dt \geq \lambda \int_a^b \left\| \frac{d(\varphi \circ C)}{dt} \right\| dt \geq \lambda r,$$

since the arc length of $\varphi(\Gamma)$ is at least r.

Therefore $d(P, Q) > 0$ if $P \neq Q$. Setting $S_P(r) = \{Q \in M, d(P, Q) \leq r\}$, we have $S_P(\lambda r) \subset \varphi^{-1}(\bar{B}_r)$, according to the above inequality. Likewise, it is possible to prove: $\varphi^{-1}(\bar{B}_r) \subset S_p(\mu r)$. Hence the topology defined by the distance d is the same as the manifold topology of M. ∎

2.2. Riemannian Connection

1.19 Definition. The Riemannian connection is the unique connection with vanishing torsion tensor, for which the covariant derivative of the metric tensor is zero.

Let us compute the expression of the Christoffel symbols in a local coordinate system. The computation gives a proof of the existence and uniqueness of the Riemannian connection.
The connection having no torsion, $\Gamma_{ij}^k = \Gamma_{ji}^k$. Moreover,

$$\begin{aligned} \nabla_k g_{ij} &= \partial_k g_{ij} - \Gamma_{ki}^l g_{jl} - \Gamma_{kj}^l g_{il} = 0, \\ \nabla_i g_{jk} &= \partial_i g_{jk} - \Gamma_{ik}^l g_{jl} - \Gamma_{ij}^l g_{kl} = 0, \\ \nabla_j g_{ik} &= \partial_j g_{ik} - \Gamma_{jk}^l g_{il} - \Gamma_{ji}^l g_{kl} = 0. \end{aligned}$$

Taking the sum of the last two equalities minus the first one, we obtain:

(3) $$\Gamma_{ij}^l = \tfrac{1}{2}[\partial_i g_{kj} + \partial_j g_{ki} - \partial_k g_{ij}]g^{kl}.$$

2.3. Sectional Curvature. Ricci Tensor. Scalar Curvature

1.20 Consider the 4-covariant tensor $R(X, Y, Z, T) = g[X, R(Z, T)Y]$ with components $R_{lkij} = g_{lm}R_{kij}^m$. For the definition of the curvature tensor see 1.13. It has the following properties: $R_{ijkl} = -R_{ijlk}$ (by definition), $R_{ijkl} = R_{klij}$, and the *Bianchi identities*: $R_{ijkl} + R_{iklj} + R_{iljk} = 0$,

(4) $$\nabla_m R_{ijkl} + \nabla_k R_{ijlm} + \nabla_l R_{ijmk} = 0.$$

1.21 Definition. $\sigma(X, Y) = R(X, Y, X, Y)$ is the *sectional curvature* of the 2-dimensional subspace of $T(M)$ defined by the vectors X and Y, which are chosen orthonormal (i.e., $g(X, X) = 1$, $g(Y, Y) = 1$, $g(X, Y) = 0$).

1.22 Definition. From the curvature tensor, only one nonzero tensor (or its negative) is obtained by contraction. It is called the *Ricci tensor*. Its components are $R_{ij} = R^k_{ikj}$. The Ricci tensor is symmetric and its contraction $R = R_{ij}g^{ij}$ is called the *scalar curvature*.
The Ricci curvature in the direction of the unit tangent vector $X = \{\xi^i\}$ is $R_{ij}\xi^i\xi^j$.

1.23 Definition. An *Einstein metric* is a metric for which the Ricci tensor and the metric tensor are proportional:

$$R_{ij}(P) = f(P)g_{ij}(P). \tag{5}$$

Contracting this equality, we obtain $f(P) = R(P)/n$, which is a constant when $n \geq 3$. Indeed, if we multiply the second Bianchi identity (4) by g^{jm}, we obtain:

$$\nabla^j R_{ijkl} + \nabla_k R_{il} - \nabla_l R_{ik} = 0,$$

which multiplied by g^{il} results in $\nabla_k R = 2\nabla^i R_{ik}$. But contracting the covariant derivative of (5) gives $\nabla_k R = n\nabla^i R_{ik}$. Hence when $n \neq 2$, the scalar curvature R must be constant.

1.24 Definition. A *normal coordinate system* at $P \in M_n$ is a local coordinate system $\{x^i\}$, for which the components of the metric tensor at P satisfy: $g_{ij}(P) = \delta^j_i$ and $\partial_k g_{ij}(P) = 0$, for all i, j, k (according to 1.19, $\partial_k g_{ij}(P) = 0$ is equivalent to $\Gamma^k_{ij}(P) = 0$).

1.25 Proposition. *At each point P, there exists a normal coordinate system.*

Proof. Let (Ω, φ) be a local chart with $\varphi(P) = 0$, and $\{x^i\}$ the corresponding coordinate system. At first we may choose in $\mathbb{R}^n$ an orthogonal frame, so that $g_{ij}(P) = \delta^j_i$. Then consider the change of coordinates defined by:

$$x^k - y^k = -\tfrac{1}{2}\Gamma^k_{ij}(P)y^iy^j.$$

In the coordinate system $\{y^k\}$, the components of the metric tensor are:

$$g'_{ij}(Q) = g_{kl}(Q)[\delta^k_i - \Gamma^k_{im}(P)y^m][\delta^l_j - \Gamma^l_{jm}(P)y^m],$$

since $\partial x^k/\partial y^i = \delta^k_i - \Gamma^k_{ij}(P)y^j$. Q is a point in the local charts corresponding to $\{x^i\}$ and $\{y^k\}$; $\{y^k\}$ is a coordinate system, according to the inverse function

theorem 3.10, the Jacobian matrix $((\partial x^k/\partial y^i))$ being equal to the unit matrix at P.
The first order in y^i of $g'_{ij}(Q) - g_{ij}(Q)$ is:

$$-[\Gamma^j_{ik}(P)y^k + \Gamma^i_{jk}(P)y^k] = -\left(\frac{\partial g_{ij}}{\partial x^k}\right)_P y^k = -\left(\frac{\partial g_{ij}}{\partial y^k}\right)_P y^k.$$

Hence $(\partial g'_{ij}/\partial y^k)_P = 0$ and all $\Gamma'^j_{ik}(P)$ are zero. ∎

2.4. Parallel Displacement. Geodesic

1.26 Definition. Let $C(t)$ be a differentiable curve. A vector field X is said to be *parallel* along C if its covariant derivative in the direction of the tangent vector to C is zero. Letting $X(t) = X(C(t))$:

$$D_{dC(t)/dt}X(t) = \frac{dC^i(t)}{dt}\nabla_i X(t)$$

$$= \frac{dC^i(t)}{dt}[\partial_i X^j(t) + \Gamma^j_{ik}(C(t))X^k(t)]\frac{\partial}{\partial x^j} = 0.$$

Thus $X(t)$ is a parallel vector field along C if, in a local chart:

$$\frac{dX^j}{dt} + \Gamma^j_{ik}X^k\frac{dC^i}{dt} = 0. \tag{6}$$

1.27 Definition. Let P and Q be two points of M_n, $C(t)$ a differentiable curve from P to Q, ($C(a) = P$, $C(b) = Q$), and X_0 a vector of $T_P(M)$.
According to Cauchy's theorem, 3.11, the initial value problem $X(a) = X_0$, of Equation (6), has a unique solution $X(t)$ defined for all $t \in [a, b]$ since (6) is linear. The vector $X(b)$ of this parallel vector field along C (with $X(a) = X_0$) is called the *parallel translate vector* of X_0 from P to Q along $C(t)$.

1.28 Definition. A differentiable curve $C(t)$ of class C^2 is a *geodesic* if its field of tangent vectors is parallel along $C(t)$. Thus $C(t)$ is a geodesic if and only if

$$\frac{d^2C^j(t)}{dt^2} + \Gamma^j_{ik}(C(t))\frac{dC^i(t)}{dt}\frac{dC^k(t)}{dt} = 0, \tag{7}$$

according to (6) with $X = dC/dt$.

Applying Cauchy's theorem, 3.11, to Equation (7) yields:

1.29 Proposition. *Given $P \in M_n$ and $X \in T_P(M)$, $X \neq 0$, there exists a unique geodesic, starting at P, such that X is its tangent vector at P. This geodesic depends differentiably on the initial conditions P and X.*

§3. Exponential Mapping

1.30 Let (Ω, φ) be a local chart related to a normal coordinate system $\{x^i\}$ at $P \in \Omega$, X a tangent vector of $T_P(M)$, $X = (\xi^1, \xi^2, \dots, \xi^n) \neq 0$, and $C^i(t)$ the coordinates of the point $C(t)$, belonging to the geodesic defined by the initial conditions $C(0) = P$, $(dC/dt)_{t=0} = X$. $C(t)$ is defined for the values of t satisfying $0 \leq t < \beta$ (β given by the Cauchy theorem). Since

$$g_{ij}(C(t)) \frac{dC^i(t)}{dt} \frac{dC^j(t)}{dt}$$

is constant along C (the covariant derivative along C of each of the three terms is zero), s, the parameter of arc length, is proportional to t: $s = \|X\| t$. $C^i(t)$ are C^∞ functions not only of t, but also of the initial conditions. We may consider $C^i(t, x^1, x^2, \dots, x^n, \xi^1, \xi^2, \dots, \xi^n)$. According to the Cauchy theorem 3.11, β may be chosen valid for initial conditions in an entire open set, for instance for $P \in \varphi^{-1}(B_r)$ and $\|X\| < \alpha$, ($B_r \subset \varphi(\Omega)$ being a ball of radius $r > 0$, and $\alpha > 0$).
It is easy to verify that $C(t, \lambda X) = C(\lambda t, X)$ for all λ, when one of the two numbers exists. Thus in all cases, if α is small enough, we may assume $\beta > 1$, without loss of generality. By Taylor's formula:

$$C^i(t, \xi^1, \xi^2, \dots, \xi^n) = x^i + t\xi^i + t^2 \psi^i(t, \xi^1, \xi^2, \dots, \xi^n).$$

1.31 Theorem. *The exponential mapping*: $\exp_P(X)$, *defined by*: $\mathbb{R}^n \supset \Theta \ni X \to C(1, P, \tilde{X}) \in M_n$ *is a diffeomorphism of* Θ *(a neighborhood of zero, where the mapping is defined) onto a neighborhood of* P. *By definition* $\exp_P(0) = P$, *and the identification of* $\mathbb{R}^n$ *with* $T_P(M)$ *is made by means of* φ_*: $\tilde{X} = (\varphi_*^{-1})_P X$, *($\varphi$ is introduced in* 1.30).

Proof. $\exp_P(X)$ is a C^∞ map of a neighborhood of $0 \in \mathbb{R}^n$ into M_n. This follows from 1.30 (β may be chosen greater than 1). At P the Jacobian matrix of this map is the unit matrix; then, according to the inverse function theorem 3.10, the exponential mapping is locally a diffeomorphism: $\xi^1, \xi^2, \dots, \xi^n$ can be expressed as functions of $C^1, C^2, \dots, C^n$. ∎

1.32 Corollary. *There exists a neighborhood Ω of P, such that every point $Q \in \Omega$ can be joined to P by a unique geodesic entirely included in Ω. $(\Omega, \exp_P^{-1})$ is a local chart and the corresponding coordinate system is called a normal geodesic coordinate system.*

Proof. Let $\{\xi^i\}$ be the coordinates of a point $Q \in \Omega$, and $C(t) = \{C^i(t)\}$ the geodesic from P to Q lying in Ω. $C^i(t) = t\xi^i$, for $t \in [0, 1]$. Since the arc length $s = \|X\|t$,

$$g_{ij}(Q)\xi^i\xi^j = \sum_{i=1}^{n} (\xi^i)^2 = \|X\|^2. \tag{8}$$

The length of the geodesic from P to Q is $\|X\|$.
Since $C(t)$ is a geodesic, by (1.28) we conclude that

$$\Gamma^i_{jk}[C(t)]\xi^j\xi^k = 0.$$

Letting $t \to 0$, we have $\Gamma^i_{jk}(P)\xi^j\xi^k = 0$ for all $\{\xi^i\}$. Thus $\Gamma^i_{jk}(P) = 0$; all Christoffel symbols are zero at P. ∎

1.33 Proposition. *Every geodesic through P is perpendicular to* $\sum_P (r)$, *the subset of the points* $Q \in \Omega$ *satisfying* $\sum_{i=1}^n (\xi^i)^2 = r^2$, *with r small enough* (ξ^i *are geodesic coordinates of Q*).

Proof. Let $Q \in \sum_P (r) \subset \Omega$. Choose an orthonormal frame of $\mathbb{R}^n$ such that the geodesic coordinates of Q are $\xi^1 = r$ and $\xi^2 = \xi^3 = \cdots = \xi^n = 0$.
We are going to prove that $g_{1i}(Q) = \delta^i_1$ for all i; thus the desired result will be established, because a vector in Q tangent to $\sum_P (r)$ has a zero first component; (if $\gamma(u)$ is a differentiable curve in $\sum_P (r)$ through Q, $\sum_{i=1}^n \gamma^i(u) \times (d\gamma^i(u)/du) = 0$, and that implies $d\gamma^1(u)/du = 0$ at Q).
Clearly, by (8), $g_{11}(Q) = 1$. Differentiation of (8) with respect to ξ^k yields:

$$\partial_k g_{ij}(Q)\xi^i\xi^j + 2g_{ik}(Q)\xi^i = 2\xi^k.$$

Hence, at Q, if $k \neq 1$:

$$r\partial_k g_{11}(r) + 2g_{1k}(r) = 0,$$

where $g_{ij}(r)$ are the components of g at the point with coordinates $\xi^1 = r$, $\xi^i = 0$ for $i > 1$. Moreover, $\Gamma^h_{ij}(r)\xi^i\xi^j = 0$ for all h leads to

$$2\partial_r g_{1k}(r) = \partial_k g_{11}(r).$$

Thus $g_{1k}(r) + r\partial_r g_{1k}(r) = 0, (\partial/\partial r)[rg_{1k}(r)] = 0$, and $rg_{1k}(r)$ is constant along the geodesic from P to Q, so

$$g_{1k}(Q) = 0, \quad \text{for } k \neq 1.$$ ∎

1.34 Definition. C is called a *minimizing curve* from P to Q if $L(C) = d(P, Q)$. See 1.17 for the definition of $L(C)$ and $d(P, Q)$.

1.35 Proposition. *A minimizing differentiable curve C from P to Q is a geodesic.*

Proof. Consider C parametrized by arc length s ($[0, \tau] \ni s \to C(s) \in C$), and suppose that $C(s)$ is of class C^2 and lies in a chart (Ω, φ).
Let $\Gamma_\lambda(s)$ be a C^2 differentiable curve from P to Q close to C, defined by $\gamma^i(s) = C^i(s) + \lambda\xi^i(s)$, with $\xi^i(0) = \xi^i(\tau) = 0$ for all i, and λ small enough. The first variation of arc length $L(\Gamma_\lambda)$ at $\lambda = 0$, $(dL_\lambda(\Gamma)/d\lambda)_{\lambda=0}$ must be zero. Now

$$\left(\frac{dL_\lambda(\Gamma)}{d\lambda}\right)_{\lambda=0} = \int_0^\tau \left[\tfrac{1}{2}\partial_k g_{ij}(C(s))\xi^k \frac{dC^i}{ds}\frac{dC^j}{ds} + g_{ij}(C(s))\frac{d\xi^i}{ds}\frac{dC^j}{ds}\right] ds.$$

Integration by parts of the second term, using (3), leads to Euler's equation:

$$\frac{d^2C^i}{ds^2} + \Gamma^i_{jk}(C(s))\frac{dC^j}{ds}\frac{dC^k}{ds} = 0. \tag{9}$$

Hence C is a geodesic. Moreover, let $S_P(r) = \{Q \in M_n, d(P, Q) \leq r\}$. According to Corollary 1.32, there exists an r_0 such that every point of $S_P(r_0)$ can be joined to P by a unique geodesic lying in $S_P(r_0)$. If we suppose $Q \in S_P(r_0)$, this unique geodesic is the minimizing curve C, and its length is $d(P, Q)$. Indeed, any other curve from P to Q does not satisfy Euler's equation (9) if it is included in $S_P(r_0)$; if it is not its length is greater than r_0, and then it cannot be minimizing. Thus we have proved that a minimizing curve C is a geodesic, because, if at a point P the Euler equation (9) were not satisfied, the above result would imply C is not minimizing. Finally, a C^1 differentiable minimizing curve must satisfy (9) as a distribution and will be C^∞ (Theorem 3.54) by induction. ∎

1.36 Theorem. *There exists $\delta(P)$, a strictly positive continuous function on M, such that every point Q satisfying $d(P, Q) \leq \delta(P)$ can be joined to P by a unique geodesic of length $d(P, Q)$. Moreover $\delta(P)$ can be chosen so that $S_P(\delta(P))$ is a convex neighborhood: every pair (Q, T) of points of $S_P(\delta(P))$ can be joined by a unique minimizing geodesic lying in $S_P(\delta(P))$.*

Proof. a) According to 1.30 and Corollary 1.32, for every $P \in M_n$ there exists $\alpha(P) > 0$, such that each $Q = \exp_P X$ with $\|X\| < \alpha(P)$ can be joined to P by a unique geodesic C included in Ω, where $(\Omega, \exp_P^{-1})$ is the chart related to the normal geodesic coordinate system. We have to prove that the length of this geodesic is $d(P, Q)$.

Let $\{\xi^i\}$ be the geodesic coordinates of Q. We suppose that $\xi^1 = r$ and $\xi^i = 0$ for $i > 1$. The equation of C is $[0, 1] \ni t \to C^i(t) = t\xi^i$, its length is r.

Consider $\gamma(t)$, $t \in [0, 1]$ a differentiable curve from P to Q lying in Ω. Its length is

$$\ell = \int_0^1 \sqrt{g_{ij}(\gamma(t)) \frac{d\gamma^i}{dt} \frac{d\gamma^j}{dt}}\, dt.$$

According to Proposition 1.33, if (ρ, θ) are geodesic polar coordinates, $ds^2 = (d\rho)^2 + \rho^2 g_{\theta_i\theta_j}\, d\theta^i\, d\theta^j$; therefore

$$g_{ij}[\gamma(t)] \frac{d\gamma^i}{dt} \frac{d\gamma^j}{dt} \geq \left(\frac{d\rho}{dt}\right)^2 = \left[\frac{d\|\gamma(t)\|}{dt}\right]^2.$$

Hence

$$\ell \geq \int_0^1 \left|\frac{d\|\gamma(t)\|}{dt}\right| dt \geq \int_0^1 \frac{d\|\gamma(t)\|}{dt}\, dt = \|\gamma(1)\| = r.$$

Consequently $r = d(P, Q)$.

Thus there exists an α such that $\exp_P X$ is a diffeomorphism of a ball with center 0 and radius α: $B_\alpha \subset \mathbb{R}^n$ onto $S_P(\alpha)$. Also, every geodesic through P is minimizing in $S_P(\alpha)$.

b) By 1.31, consider the following differentiable map ψ defined on a neighborhood of $(P, 0) \in T(M)$:

$$T(M) \ni (Q, \tilde{X}) \to (Q, \exp_Q \varphi_* \tilde{X}) \in M \times M.$$

The Jacobian matrix of ψ at $(P, 0)$ is the unit matrix; thus, by the inverse function Theorem 3.10, the restriction of ψ to a neighborhood Θ of $(P, 0)$ in $T(M)$ is a diffeomorphism onto $\psi(\Theta)$. This result allows us to choose $\delta(P)$ to be continuous. Moreover, we choose Θ as $S_P(\beta) \times B_\beta$, with β small enough (in particular $\beta \leq \alpha/2$).

Pick λ small enough so that $S_P(\lambda) \times S_P(\lambda) \subset \psi(\Theta)$. Then $S_P(\lambda)$ is a neighborhood of P such that every pair (Q, T) of points belonging to $S_P(\lambda)$ can be joined by a geodesic.

Since $\lambda \leq \beta \leq \alpha/2$, the length of this geodesic is not greater than β. Thus it is included in $S_P(\alpha)$, and is minimizing and unique.

c) Let us prove that this geodesic γ is included in $S_P(\lambda)$ for λ small enough. Denote by R, the (or a) point of γ, whose distance to P is maximum. If R is not Q or T, $\|\gamma(t)\|^2 = \sum_{i=1}^n [\gamma^i(t)]^2$ has a maximum at R for $t = t_0$. Thus its second derivative at t_0 is less than or equal to zero:

$$\sum_{i=1}^n \gamma^i(t_0) \frac{d^2\gamma^i}{dt^2}(t_0) + \sum_{i=1}^n \left(\frac{d\gamma^i}{dt}(t_0)\right)^2 \leq 0.$$

Since γ is a geodesic,

$$\frac{d^2\gamma^i(t_0)}{dt^2} + \Gamma^i_{jk}(R)\frac{d\gamma^j(t_0)}{dt}\frac{d\gamma^k(t_0)}{dt} = 0.$$

Multiplying by $\gamma^i(t_0)$ and summing over i leads to:

$$\left[g_{jk}(P) - \sum_{i=1}^{n} \gamma^i(t_0)\Gamma^i_{jk}(R)\right]\frac{d\gamma^j}{dt}\frac{d\gamma^k}{dt} \leq 0,$$

since $\sum_{i=1}^n (d\gamma^i/dt)^2 = g_{jk}(d\gamma^j/dt)(d\gamma^k/dt)$.
But this inequality is impossible, if λ is small enough, because when $\lambda \to 0$, $R \to P$, $\gamma^i(t_0) \to 0$, and $\Gamma^i_{jk}(R) \to 0$. Hence, for λ small enough, R is Q or T, and $\gamma \subset S_P(\lambda)$. ∎

§4. The Hopf–Rinow Theorem

1.37. *The following four propositions are equivalent*:

- (*a*) *The Riemannian manifold M is complete as a metric space.*
- (b) *For some point $P \in M$, all geodesics from P are infinitely extendable.*
- (c) *All geodesics are infinitely extendable.*
- (d) *All bounded closed subsets of M are compact.*

Moreover, we also have the following:

1.38 Theorem. *If M is connected and complete, then any pair (P, Q) of points of M can be joined by a geodesic arc whose length is equal to $d(P, Q)$.*

Proof. a) $\Rightarrow$ b) and c).
Let $P \in M$ and a geodesic $C(s)$ through P be defined for $0 \leq s < L$, where s is the canonical parameter of arc length. Consider s_p, an increasing sequence converging to L, and set $x_p = C(s_p)$. We have $d(x_p, x_q) \leq |s_p - s_q|$. Hence $\{x_p\}$ is a Cauchy sequence in M, and it converges to a point, say Q, which does not depend on the sequence $\{s_p\}$.
Applying Theorem 1.31 at Q, we prove that the geodesic can be extended for all values of s such that $L \leq s < L + \varepsilon$ for some $\varepsilon > 0$.

Proof. b) $\Rightarrow$ d) and Theorem 1.38.
Denote by $E_P(r)$ the subset of the points $Q \in S_P(r)$, such that there exists a minimizing geodesic from P to Q. Recall $S_P(r) = \{Q \in M, d(P, Q) \leq r\}$.
We are going to prove that $E(r) = E_P(r)$ is compact and is the same as $S(r) = S_P(r)$.

Let $\{Q_i\}$ be a sequence of points in $E(r)$, $\tilde{X}_i$ (with $\|X_i\| = 1$ (recall $X_i = \varphi_* \tilde{X}_i$)) the corresponding tangent vectors at P to the minimizing geodesic (or one of them) from P to Q_i, and $s_i = d(P, Q_i)$. Since the sphere $\mathbb{S}_{n-1}(1)$ is compact and the sequence $\{s_i\}$ bounded, there exists a subsequence $\{Q_j\}$ of $\{Q_i\}$ such that $\{X_j\}$ converges to a unit vector $X_0 \in \mathbb{S}_{n-1}(1)$ and $s_j \to s_0$.
Assuming b), $Q_0 = \exp_P s_0 X_0$ exists. It follows that $Q_j \to Q_0$ and $d(P, Q_0) = s_0 \leq r$. Hence $E(r)$ is compact. Indeed, $\exp_P$ is continuous: We have only to consider a finite covering of the geodesic, from P to Q_0 by open balls, where we can apply Proposition 1.29.
According to Theorem 1.36, $E(r) = S(r)$ for $0 < r < \delta(P)$. Suppose $E(r) = S(r)$ for $0 < r < r_0$ and let us prove first, that equality occurs for $r = r_0$, then for $r > r_0$. Let $Q \in S(r_0)$ and $\{Q_i\}$ be a sequence, which converges to Q, such that $d(P, Q_i) < r_0$. Such a sequence exists because P and Q can be joined by a differentiable curve whose length is as close as one wants to r_0. $Q_i \in E(r_0)$, which is compact; hence $E(r_0) = S(r_0)$. By Theorem 1.36, $\delta(Q)$ is continuous. It follows that there exists a $\delta_0 > 0$ such that $\delta(Q) \geq \delta_0$ when $Q \in E(r_0)$, since $E(r_0)$ is compact.

Let us prove that $E(r_0 + \delta_0) = S(r_0 + \delta_0)$.
Pick $Q \in S(r_0 + \delta_0)$, $Q \notin S(r_0)$. For every $k \in \mathbb{N}$, there exists C_k, a differentiable curve from P to Q, whose length is smaller than $d(P, Q) + 1/k$. Denote by T_k the last point on C_k, which belongs to $E(r_0)$. After possibly passing to a subsequence, since $E(r_0)$ is compact, T_k converges to a point T. Clearly, $d(P, T) = r_0$, $d(T, Q) \leq \delta_0 \leq \delta(T)$, and

$$d(P, T) + d(T, Q) = d(P, Q), \quad \text{since } d(P, T_k) + d(T_k, Q) < d(P, Q) + 1/k.$$

There exists a minimizing geodesic from P to T and another from T to Q. The union of these two geodesics is a piecewise differentiable curve from P to Q, whose length is $d(P, Q)$. Hence it is a minimizing geodesic from P to Q.

This proves d) and Theorem 1.38, any bounded subset of M being included in $S(r)$ for r large enough, and $S(r) = E(r)$ being compact.

Finally, d) $\Rightarrow$ a), obviously. ∎

1.39 Definition. Cut-locus of a point P on a complete Riemannian manifold. According to Theorem 1.37, $\exp_P(rX)$ with $\|X\| = 1$ is defined for all $r \in \mathbb{R}$ and $X \in \mathbb{S}_{n-1}(1)$. Moreover the exponential mapping is differentiable.

Consider the following map $\mathbb{S}_{n-1}(1) \ni X \to \mu(X) \in]0, +\infty]$, $\mu(X)$ being the upper bound of the set of the r, such that the geodesic $[0, r] \ni s \to C(s) = \exp_P sX$ is minimizing. It is obvious that, for $0 < r \leq \mu(X)$, the geodesic $C(s)$ is minimizing.

The set of the points $\exp_P[\mu(X)\, X]$, when X varies over $\mathbb{S}_{n-1}(1)$, is called the *cut-locus* of P.

It is possible to show that $\mu(X)$ is a continuous function on $\mathbb{S}_{n-1}(1)$ with value in $]0, \infty]$ (Bishop and Crittenden [53]). Thus the cut-locus is a closed subset of M. So when M is complete, $\exp_P$, which is defined and differentiable on the whole $\mathbb{R}^n$, is a diffeomorphism of

$$\Theta = \{rX \in \mathbb{R}^n | 0 \leq r < \mu(X)\} \quad \text{onto } \Omega = \exp_P \Theta.$$

M is the union of the two disjoint sets: Ω and the cut-locus of P.

1.40 Definition. Let $\mu(X)$ be as above and $\delta_P = \inf \mu(X)$, $X \in \mathbb{S}_{n-1}(1)$. δ_P is called the *injectivity radius* at P. Clearly $\delta_P > 0$. The *injectivity radius* δ of a manifold M is the greatest real number such that $\delta \leq \delta_P$ for all $P \in M$. Clearly δ may be zero. But according to Theorem 1.36, δ is strictly positive if the manifold is compact.

§5. Second Variation of the Length Integral

5.1. Existence of Tubular Neighborhoods

1.41 Let $C(s)$ be an imbedded geodesic $[a, b] \ni s \to C(s) \in M$. At $P = C(a)$, fix an orthonormal frame of $T_P(M)$, $\{e_i\}$, $(i = 1, 2, \ldots, n)$ with $e_1 = (dC/ds)_{s=a}$, s being the parameter of arc length. Consider $e_i(s)$, the parallel translate vector of e_i from P to $C(s)$ (see Definition 1.27).
$\{e_i(s)\}$ forms an orthonormal frame of $T_{C(s)}(M)$ with $e_1(s) = dC(s)/ds$, since $g_{C(s)}(e_i(s), e_j(s))$ is constant along C.
Consider the following map Γ defined on an open subset of $\mathbb{R}^n$: $\mathbb{R} \times \mathbb{R}^{n-1} \ni (s, \tilde{\xi}) \to \exp_{C(s)} \xi$. To define Γ, associate to $\tilde{\xi} \in \mathbb{R}^{n-1}$ the vector $\xi \in \mathbb{R}^n$, whose first component ξ^1 is zero. According to Cauchy's theorem (see Proposition 1.29), Γ is differentiable. Moreover, by 1.30, the differential of Γ at each point $C(s)$ is the identity map of $\mathbb{R}^n$ if we identify the tangent space with $\mathbb{R}^n$; thus Γ is locally invertible in a neighborhood of C, by the inverse function theorem, 3.10.

For $\mu > 0$, define $T_\mu = \{$the set of the $\Gamma(s, \tilde{\xi})$ with $s \in [a, b]$ and $\|\tilde{\xi}\| < \mu\}$. T_μ is called a tubular neighborhood of C. The restriction Γ_μ of Γ to $[a, b] \times B_\mu \subset \mathbb{R}^n$ is a diffeomorphism onto T_μ, provided μ is small enough. Indeed, it is sufficient to show that for μ small enough Γ_μ is one-to-one. Suppose the contrary: there exists a sequence $\{Q_i\}$ of points belonging to $T_{1/i}$, such that $Q_i = \Gamma(s_i, X_i) = \Gamma(\sigma_i, Y_i)$ with $(s_i, X_i) \neq (\sigma_i, Y_i)$ and $\|X_i\| \leq \|Y_i\| < 1/i$. After possibly passing to a subsequence Q_j, when $j \to \infty$, Q_j converges to a point of C, say $C(s_0)$. Accordingly, $s_j \to s_0$ and $\sigma_j \to s_0$. This yields the desired contradiction, since Γ is locally invertible at $C(s_0)$, as proved above.

5.2. Second Variation of the Length Integral

1.42 Let C be a geodesic from P to Q, $[0, r] \ni s \to C(s) \in M$ being injective. Choose μ small enough so that Γ_μ is injective (for the definition of Γ_μ see 1.41).

On T_μ, the tubular neighborhood of C, $(s, \tilde{\xi})$ forms a coordinate system (called *Fermi coordinates*), which is normal at each point of C, as it is possible to show. We are going to compute the second variation of arc length in this chart (T_μ, Γ_μ^{-1}). Set $x^1 = s$ and $x^i = \xi^i$, for $i > 1$.
Let $\{C_\lambda\}$ be a family of curves close to C, defined by the C^2 differentiable mappings: $[0, r] \times]-\varepsilon, +\varepsilon[\ni (s, \lambda) \to x^i(s, \lambda)$, the coordinates of the point $Q(s, \lambda) \in C_\lambda$. In addition, suppose that $Q(s, 0) = C(s)$, $x^1(s, \lambda) = s$, and that $\varepsilon > 0$ is chosen small enough so that C_λ is included in T_μ for all $\lambda \in]-\varepsilon, +\varepsilon[$.
The first variation of the length integral

$$L(\lambda) = \int_0^r \sqrt{g_{ij}[Q(s, \lambda)] \frac{\partial x^i}{\partial s} \frac{\partial x^j}{\partial s}}\, ds$$

is zero at $\lambda = 0$, since $C_0 = C$ is a geodesic. A straightforward calculation leads to

$$(10) \qquad I = \left(\frac{\partial^2 L(\lambda)}{\partial \lambda^2}\right)_{\lambda=0} = \int_0^r \left[\sum_{i=2}^n \left(\frac{dy^i}{ds}\right)^2 - R_{1i1j}(C(s))y^i(s)y^j(s)\right] ds,$$

where $y^i(s) = [\partial x^i(s, \lambda)/\partial \lambda]_{\lambda=0}$. Indeed, by 1.13, $R_{1i1j} = -\frac{1}{2}\partial_{ij} g_{11}$ on C. Recall that on C, $g_{ij} = \delta_i^j$ and $\partial_k g_{ij} = 0$.

5.3. Myers' Theorem

1.43 *A connected complete Riemannian manifold M_n with Ricci curvature $\geq (n-1)k^2 > 0$ is compact and its diameter is $\leq \pi/k$.*

Proof. Let P and Q be two points of M_n and let C be the (or a) minimizing geodesic from P to Q, r its length.
Consider the second variation I_j $(j \geq 2)$ related to the family C_λ defined by $x^j(s, \lambda) = \lambda \sin(\pi s/r)$ and $x^i(s, \lambda) = 0$ for all $i > 1$, $i \neq j$. According to (10):

$$I_j = \int_0^r \left[\frac{\pi^2}{r^2} \cos^2 \frac{\pi s}{r} - R_{1j1j}(s) \sin^2 \frac{\pi s}{r}\right] ds.$$

Adding these equations and using the hypothesis $R_{11} \geq (n-1)k^2$, it follows that

$$\sum_{j=2}^n I_j = \int_0^r \left[(n-1)\frac{\pi^2}{r^2} \cos^2 \frac{\pi s}{r} - R_{11}(s) \sin^2 \frac{\pi s}{r}\right] ds \leq (n-1)\frac{r}{2}\left(\frac{\pi^2}{r^2} - k^2\right).$$

If $r > \pi/k$, this expression will be negative and at least one of the I_j must be negative. It follows that C is not minimizing, since there exists a curve from P to Q with length smaller than r. Hence $d(P, Q) \leq \pi/k$ for all pair of points P and Q. By Theorem 1.37, M is compact. ■

§6. Jacobi Field

1.44 Definition. A vector field $Z(s)$, along a geodesic C, is a Jacobi field if its components $\xi^i(s)$ satisfy the equations:

$$(\xi^i)''(s) = -R^i_{1j1}(s)\xi^j(s) \tag{11}$$

in a Fermi coordinate system (see 1.42).

The set of the Jacobi fields along C forms a vector space of dimension $2n$, because by Cauchy's Theorem, 3.11, there is a unique Jacobi field which satisfies $Z(s_0) = Z_0$ and $Z'(s_0) = Y_0$, $s_0 \in [0, r]$, when Z_0 and Y_0 belong to $T_{C(s_0)}(M)$. The subset of the Jacobi fields which vanish at a fixed s_0 forms a vector subspace of dimension n. Those, which are in addition, orthogonal to C, form a vector subspace of dimension $(n - 1)$. Indeed, if $\xi^1(s_0) = 0$ and $(\xi^1)'(s_0) = 0$, $\xi^1(s) = 0$ for all $s \in [0, r]$, since $(\xi^1)''(s) = 0$, for all s (by definition 1.44).

1.45 Definition. If there exists a non-identically-zero Jacobi field which vanishes at P and Q, two points of C, then Q is called a *conjugate point* to P.

1.46 Theorem. $\exp_P X$ *is singular at* X_0 *if and only if* $Q = \exp_P X_0$ *is a conjugate point to* P.

Proof. $\exp_P X$ is singular at X_0 if and only if there exists a vector $Y \neq 0$ orthogonal to X_0 such that

$$\left(\frac{\partial \exp_P(X_0 + \lambda Y)}{\partial \lambda}\right)_{\lambda=0} = 0. \tag{12}$$

Consider the family $\{C_\lambda\}$ of geodesics through P, defined by $[0, r] \ni s \to Q_\lambda(s) = \exp_P[(s/r)(X_0 + \lambda Y)] \in C_\lambda$, with $r = \|X_0\|$.
In a Fermi coordinate system (see 1.41) on a tubular neighborhood of C_0, the coordinates $x^i(s, \lambda)$ of $Q_\lambda(s)$ satisfy:

$$\frac{\partial^2 x^i(s, \lambda)}{\partial s^2} = -\Gamma^i_{jk}(Q_\lambda(s)) \frac{\partial x^j}{\partial s}\frac{\partial x^k}{\partial s}; \tag{13}$$

for λ small enough, $\lambda \in]-\varepsilon, +\varepsilon[$, by (7), since C_λ is a geodesic.
The first order term in λ of (13) leads to

$$\frac{d^2 y^i(s)}{ds^2} = -\partial_j \Gamma^i_{11}(Q_0(s))y^j(s) = -R_{1i1j}(Q_0(s))y^j(s),$$

where $y^i(s) = (\partial x^i(s, \lambda)/\partial\lambda)_{\lambda=0}$ (recall that Christoffel's symbols are zero on C_0).

Hence $\{y^i(s)\}$ are the components of a Jacobi field $Z(s)$ along C_0, orthogonal to C_0.
If (12) holds, the preceding Jacobi field $Z(s)$ vanishes at P and Q, and it is not identically zero, since $Z'(0) = Y/r$. Conversely, if there exists a Jacobi field $Z(s) \not\equiv 0$, which vanishes at P and Q, then (12) holds with $Y = rZ'(0) \neq 0$. $\mathbb{R}^n$ and $T_P(M)$ are identified by $(\Gamma_\mu)_*$ (for the definition of $(\Gamma_\mu)_*$ see 1.8 and 1.41). ∎

1.47 Theorem. *If Q belongs to the cut-locus of P, then one at least of the following two situations occurs:*

(a) *Q is a conjugate point to P;*
(b) *There exist at least two minimizing geodesics from P to Q.*

For the proof see Kobayashi and Nomizu [167].

1.48 Theorem. *On a complete Riemannian manifold with nonpositive curvature, two points are never conjugate.*

Proof. Let $\{y^i(s)\}$ be the components of $Z(s) \not\equiv 0$, a Jacobi field which vanishes at P, as above. Then

$$\frac{1}{2}\left[\sum_{i=2}^{n}(y^i)^2\right]'' = \sum_{i=2}^{n}[(y^i)']^2 + \sum_{i=2}^{n}(y^i)(y^i)''$$

$$= \sum_{i=2}^{n}[(y^i)']^2 - R_{1i1j}(s)y^i(s)y^j(s) \geq \sum_{i=2}^{n}[(y^i)']^2.$$

Now $f(s) = \|Z(s)\|^2 = \sum_{i=2}^{n}[y^i(s)]^2$ cannot be zero for $s > 0$, since $f(0) = f'(0) = 0$ and $f''(0) > 0$, with $f''(s) \geq 0$ for all $s > 0$. ∎

§7. The Index Inequality

1.49 Proposition. *Let Y and Z be two Jacobi fields along (C), as in* 1.44. *Then $g(Y, Z') - g(Y', Z)$ is constant along (C). In particular, if Y and Z vanish at P, then $g(Y, Z') = g(Y', Z)$.*

Indeed, $[\sum_{i=1}^{n}(y^i z'^i - y'^i z^i)]' = 0$.

1.50 Definition (The Index Form). Let Z be a differentiable (or piecewise differentiable) vector field along a geodesic (C): $[0, r] \ni t \to C(t) \in M$. For Z orthogonal to dC/dt, the *index form* is

$$I(Z) = \int_0^r \left\{g(Z'(t), Z'(t)) + g\left[R\left(\frac{dC}{dt}, Z\right)\frac{dC}{dt}, Z\right]\right\} dt. \tag{14}$$

1.51 Theorem (The Index Inequality). *Let P and Q be two points of M_n, and let (C) be a geodesic from P to Q: $[0, r] \ni s \to C(s) \in M$ such that P admits no conjugate point along (C). Given a differentiable (or piecewise differentiable) vector field Z along (C), orthogonal to dC/dt and vanishing at P, consider the Jacobi field Y along (C) such that $Y(0) = 0$ and $Y(r) = Z(r)$. Then $I(Y) \leq I(Z)$. Equality occurs if and only if $Z = Y$.*

Proof. First of all, such a Jacobi field exists. Indeed, by 1.44, the Jacobi fields V, vanishing at P and orthogonal to dC/dt, form a vector space $\mathscr{V}$ of dimension $n-1$.
Since P has no conjugate point on (C), the map $V'(0) \to V(r)$ is one-to-one, from the orthogonal complement of dC/dt in $T_P(M)$ to that of dC/dt in $T_Q(M)$. Thus this map is onto. And given $Z(r)$, Y exists.
Let $\{V_i\}$ $(i = 2, 3, \ldots, n)$ be a basis of $\mathscr{V}$. For the same reason as above, $\{V_i(s)\}$ $(2 \leq i \leq n)$ and dC/ds form a basis of $T_{C(s)}(M)$. Hence there exist differentiable (or piecewise differentiable) functions $f_i(s)$, such that $Z(s) = \sum_{i=2}^n f_i(s) V_i(s)$.
Furthermore, set $W(s) = \sum_{i=2}^n f'_i(s) V_i(s)$ and $e_1 = dC/ds$. Then by (11), $g[R(e_1, Z)e_1, Z] = \sum_{i=2}^n f_i g[R(e_1, V_i)e_1, Z] = \sum_{i=2}^n f_i g(V''_i, Z)$. Thus:

$$I(Z) = \int_0^r \Big[g(W, W) + \sum_{i,j} g(f_i V'_i, f_j V'_j) + \sum_{i,j} g(f_i V'_i, f'_j V_j) + \sum_{i,j} g(f'_i V_i, f_j V'_j) + \sum_{i,j} g(f_i V''_i, f_j V_j) \Big] \, dS.$$

By virtue of Proposition 1.49, $g(V_i, V'_j) = g(V'_i, V_j)$. Thus, integrating the last term of $I(Z)$ by parts gives

$$I(Z) = \int_0^r g(W, W) \, ds + g[Y'(r), Y(r)],$$

because $Y(s) = \sum_{i=2}^n f_i(r) V_i(s)$ and $Y'(s) = \sum_{i=2}^n f_i(r) V'_i(s)$.
If f_i are constant for all i, we find:

$$I(Y) = g[Y'(r), Y(r)]. \tag{15}$$

Hence $I(Z) \geq I(Y)$ and equality occurs if and only if $W = 0$, which is equivalent to $f'_i = 0$ for all i, that is to say, if $Y = Z$. ■

1.52 Proposition. *Let b^2 be an upper bound for the sectional curvature of M and δ its injectivity radius. Then the ball $S_P(r)$ is convex, if r satisfies $r < \delta/2$ and $r \leq \pi/4b$.*

Proof. Let $Q \in S_P(r)$ with $d(P, Q) = r$, and (C) the minimizing geodesic from P to Q. In a tubular neighborhood of (C), we consider a Fermi coordinate system, (see 1.42).

Given a geodesic γ through Q orthogonal to (C) at Q, so that $]-\varepsilon, +\varepsilon[\ni \lambda \to \gamma(\lambda) \in M$, with $\gamma(0) = Q$, set $Y_0 = (d\gamma/d\lambda)_{\lambda=0}$. The first coordinate of $\gamma(\lambda)$ is equal to r, for all λ.

By (10), the second variation of $d(P, \gamma(\lambda))$ at $\lambda = 0$ is $I(Y)$, where Y is the Jacobi field along (C) satisfying $Y(P) = 0$, $Y(Q) = Y_0$. But

$$I(Y) \geq \int_0^r [g(Y', Y') - b^2 g(Y, Y)]\, ds = I_b(Y);$$

$I_b(Y)$ is the index form (14) on a manifold with constant sectional curvature b^2.

On such a manifold, the solutions of (11) vanishing at $s = 0$ are of the type $\xi^i = \beta^i \sin bs$, for $i \geq 2$, where β^i are some constants. If $br < \pi$, a solution does not vanish for some $s \in]0, r]$, without being identically zero. In that case, according to Theorem 1.51, and by (15):

$$I_b(Y) \geq I_b\left(\frac{\sin bs}{\sin br} Y_0\right) = b \cot br\, g(Y_0, Y_0).$$

If $r < \pi/2b$, then $I(Y) > 0$ and for ε small enough, the points of γ, except Q, lie outside $S_P(r)$. Henceforth suppose $r < \delta/2$ and $r \leq \pi/4b$.

Consider Q_1 and Q_2, two points of $S_P(r)$, and γ a minimizing geodesic from Q_1 to Q_2 (see Theorem 1.38). Since $d(Q_1, Q_2) \leq 2r < \delta$, γ is unique and included in $S_P(2r)$. Let T be the (or a) point of γ, whose distance to P is maximum. Since $d(P, T) < 2r \leq \pi/2b$, T is one end point of γ. Indeed, if T is not Q_1 or Q_2, γ is orthogonal at T to the geodesic from P to T and by virtue of the above result, γ is not included in $S_P(d(P, T))$ and that contradicts the definition of T. ∎

§8. Estimates on the Components of the Metric Tensor

1.53 Theorem. *Let M_n be a Riemannian manifold whose sectional curvature K satisfies the bounds $-a^2 \leq K \leq b^2$, the Ricci curvature being greater than $a' = (n-1)\alpha^2$. Let $S_P(r_0)$ be a ball of M with center P and radius $r_0 < \delta_P$ the injectivity radius at P. Consider $(S_P(r_0), \exp_P^{-1})$, a normal geodesic coordinate system. Denote the coordinates of a point $Q = (r, \theta) \in [0, r_0] \times \mathbb{S}_{n-1}(1)$, locally by $\theta = \{\theta^i\}$, $(i = 1, 2, \ldots, n-1)$. The metric tensor g can be expressed by*

$$ds^2 = (dr)^2 + r^2 g_{\theta^i\theta^j}(r, \theta)\, d\theta^i\, d\theta^j.$$

For convenience let $g_{\theta\theta}$ be one of the components $g_{\theta^i\theta^i}$ and $|g| = \det((g_{\theta^i\theta^j}))$. *Then $g_{\theta\theta}$ and $|g|$ satisfy the following inequalities*:

(α) $\partial/\partial r \log \sqrt{g_{\theta\theta}(r, \theta)} \geq \partial/\partial r \log[\sin(br)/r]$, $g_{\theta\theta}(r, \theta) \geq [\sin(br)/br]^2$ *when* $br < \pi$;

(β) $\partial/\partial r \log \sqrt{g_{\theta\theta}(r, \theta)} \leq \partial/\partial r \log[\sinh(ar)/r]$, $g_{\theta\theta}(r, \theta) \leq [\sinh(ar)/ar]^2$;

(γ) $\partial/\partial r \log \sqrt{|g(r, \theta)|} \leq (n-1)(\partial/\partial r)\log[\sin(\alpha r)/r] \leq -a'r/3$,

(16)

$$\sqrt{|g(r, \theta)|} \leq \left[\frac{\sin(\alpha r)}{\alpha r}\right]^{n-1};$$

(δ) $\partial/\partial r \log \sqrt{|g(r, \theta)|} \geq (n-1)(\partial/\partial r)\log[\sin(br)/r]$,

$$\sqrt{|g(r, \theta)|} \geq \left[\frac{\sin(br)}{br}\right]^{n-1} \quad \textit{when } br < \pi.$$

As usual, if $a' = \alpha = 0$, *we set* $\sin(\alpha r)/\alpha = r$, *while if* $(n-1)\alpha^2 = a' < 0$, $\sinh i\alpha r = i \sin \alpha r$ *and* $\cosh i\alpha r = \cos \alpha r$.

Proof. Let Y be a Jacobi field along (C), the minimizing geodesic from P to Q, $[0, r] \ni s \to C(s) \in M$, Y satisfying $Y(0) = 0$ and $Y \not\equiv 0$. When $br < \pi$, according to the proof of Proposition 1.52, and using (15):

$$g[Y'(r), Y(r)] = I(Y) \geq I_b(Y) \geq b \cot br\, g[Y(r), Y(r)],$$

where $I_b(Y)$ is the index form (14) on a manifold with constant sectional curvature b^2.
Moreover, according to the proof of Theorem 1.46,

$$Y(r) = \left(\frac{\partial \exp_P(X_0 + \lambda Y'(0))}{\partial \lambda}\right)_{\lambda=0},$$

where $X_0 = \exp_P^{-1} Q$ and we identify $T_P(M)$ with $\mathbb{R}^n$.
Thus $g[Y(r), Y(r)] = r^2 g_{\theta\theta}(r, \theta)\|Y'(0)\|^2$, θ being in the direction defined by $Y(r)$. Differentiating this equality, we obtain:

$$(\partial/\partial r)\log \sqrt{g_{\theta\theta}(r, \theta)} \geq g[Y'(r), Y(r)]/g[Y(r), Y(r)] - 1/r \geq b \cot br - 1/r.$$

The inequality α) follows, since $g_{\theta\theta}$ is equal to 1 at P and $\lim_{r\to 0}[\sin(br)/br] = 1$. To establish β), let us use the index inequality, Theorem 1.51:

$$g(Y'(r), Y(r)) = I(Y) \le I\left[\frac{\sinh as}{\sinh ar} Y(r)\right]$$

$$\le \left[a^2 \int_0^r \left(\frac{\cosh as}{\sinh ar}\right)^2 ds + a^2 \int_0^r \left(\frac{\sinh as}{\sinh ar}\right)^2 ds\right] g[Y(r), Y(r)]$$

$$= a \coth ar\, g[Y(r), Y(r)].$$

Thus $(\partial/\partial r)\log\sqrt{g_{\theta\theta}(r, \theta)} \le a \coth ar - 1/r$.

Let us now prove γ). Consider $\{e_i(s)\}$, an orthonormal frame on $T_{C(s)}(M)$, as in 1.41. Denote by $Y_2, Y_3, \ldots, Y_n$, the Jacobi fields along (C), such that $Y_i(0) = 0$ and $Y_i(r) = e_i(r)$, for $2 \le i \le n$. Using the index inequality, Theorem 1.51, yields:

$$g(Y_i'(r), Y_i(r)) = I(Y_i) \le I\left(\frac{\sin \alpha s}{\sin \alpha r} e_i(r)\right).$$

The possibility that $\sin \alpha r = 0$ for some $r > 0$ does not occur, even if $\alpha^2 > 0$, since $r < \delta_P \le \pi/\alpha$ (Myers' Theorem, 1.43).
Adding these inequalities leads to:

$$\sum_{i=2}^n g(Y_i'(r), Y_i(r)) \le (n-1)\alpha^2 \int_0^r \left(\frac{\cos \alpha s}{\sin \alpha r}\right)^2 ds$$

$$- \sum_{i-2}^n \int_0^r R_{1i1i}(s)\left(\frac{\sin \alpha s}{\sin \alpha r}\right)^2 ds.$$

Since

$$\sum_{i=2}^n R_{1i1i}(s) = R_{11}(s) \ge (n-1)\alpha^2,$$

$$\sum_{i=2}^n g(Y_i'(r), Y_i(r)) \le (n-1)\alpha \cot \alpha r.$$

Therefore $(\partial/\partial r)\log\sqrt{|g(r, \theta)|} = \sum_{i=2}^n g(Y_i'(r), Y_i(r)) - (n-1)/r \le (n-1) \times (\partial/\partial r)\log[\sin(\alpha r)/r]$ and the properties of $\cot u$ give the second inequality for γ). The third inequality follows by integrating the first, since $|g| = 1$ at P. To prove δ), we have only to add $n-1$ inequalities α) in the $n-1$ directions $e_i(r)$, $i = 2, \ldots, n$. ∎

§9. Integration over Riemannian Manifolds

1.54 Definition. A differentiable manifold is said to be orientable if there exists an atlas all of whose changes of coordinate charts have positive Jacobian.

Given two charts of the atlas, (Ω, φ) and (Θ, ψ), with $\Omega \cap \Theta \neq \emptyset$, denote by $\{x^i\}$ the coordinates corresponding to (Ω, φ) and by $\{y^\alpha\}$ those corresponding to (Θ, ψ). In $\Omega \cap \Theta$, let $A_i^\alpha = \partial y^\alpha/\partial x^i$ and $B_\beta^j = \partial x^j/\partial y^\beta$, the Jacobian matrix $A = ((A_i^\alpha)) \in GL(\mathbb{R}^n)^+$, the subgroup of $GL(\mathbb{R}^n)$ consisting of those matrices A for which $\det A = |A| > 0$.

1.55 Theorem. *A differentiable manifold M_n is orientable if and only if there exists an exterior differential n-form, everywhere nonvanishing.*

Proof. Suppose M_n orientable. Let $(\Omega_i, \varphi_i)_{i \in I}$ be an atlas, all of whose changes of charts have positive Jacobian, and $\{\alpha_i\}$ a partition of unity subordinated to the covering $\{\Omega_i\}$.
Consider the differential n-forms $\omega_i = \alpha_i \, dx^1 \wedge dx^2 \wedge \cdots \wedge dx^n$ ($x^1, x^2, \ldots, x^n$ being the coordinates on Ω_i). It is easy to verify that the differential n-form $\omega = \sum_{i \in I} \omega_i$ is nowhere zero.

Conversely, let ω be a nonvanishing differentiable n-form, and $\mathscr{A} = (\Omega_i, \varphi_i)_{i \in I}$ an atlas such that all Ω_i are connected. On Ω_i there exists f_i, a nonvanishing function, such that $\omega = f_i \, dx^1 \wedge dx^2 \wedge \cdots \wedge dx^n$. Since Ω_i is connected, f_i has a fixed sign. If f_i is positive, we keep the chart (Ω_i, φ_i). In that case set $\tilde{\varphi}_i = \varphi_i$. Otherwise, whenever f_j is negative, we consider $\tilde{\varphi}_j$, the composition of φ_j with the transformation $(x^1, x^2, \ldots, x^n) \to (-x^1, x^2, \ldots, x^n)$ of $\mathbb{R}^n$. So from $\mathscr{A}$, we construct an atlas $\mathscr{A}_0$.
The charts of $\mathscr{A}_0$ are (Ω_i, φ_i) or $(\Omega_j, \tilde{\varphi}_j)$, depending on whether $f_i > 0$ or $f_j < 0$. Set $\tilde{f}_j = -f_j \circ \varphi_j \circ \tilde{\varphi}_j^{-1}$. All changes of charts of $\mathscr{A}_0$ have positive Jacobian. Indeed, at $x \in \Omega_i \cap \Omega_j$, denoting by $|A|$ the determinant of the Jacobian of $\tilde{\varphi}_j \circ \tilde{\varphi}_i^{-1}$, we have $\tilde{f}_j |A| = \tilde{f}_i$. Since $\tilde{f}_j$ and $\tilde{f}_i$ are positive, $|A| > 0$.

1.56 Definition. Let M be a connected orientable manifold. On the set of nonvanishing differentiable n-forms, consider the equivalence relation: $\omega_1 \sim \omega_2$ if there exists $f > 0$ such that $\omega_1 = f\omega_2$. There are two equivalence classes. Choosing one of them defines an *orientation* of M; then M is called *oriented*. There are two possible orientations of an orientable connected manifold.

Some examples of nonorientable manifolds: Möbius' band, Klein's bottle, the real projective space $\mathbb{P}_{2m}$ of even dimension $2m$.

Some examples of orientable manifolds: the sphere $\mathbb{S}_n$, the tangent space of any manifold, the complex manifolds.

1.57 Definition. Let M_n be a differentiable oriented manifold. We define *the integral of* ω, a differentiable n-form with compact support, as follows: Let $(\Omega_i, \varphi_i)_{i\in I}$ be an atlas compatible with the orientation chosen, and $\{\alpha_i\}_{i\in I}$ a partition of unity subordinate to the covering $\{\Omega_i\}_{i\in I}$. On Ω_i, ω is equal to $f_i(x)\, dx^1 \wedge \cdots \wedge dx^n$. By definition

$$\int_M \omega = \sum_{i\in I} \int_{\varphi_i(\Omega_i)} [\alpha_i(x) f_i(x)] \circ \varphi_i^{-1}\, dx^1 \wedge dx^2 \wedge \cdots \wedge dx^n.$$

One may verify that the definition makes sense. The integral does not depend on the partition of unity (see 1.73) and the sum is finite.

1.58 Theorem. *If M_n is nonorientable, there exists a covering manifold $\tilde{M}$ of M with two sheets, such that $\tilde{M}$ is orientable.*

For the proof see Narasimhan [212].

1.59 Definition. $\tilde{M}$ is called a *covering manifold* of M, if there exists $\pi: \tilde{M} \to M$, a differentiable map, such that for every $P \in M$:

α) $\pi^{-1}(P)$ is a discrete space, F;

β) there exists a neighborhood Ω of P, such that $\pi^{-1}(\Omega)$ is diffeomorphic to $\Omega \times F$. Each point $P' \in \pi^{-1}(P)$ has a neighborhood $\Omega' \subset \tilde{M}$, such that the restriction π' of π to Ω' is a diffeomorphism of Ω' onto Ω.
The map π is a 2-sheeted covering, if F consists of two points.

If (M, g) is a Riemannian manifold, on a covering manifold $\tilde{M}$ of M, we can consider the Riemannian metric $\tilde{g} = \pi^* g$. We call $(\tilde{M}, \tilde{g})$ a Riemannian covering of M.

1.60 Theorem. *If M is simply connected, then M is orientable.*

For the proof see Narasimhan [212].

1.61 Definition. Let E be the half-space of $\mathbb{R}^n$ $(x^1 < 0)$, x^1 the first coordinate of $\mathbb{R}^n$. Consider $\bar{E} \subset \mathbb{R}^n$ with the induced topology. We identify the hyperplane of $\mathbb{R}^n$, $x^1 = 0$, with $\mathbb{R}^{n-1}$.
Letting Ω and θ be two open sets of $\bar{E}$, and $\varphi: \Omega \to \Theta$ a homeomorphism, it is possible to prove that the restriction of φ to $\Omega \cap \mathbb{R}^{n-1}$ is a homeomorphism of $\Omega \cap \mathbb{R}^{n-1}$ onto $\Theta \cap \mathbb{R}^{n-1}$. B will denote B_1, the unit ball with center 0 in $\mathbb{R}^n$, and we set $D = B \cap \bar{E}$.

§10. Manifold with Boundary

1.62 Definition. M_n is a manifold with boundary if each point of M_n has a neighborhood homeomorphic to an open set of $\bar{E}$.

The points of M_n which have a neighborhood homeomorphic to $\mathbb{R}^n$ are called interior points. They form the inside of M_n. The other points are called boundary points. We denote the set of boundary points by ∂M.

As in 1.4, we define a C^k-differentiable manifold with boundary. By definition, a function is C^k-differentiable on $\bar{E}$, if it is the restriction to $\bar{E}$, of a C^k-differentiable function on $\mathbb{R}^n$.

1.63 Theorem. *Let M_n be a (C^k-differentiable) manifold with boundary. If ∂M is not empty, then ∂M is a (C^k-differentiable) manifold of dimension $(n - 1)$, without boundary*: $\partial(\partial M) = \emptyset$.

Proof. If $Q \in \partial M$, there exists a neighborhood Ω of Q homeomorphic by φ, to an open set $\Theta \subset \bar{E}$. The restriction $\tilde{\varphi}$ of φ to $\tilde{\Omega} = \Omega \cap \partial M$ is a homeomorphism of a neighborhood $\tilde{\Omega}$ of $Q \in \partial M$ onto an open set $\tilde{\Theta} \subset \mathbb{R}^{n-1}$. Thus ∂M is a manifold (without boundary) of dimension $(n - 1)$ (Definition 1.1). If M_n is C^k-differentiable, let $(\Omega_i, \varphi_i)_{i \in I}$ be a C^k-atlas. Clearly, $(\tilde{\Omega}_i, \tilde{\varphi}_i)_{i \in I}$ form a C^k-atlas for ∂M. ■

1.64 Definition. By $\overline{W}_n$ a *compact Riemannian manifold with boundary* of *class* C^k, we understand the following: $\overline{W}_n$ is a C^k-differentiable manifold with boundary and $\overline{W}_n$ is a compact subset of M_n, a C^∞ Riemannian manifold. We set $W = \mathring{\overline{W}}$. We always suppose that the boundary is C^1, or at least Lipschitzian (Remark 2.35).

1.65 Theorem. *If M_n is a C^k-differentiable oriented manifold with boundary, ∂M is orientable. An orientation of M_n induces a natural orientation of ∂M.*

Proof. Let $(\Omega_j, \varphi_j)_{j \in I}$ be an allowable atlas with the orientation of M_n, and $(\tilde{\Omega}_j, \tilde{\varphi}_j)_{j \in I}$ the corresponding atlas of ∂M, as above. Set $\mathrm{i}\colon \partial M \to M$, the canonical imbedding of ∂M into M. We identify Q with $\mathrm{i}(Q)$, and $X \in T_Q(\partial M)$ with $\mathrm{i}_*(X) \in T_Q(M)$. Given $Q \in \partial M$, pick $e_1 \in T_Q(M)$, $e_1 \notin T_Q(\partial M)$, e_1 being oriented to the outside, namely, $e_1(f) \geq 0$ for all functions differentiable on a neighborhood of Q, which satisfy $f \leq 0$ in M_n, $f(Q) = 0$. We choose a basis of $T_Q(\partial M) = \{e_2, e_3, \ldots, e_n\}$, such that the basis of $T_Q(M)$: $\{e_1, e_2, \ldots, e_n\}$, belongs to the positive orientation given on M_n. ■

This procedure defines a canonical orientation on ∂M, as one can see.

10.1. Stokes' Formula

1.66 *Let M_n be a C^k-differentiable oriented compact manifold with boundary, and ω a differentiable $(n-1)$-form on M_n; then*

$$\int_M d\omega = \int_{\partial M} \omega, \tag{17}$$

where ∂M is oriented according to the preceding theorem. For convenience we have written $\int_{\partial M} \omega$ instead of $\int_{\partial M} \mathrm{i}^*\omega$, (for the definition of i^* *see* (1.8)).

Proof Let $(\Omega_i, \varphi_i)_{i\in I}$ be a finite atlas compatible with the orientation of M_n; such an atlas exists, because M_n is compact. Set $\Theta_i = \varphi_i(\Omega_i)$. Consider $\{\alpha_i\}$, a C^k-partition of unity subordinate to $\{\Omega_i\}$. By definition $\int_M d\omega = \sum_{i\in I} \int_{\Theta_i} d(\alpha_i\omega)$. Thus we have only to prove that $\int_{\Theta_i} d(\alpha_i\omega) = \int_{\tilde{\Theta}_i} \alpha_i\omega$, where we recall that $\tilde{\Omega}_i = \Omega_i \cap \partial V$ and have set $\tilde{\Theta}_i = \varphi_i(\tilde{\Omega}_i) = \Theta_i \cap \mathbb{R}^{n-1}$. In (Ω_i, φ_i), $\alpha_i\omega = \sum_{j=1}^n f_j(x)\, dx^1 \wedge \cdots \wedge \widehat{dx_j} \wedge \cdots \wedge dx^n$, $f_j(x)$ are C^k-differentiable functions with compact support included in $\varphi_i(\Omega_i)$; $\widehat{dx_j}$ means: this term is missing. Now,

$$d(\alpha_i\omega) = \left[\sum_{j=1}^n (-1)^{j-1} \frac{\partial f_j(x)}{\partial x^j}\right] dx^1 \wedge dx^2 \wedge \cdots \wedge dx^n,$$

by Definition 1.10. According to Fubini's theorem:

$$\int_{\Theta_i} d(\alpha_i\omega) = \int_{\tilde{\Theta}_i} f_1(x)\, dx^2 \wedge dx^3 \wedge \cdots \wedge dx^n = \int_{\tilde{\Theta}_i} \alpha_i\omega. \qquad \blacksquare$$

§11. Harmonic Forms

11.1. Oriented Volume Element

1.67 Definition. Let M_n be an oriented Riemannian manifold, and $\mathscr{A}$ an atlas compatible with the orientation. In the coordinate system $\{x^i\}$ corresponding to $(\Omega, \varphi) \in \mathscr{A}$, define the differential n-form η by:

$$\eta = \sqrt{|g|}\, dx^1 \wedge dx^2 \wedge \cdots \wedge dx^n, \tag{18}$$

where $|g|$ is the determinant of the metric matrix $((g_{ij}))$. η is a global differentiable n-form, called *oriented volume element*, and is nowhere zero.

Indeed, in another chart $(\Theta, \psi) \in \mathscr{A}$, such that $\Theta \cap \Omega \neq \emptyset$, consider the differentiable n-form: $\eta' = \sqrt{|g'|}\, dy^1 \wedge dy^2 \wedge \cdots \wedge dy^n$. But $g'_{\alpha\beta} = B^i_\alpha B^j_\beta g_{ij}$, hence $\sqrt{|g'|} = \sqrt{|B|^2}\sqrt{|g|}$ (for the definition of the matrices A and B see 1.54). Thus on $\Theta \cap \Omega$:

$$\eta' = \sqrt{|B|^2}\sqrt{|g|}|A|\, dx^1 \wedge dx^2 \wedge \cdots \wedge dx^n = \eta,$$

since $|A| > 0$ and $|A||B| = 1$. Moreover, η does not vanish.

1.68 Definition (Adjoint operator $*$). Let M_n be a Riemannian oriented manifold and η its oriented volume element. We associate to a p-form α, an $(n-p)$-form $*\alpha$, called the *adjoint* of α, defined as follows:
In a chart $(\Omega, \varphi) \in \mathscr{A}$, the components of $*\alpha$ are

$$(*\alpha)_{\lambda_{p+1}, \lambda_{p+2}, \ldots, \lambda_n} = \frac{1}{p!} \eta_{\lambda_1, \lambda_2, \ldots, \lambda_n} \alpha^{\lambda_1, \lambda_2, \ldots, \lambda_p}. \tag{19}$$

We can verify that:

$$*1 = \eta, \qquad **\alpha = (-1)^{p(n-p)}\alpha, \qquad \alpha \wedge (*\beta) = (\alpha, \beta)\eta, \tag{20}$$

where β is a p-form, and (α, β) denotes the scalar product of α and β:

$$(\alpha, \beta) = \alpha_{\lambda_1 \lambda_2, \ldots, \lambda_p} \beta^{\lambda_1 \lambda_2, \ldots, \lambda_p}.$$

Note that the adjoint operator is an isomorphism between the spaces $\Lambda^p(M)$ and $\Lambda^{n-p}(M)$.

11.2. Laplacian

1.69 Definition. (Co-differential δ, Laplacian Δ). Let $\alpha \in \Lambda^p(M)$. We define $\delta\alpha$, by its components in a chart $(\Omega, \varphi) \in \mathscr{A}$, as follows:

$$(\delta\alpha)_{\lambda_1, \ldots, \lambda_{p-1}} = -\nabla^\nu \alpha_{\nu\lambda_1, \ldots, \lambda_{p-1}}. \tag{21}$$

The differentiable $(p-1)$-form $\delta\alpha$ is called the *co-differential* of α and has the properties:

$$\delta = (-1)^p *^{-1} d\, *, \quad \delta\delta = -*^{-1} dd\, *, \quad \text{hence } \delta\delta = 0. \tag{22}$$

The *Laplacian operator* Δ is defined by:

$$\Delta = d\delta + \delta d. \tag{23}$$

If $\alpha \in \Lambda^p(M)$, $\Delta\alpha \in \Lambda^p(M)$.

The Laplacian commutes with the adjoint operator:

(24) $$* \Delta = \Delta *.$$

For a function φ, $\delta\varphi = 0$, and

(25) $$\Delta\varphi = \delta d\varphi = -\nabla^\nu\nabla_\nu\varphi.$$

α is said to be *closed* if $d\alpha = 0$, co-closed if $\delta\alpha = 0$, harmonic if $\Delta\alpha = 0$.
α is said to be *exact* if there exists a differential form β, such that $\alpha = d\beta$.
α is said to be *co-exact* if there exists a differential form γ, such that $\alpha = \delta\gamma$.

Two p-forms are homologous if their difference is exact.

1.70 Definition (Global scalar product). On a compact oriented Riemannian manifold, we define the global scalar product $\langle\alpha, \beta\rangle$ of two p-forms α and β, as follows:

$$\langle\alpha, \beta\rangle = \int_M (\alpha, \beta)\eta.$$

Recall that $(\alpha, \beta) = \alpha_{\lambda_1, \lambda_2, \ldots, \lambda_p}\beta^{\lambda_1, \lambda_2, \ldots, \lambda_p}$.

The name of the operator δ comes from the formula:

(26) $$\langle d\alpha, \gamma\rangle = \langle\alpha, \delta\gamma\rangle \quad \text{for all } \gamma \in \Lambda^{p+1}(M) \quad \text{and } \alpha \in \Lambda^p(M).$$

Let us verify this. Using (1.10) we have:

(27) $$d(\alpha \wedge *\gamma) = d\alpha \wedge (*\gamma) + (-1)^p\alpha \wedge d(*\gamma),$$

while by (22), $*\delta\gamma = (-1)^{p+1}d(*\gamma)$. According to Stokes' formula (17), integrating (27) over M leads to:

$$0 = \int_M d\alpha \wedge (*\gamma) + (-1)^p \int_M \alpha \wedge d(*\gamma).$$

That is the equality (26) (see (20)). By (26), obviously, α and β being any p-forms:

(28) $$\langle\Delta\alpha, \beta\rangle = \langle\delta\alpha, \delta\beta\rangle + \langle d\alpha, d\beta\rangle = \langle\alpha, \Delta\beta\rangle.$$

Δ is an elliptic selfadjoint differential operator (for the definition see 3.51). If $\varphi \in C^2(M)$:

(29) $$\langle\Delta\varphi, \varphi\rangle = \int_M \nabla^\nu\varphi\nabla_\nu\varphi \, dV.$$

1.71 Theorem. *On a compact oriented Riemannian manifold, any harmonic form is closed and co-closed. A harmonic function is necessarily a constant.*

Proof. By (28), if α is harmonic:

$$0 = \langle \Delta\alpha, \alpha \rangle = \langle \delta\alpha, \delta\alpha \rangle + \langle d\alpha, d\alpha \rangle.$$

Thus $\delta\alpha = 0$ and $d\alpha = 0$. For a harmonic function φ, this implies $d\varphi = 0$, $\varphi = \text{const}$. ∎

11.3. Hodge Decomposition Theorem

1.72 *Let M_n be a compact and orientable Riemannian manifold. A p-form α may be uniquely decomposed into the sum of three p-forms:*

$$\alpha = d\lambda + \delta\mu + H\alpha,$$

where $H\alpha$ is a harmonic p-form.
Uniqueness comes from the orthogonality of the three spaces for the global scalar product

$$\langle \alpha, \alpha \rangle = \langle d\lambda, d\lambda \rangle + \langle \delta\mu, \delta\mu \rangle + \langle H\alpha, H\alpha \rangle.$$

For the proof see De Rham [106].

The dimension of $H_p(M_n)$, the space of harmonic p-forms, is called the pth Betti number of M. It is finite. By (24), $\Delta * = * \Delta$, $*$ defines an isomorphism between the spaces $H_p(M_n)$ and $H_{n-p}(M_n)$. Hence $b_p(M) = b_{n-p}(M)$. Clearly, $b_0(M_n) = b_n(M_n) = 1$ (Theorem 1.71). We set $\chi(M_n) = \sum_{p=0}^{n} (-1)^p b_p$.

1.73 Definition (The *Lebesgue Integral*). Let M_n be a Riemannian manifold and (Ω, φ) a local chart, with $\{x^i\}$ the associated coordinate system. We set:

$$\text{(30)} \qquad \int_M f\,dV = \int_{\varphi(\Omega)} (\sqrt{|g|}\,f) \circ \varphi^{-1}\, dx^1\, dx^2 \cdots dx^n$$

for the continuous functions f on M_n with compact support lying in Ω.

Let (Θ, ψ) be another chart, $\{y^\alpha\}$ the associated coordinate system. We check that the definition (30) makes sense. Suppose $\operatorname{supp} f \subset \Omega \cap \Theta$; set $\partial y^\alpha/\partial x^i = A_i^\alpha$ and $\partial x^j/\partial y^\beta = B_\beta^j$ (see [1.54]). Then,

$$\int_{\varphi(\Omega\cap\Theta)} (\sqrt{|g|}\,f) \circ \varphi^{-1}\, dx^1\, dx^2 \cdots dx^n$$
$$= \int_{\psi(\Omega\cap\Theta)} (\sqrt{|g'|}\,f) \circ \psi^{-1}\, dy^1\, dy^2 \cdots dy^n.$$

Indeed $|g'| = |B|^2 |g|$ and $dy^1\, dy^2 \cdots dy^n = ||A||\, dx^1\, dx^2 \cdots dx^n$.

Consider $\{\Omega_i, \varphi_i\}_{i\in I}$ and $\{\Theta_j, \psi_j\}_{j\in J}$, two atlases, and $\{\alpha_i\}_{i\in I}$ (respectively, $\{\beta_j\}_{j\in J}$), a partition of unity subordinate to the covering $\{\Omega_i\}_{i\in I}$, (respectively, $\{\Theta_j\}_{j\in J}$). Since only a finite number of terms are nonzero, we have

$$\sum_{i\in I}\int_M \alpha_i f\,dV = \sum_{i\in I}\sum_{j\in J}\int_M (\alpha_i\beta_j)f\,dV = \sum_{j\in J}\int_M \beta_j f\,dV.$$

Thus $f \to \int_M f\,dV = \sum_{i\in I}\int_M \alpha_i f\,dV$ defines a positive Radon measure and the theory of the Lebesgue integral can be applied.

1.74 Definition. $dV = \sqrt{|g|}\,dx^1\,dx^2 \cdots dx^n$ is called the *Riemannian volume element*.

1.75 Proposition. *Let M_n be a compact Riemannian manifold, and ω a 1-form. Then $\int_M \delta\omega\,dV = 0$. In particular if $f\in C^2(M)$, $\int_M \Delta f\,dV = 0$.*

Proof. Consider $\tilde{M}$ an orientable Riemannian covering manifold of M with two sheets, Theorem (1.58) and Definition 1.59. Let π be the covering map: $\tilde{M} \to M$ and let $\tilde{\omega} = \pi^*\omega$. Since $\tilde{M}$ is orientable, let $\tilde{\eta}$ be one of its two oriented volume elements.
According to (26):

$$\int_M \tilde{\delta}\tilde{\omega}\tilde{\eta} = \int_{\tilde{M}} (\tilde{\delta}\tilde{\omega}, 1)\tilde{\eta} = \int_{\tilde{M}} (\tilde{\omega}, d1)\tilde{\eta} = 0.$$

Moreover, from Definition 1.59 it follows that

$$\left|\int_{\tilde{M}} \tilde{\delta}\tilde{\omega}\tilde{\eta}\right| = 2\left|\int_M \delta\omega\,dV\right|.$$

If $f\in C^2(M)$, $\Delta f = \delta df$ and the preceding result applied to $\omega = df$ gives $\int_M \Delta f\,dV = 0$. Or else, by using the Stokes' formula (17), (20), and (24):

$$d\tilde{\delta}(*\tilde{f}) = \tilde{\Delta}(*\tilde{f}) = *\tilde{\Delta}\tilde{f} = (\tilde{\Delta}\tilde{f})\tilde{\eta},$$

and $\int_{\tilde{M}} (\tilde{\Delta}\tilde{f})\tilde{\eta} = 0$, where $\tilde{f} = f\circ\pi$. ∎

1.76 Theorem. *Let M be a compact Riemannian manifold with strictly positive Ricci curvature. Then $b_1(M)$, the first Betti number of M, is zero.*

Proof. Let α be a harmonic 1-form. Then, by Theorem 1.71, $d\alpha = 0$ and $\delta\alpha = 0$. That is to say, in a local coordinate system with $\alpha = \alpha_i\,dx^i$ we have $\nabla_j\alpha_i = \nabla_i\alpha_j$ and $\nabla^i\alpha_i = 0$.

Contracting (1), ($i = l$), with $Z^i = \alpha^i$, gives

$$(31) \qquad R_{ij}\alpha^i = \nabla_i(\nabla_j\alpha^i) - \nabla_j(\nabla_i\alpha^i).$$

Multiplying (31) by α^j and integrating over M lead to:

$$(32) \qquad \int_M R_{ij}\alpha^i\alpha^j\,dV = \int_M \nabla_i[\alpha^j(\nabla_j\alpha^i)]\,dV - \int_M (\nabla_i\alpha^j)(\nabla_j\alpha^i)\,dV.$$

According to Proposition 1.75,

$$\int_M \nabla_i[\alpha^j(\nabla_j\alpha^i)]\,dV = -\int_M \delta[\alpha^j(\nabla_j\alpha^i)]\,dV = 0.$$

Hence, if α does not vanish everywhere, the first member of (32) will be strictly positive, while the second member is ≤ 0, since $\nabla_j\alpha_i = \nabla_i\alpha_j$. ∎

11.4. Spectrum

1.77 Definition. Let M be a compact Riemannian manifold. $Sp(M) = \{\lambda \in \mathbb{R}$, such that there exists $f \in C^2(M)$, $f \not\equiv 0$, satisfying $\Delta f = \lambda f\}$ is called the *spectrum* of M. λ is called an *eigenvalue* of the Laplacian and f an *eigenfunction.*

If $\lambda \in Sp(M)$, $\lambda \ge 0$, because

$$\lambda \int_M f^2\,dV = \int_M f\,\Delta f\,dV = \int_M \nabla^\nu f\,\nabla_\nu f\,dV \ge 0.$$

The eigenvalues of the Laplacian form an infinite sequence $0 = \lambda_0 < \lambda_1 < \lambda_2, \ldots$ going to $+\infty$. And for each eigenvalue λ_i, the set of the corresponding eigenfunctions forms a vector space of finite dimension (Fredholm's theorem (3.24). For λ_0 the vector space has one dimension.

1.78 Lichnerowicz's theorem. *If the Ricci tensor of M_n, a compact Riemannian manifold, is such that the 2-tensor $R_{ij} - kg_{ij}$ is non-negative for some $k > 0$, then $\lambda_1 \ge nk/(n-1)$.*

Proof. Let f be an eigenfunction: $\Delta f = \lambda f$ with $\lambda > 0$. Multiplying formula (31), with $\alpha = df$, by $\nabla^j f$, and integrating over M_n lead to:

$$\lambda \int_M \nabla^i f\,\nabla_i f\,dV - \int_M \nabla_i\nabla_j f\,\nabla^i\nabla^j f\,dV = \int_M R_{ij}\nabla^i f\,\nabla^j f\,dV.$$

As $(\nabla_i\nabla_j f + (1/n)\Delta f g_{ij})(\nabla^i\nabla^j f + (1/n)\Delta f g^{ij}) \ge 0$, it follows that $\nabla_i\nabla_j f\,\nabla^i\nabla^j f \ge (1/n)(\Delta f)^2$, hence $\lambda(1 - 1/n) \ge k$. ∎

Chapter 2

Sobolev Spaces

§1. First Definitions

2.1 We are going to define *Sobolev spaces* of integer order on a Riemannian manifold. First we shall be concerned with density problems. Then we shall prove the Sobolev imbedding theorem and the Kondrakov theorem. After that we shall introduce the notion of best constant in the Sobolev imbedding theorem. Finally, we shall study the exceptional case of this theorem (i.e., H_1^n on n-dimensional manifolds).

For Sobolev spaces on the open sets in n-dimensional, real Euclidean space $\mathbb{R}^n$, we recommend the very complete book of Adams [1].

2.2 Definitions. Let (M_n, g) be a smooth Riemannian manifold of dimension n (smooth means C^∞). For a real function φ belonging to $C^k(M_n)$ ($k \geq 0$ an integer), we define:

$$|\nabla^k \varphi|^2 = \nabla^{\alpha_1}\nabla^{\alpha_2}\cdots\nabla^{\alpha_k}\varphi\nabla_{\alpha_1}\nabla_{\alpha_2}\cdots\nabla_{\alpha_k}\varphi$$

In particular, $|\nabla^0\varphi| = |\varphi|$, $|\nabla^1\varphi|^2 = |\nabla\varphi|^2 = \nabla^\nu\varphi\nabla_\nu\varphi$. $\nabla^k\varphi$ will mean any kth covariant derivative of φ.

Let us consider the vector space $\mathfrak{C}_k^p$ of C^∞ functions φ, such that $|\nabla^\ell\varphi| \in L_p(M_n)$, for all ℓ with $0 \leq \ell \leq k$, where k and ℓ are integers and $p \geq 1$ is a real number.

2.3 Definitions. The *Sobolev space* $H_k^p(M_n)$ is the completion of $\mathfrak{C}_k^p$ with respect to the norm

$$\|\varphi\|_{H_k^p} = \sum_{\ell=0}^{k} \|\nabla^\ell\varphi\|_p.$$

$\overset{\circ}{H}{}_k^p(M_n)$ is the closure of $\mathscr{D}(M_n)$ in $H_k^p(M_n)$. $\mathscr{D}(M_n)$ is the space of C^∞ functions with compact support in M_n and $H_0^p = L_p$.

It is possible to consider some other norms which are equivalent; for instance, we could use

$$\left[\sum_{\ell=0}^{k} \|\nabla^{\ell}\varphi\|_p^p\right]^{1/p}.$$

When $p = 2$, H_k^2 is a Hilbert space, and this norm comes from the inner product. For simplicity we will write H_k for the Hilbert space H_k^2.

§2. Density Problems

2.4 Theorem. *$\mathscr{D}(\mathbb{R}^n)$ is dense in $H_k^p(\mathbb{R}^n)$.*

Proof. Let $f(t)$ be a C^∞ decreasing function on $\mathbb{R}$, such that $f(t) = 1$ for $t \leq 0$ and $f(t) = 0$ for $t \geq 1$.

It is sufficient to prove that a function $\varphi \in C^\infty(\mathbb{R}^n) \cap H_k^p(\mathbb{R}^n)$ can be approximated in $H_k^p(\mathbb{R}^n)$ by functions of $\mathscr{D}(\mathbb{R}^n)$. We claim that the sequence of functions $\varphi_j(x) = \varphi(x)f(\|x\| - j)$, of $\mathscr{D}(\mathbb{R}^n)$, converges to $\varphi(x)$ in $H_k^p(\mathbb{R}^n)$.

Let us verify this for the functions and the first derivatives, that is, in the case of $H_1^p(\mathbb{R}^n)$. When $j \to \infty$, $\varphi_j(x) \to \varphi(x)$ everywhere and $|\varphi_j(x)| \leq |\varphi(x)|$, which belongs to L_p. So by the Lebesgue dominated convergence theorem $\|\varphi_j - \varphi\|_p \to 0$. Moreover, when $j \to \infty$, $|\nabla\varphi_j(x)| \to |\nabla\varphi(x)|$ everywhere, and $|\nabla\varphi_j(x)| \leq |\nabla\varphi(x)| + |\varphi(x)| \sup_{t\in[0,1]} |f'(t)|$ which belongs to L_p. Thus $\|\nabla(\varphi_j - \varphi)\|_p \to 0$.

This proves the density assertion for $H_1^p(\mathbb{R}^n)$. For $k > 1$, we have to use Leibnitz's formula. ∎

2.5 Remark. The preceding theorem is not true for a bounded open set Ω in Euclidean space. Indeed, let us verify that $\mathring{H}_1^2(\Omega)$ is strictly included in $H_1^2(\Omega)$. For this purpose consider the inner product

$$\langle\varphi, \psi\rangle = \int_\Omega \varphi\psi\, dx + \sum_{i=1}^{n} \int_\Omega \partial_i\varphi\partial_i\psi\, dx.$$

For $\psi \in C^\infty(\Omega) \cap H_1^2(\Omega)$ and $\varphi \in \mathscr{D}(\Omega)$,

$$\langle\varphi, \psi\rangle = \int_\Omega \left(\psi - \sum_{i=1}^{n} \partial_{ii}\psi\right)\varphi\, dx.$$

If $\psi \not\equiv 0$ satisfies $\psi = \sum_{i=1}^n \partial_{ii}\psi$, then for all $\varphi \in \mathscr{D}(\Omega)$, $\langle\varphi, \psi\rangle = 0$, so that $\psi \notin \mathring{H}_1^2(\Omega)$.

Such a function ψ exists on a bounded open set Ω; for instance, $\psi = \sinh x_1$ (x_1 the first coordinate of x), $\int_\Omega |\sinh x_1|^p\, dx$, and $\int_\Omega |\cosh x_1|^p\, dx$ are finite.

For this reason we only try to prove the following theorem for complete Riemannian manifolds.

2.6 Theorem. *For a complete Riemannian manifold* $\overset{\circ}{H}{}_1^p(M_n) = H_1^p(M_n)$.

Proof. It is not useful to consider a function $f \in C^\infty$ on $\mathbb{R}$, as in the proof of the preceding theorem, because for a Riemannian manifold $[d(P, Q)]^2$ is only a Lipschitz function in $Q \in M_n$, P being a fixed point of M_n. So let us consider the function $f(t)$ on $\mathbb{R}$, defined by $f(t) = 1$ for $t \leq 0$, $f(t) = 1 - t$ for $0 < t < 1$, and $f(t) = 0$ for $t > 1$.

Let $\varphi(Q)$ be a C^∞ function belonging to $H_1^p(M_n)$, and P a fixed point of M_n. The sequence of functions $\varphi_j(Q) = \varphi(Q)f[d(P, Q) - j]$ belongs to $H_1^p(M_n)$, because the gradient exists almost everywhere, is bounded, and equals zero outside a compact set. One proves that the sequence $\varphi_j(Q)$ converges to $\varphi(Q)$ in $H_1^p(M_n)$, as in the proof of Theorem 2.4. Now we consider a regularization of $\varphi_j(Q)$.

Let K be the support of φ_j, and $\{\Theta_i\}$ be a finite covering of K such that Θ_i is homeomorphic to the open unit ball B of $\mathbb{R}^n$, (Θ_i, ψ_i) being the corresponding chart. Let $\{\alpha_i\}$ be a partition of unity of K subordinate to the covering $\{\Theta_i\}$. We approximate each function $\alpha_i \varphi_j$.
A function like $h = (\alpha_i \varphi_j) \circ \psi_i^{-1}$ has its support in B and is a uniformly Lipschitz function. Thus there exists a sequence $h_k \in \mathscr{D}(B)$, such that $h_k \to h$ in $H_1^p(B)$ (the usual regularization; for instance, see Adams [1]). It is now easy to show that $h_k \circ \psi_i$ converges, when $k \to \infty$, to $\alpha_i \varphi_j$ in $H_1^p(M_n)$.

2.7 Remark. It is possible to prove that if the manifold has an injectivity radius $\delta_0 > 0$, and if the curvature is bounded, then $\overset{\circ}{H}{}_2^p(M_n) = H_2^p(M_n)$. But for the proof of $\overset{\circ}{H}{}_k^p(M_n) = H_k^p(M_n)$, $(k > 2)$, besides these assumptions we need some hypothesis on the covariant derivatives of the components of the curvature tensor. (See Aubin [17] p. 154).

If there exists an injectivity radius $\delta_0 > 0$, it is simpler to consider, instead of the preceding functions $\varphi_j(Q)$, the following C^∞ functions $\tilde{\varphi}_j(Q)$, that tend to $\varphi(Q)$ in H_1^p or in H_k^p the space considered, under some conditions. Let T be a point of M_n, $B_T(q)$ the set of the points $Q \in M_n$ such that $d(T, Q) < q$ with $q \in \mathbb{N}$, and $\chi_q(Q)$ the characteristic function of $B_T(q)$. Let us consider $\gamma(t)$, a C^∞ decreasing function, which is equal to 1 for $t \leq 0$ and to zero for $t \geq \delta$, $(0 < \delta < \delta_0)$ and $\psi(P, Q) = \gamma[d(P, Q)]$. Now define the functions

$$h_q(P) = \int_{M_n} \psi(P, Q)\chi_q(Q)\, dV(Q) \Big/ \int_{M_n} \psi(P, Q)\, dV(Q).$$

These functions are C^∞, equal to 1 when $d(T, P) < q - \delta$, and equal to zero when $d(T, P) > q + \delta$. The functions $\tilde{\varphi}_j(P) = \varphi(P)h_j(P)$ have the desired property.

2.8 Theorem. *$C^\infty(\bar{E})$ is dense in $H_k^p(E)$, where E is the half-space:*

$$E = \{x \in \mathbb{R}^n / x_1 < 0\}.$$

By definition $C^\infty(\bar{E})$ is the set of functions that are restrictions to $\bar{E}$ of C^∞ functions on $\mathbb{R}^n$.

Proof. Let f belong to $C^\infty(E) \cap H_k^p(E)$. Consider the sequence of functions f_m, which are the restrictions to $\bar{E}$ of the functions $f(x_1 - 1/m, x_2, \ldots, x_n)$. It is obvious that $f_m \in C^\infty(\bar{E})$ and it is well known that, if $f \in L_p(E)$, $f_m \to f$ in $L_p(E)$. The same result holds for the derivatives of order $\leq k$.

For manifolds, we have the following theorem:

2.9 Theorem. *Let $\overline{W}_n$ be a compact Riemannian manifold with boundary of class C^r. Then $C^r(\overline{W})$ is dense in $H_k^p(W)$ for $k \leq r$.*

Proof. Let (Ω_i, φ_i) be a finite C^r atlas of $\overline{W}$, each Ω_i being homeomorphic either to a ball B of $\mathbb{R}^n$, or to a half ball $D \subset \bar{E}(D = B \cap \bar{E})$.
$C^r(\overline{W})$ is the set of functions belonging to $C^r(W) \cap C^0(\overline{W})$, whose derivatives of order $\leq r$, in each Ω_i, can be extended to continuous functions on $\overline{W} \cap \Omega_i$.

Consider a C^∞ partition of unity $\{\alpha_i\}$ subordinate to the covering $\{\Omega_i\}$ of $\overline{W}$. Let $f \in H_k^p(W) \cap C^\infty(W)$. We have to prove that each function $\alpha_i f$ can be approximated in $H_k^p(W)$ by functions of $C^r(\overline{W})$. There is only a problem for the Ω_i homeomorphic to D. Let Ω_i be one of them.
The sequence of functions h_m defined, for m sufficiently large, as the restriction to D of $[(\alpha_i f) \circ \varphi_i^{-1}](x_1 - 1/m, x_2, \ldots, x_n)$ converges to $(\alpha_i f) \circ \varphi_i^{-1}$ in $H_k^p(D)$, where D has the Euclidean metric. Since the metric tensor, and all its derivatives are bounded on Ω_i (by a proper choice of the Ω_i, without loss of generality), $h_m \circ \varphi_i \in C^r(\overline{W})$ and converges to $\alpha_i f$ in $H_k^p(W)$ for $k \leq r$, when $m \to \infty$. ■

§3. Sobolev Imbedding Theorem

2.10 First part of the theorem.
Let k and ℓ be two integers ($k > \ell \geq 0$), p and q two real numbers ($1 \leq q < p$) satisfying $1/p = 1/q - (k - \ell)/n$. The Sobolev imbedding theorem asserts that for $\mathbb{R}^n$, $H_k^q \subset H_\ell^p$ and that the identity operator is continuous.

Second part.
If $(k - r)/n > 1/q$, $H_k^q \subset C_B^r$ and the identity operator is continuous. Here $r \geq 0$ is an integer and C_B^r is the space of C^r functions which are bounded as well as their derivatives of order $\leq r$, ($\|u\|_{C^r} \equiv \max_{0 \leq \ell \leq r} \sup |\nabla^\ell u|$).
If $(k - r - \alpha)/n \geq 1/q$, $H_k^q \subset C^{r+\alpha}$, where α is a real number satisfying $0 < \alpha < 1$ and $C^{r+\alpha}$ the space of the C^r functions, the rth derivatives of

which satisfy a Hölder condition of exponent α. *Furthermore, the identity operator* $H_k^q \subset C^\alpha$ *is continuous. The norm of* C^α *is*:

$$\|u\|_{C^\alpha} = \sup|u| + \sup_{P \neq Q}\{|u(P) - u(Q)|[d(P, Q)]^{-\alpha}\}.$$

We shall also denote the space $C^{r+\alpha}$, by $C^{r,\alpha}$ when $0 < \alpha \leq 1$. $C^{r,0} = C^r$. We will mainly discuss the first part of the theorem, because the other part concerns local properties (except the continuity of the imbedding) and so there is no difference in the case of manifolds. One will find the complete proof in Theorem 2.21.

But first of all, let us prove that the first part of the Sobolev imbedding theorem holds for all k, assuming it is true for $k = 1$.

2.11 Proposition. *Let* M_n *be a* C^∞ *Riemannian manifold. If* $H_1^{q_0}(M_n)$ *is imbedded in* $L_{p_0}(M_n)$, *with* $1/p_0 = 1/q_0 - 1/n$ $(1 \leq q_0 < n)$, *then* $H_k^q(M_n)$ *is imbedded in* $H_\ell^{p_\ell}(M_n)$, *with* $1/p_\ell = 1/q - (k - \ell)/n > 0$.

Proof. Let r be an integer and let $\psi \in C^{r+1}$. Then

$$|\nabla|\nabla^r\psi|| \leq |\nabla^{r+1}\psi|. \tag{1}$$

To establish this inequality, it is sufficient to develop

$$\begin{aligned}(\nabla_\nu\nabla_{\alpha_1}\cdots\nabla_{\alpha_r}\psi\nabla_{\beta_1}\cdots\nabla_{\beta_r}\psi - \nabla_\nu\nabla_{\beta_1}\cdots\nabla_{\beta_r}\psi\nabla_{\alpha_1}\cdots\nabla_{\alpha_r}\psi)\\ \times g^{\nu\mu}g^{\alpha_1\lambda_1}g^{\alpha_2\lambda_2}\cdots g^{\alpha_r\lambda_r}g^{\beta_1\gamma_1}\cdots g^{\beta_r\gamma_r}(\nabla_\mu\nabla_{\lambda_1}\cdots\nabla_{\lambda_r}\psi\nabla_{\gamma_1}\cdots\nabla_{\gamma_r}\psi\\ - \nabla_\mu\nabla_{\gamma_1}\cdots\nabla_{\gamma_r}\psi\nabla_{\lambda_1}\cdots\nabla_{\lambda_r}\psi) \geq 0.\end{aligned}$$

We find $4|\nabla^{r+1}\psi|^2|\nabla^r\psi|^2 - |\nabla|\nabla^r\psi|^2|^2 \leq 0$.
Since $H_1^{q_0}(M_n)$ is imbedded in $L_{p_0}(M_n)$, there exists a constant A, such that for all $\varphi \in H_1^{q_0}(M_n)$:

$$\|\varphi\|_{p_0} \leq A(\|\nabla\varphi\|_{q_0} + \|\varphi\|_{q_0}).$$

Let us apply this inequality with $\varphi = |\nabla^r\psi|$, assuming φ belongs to $H_1^{q_0}$:

$$\begin{aligned}\|\nabla^r\psi\|_{p_0} &\leq A(\|\nabla|\nabla^r\psi|\|_{q_0} + \|\nabla^r\psi\|_{q_0}).\\ &\leq A(\|\nabla^{r+1}\psi\|_{q_0} + \|\nabla^r\psi\|_{q_0}).\end{aligned} \tag{2}$$

Now let $\psi \in H_k^q(M_n) \cap C^\infty(M_n)$. Applying inequalities (1) and (2) with $q = q_0$ and $r = k - 1, k - 2, \ldots$, we find:

$$\|\nabla^{k-1}\psi\|_{p_{k-1}} \leq A(\|\nabla^k\psi\|_q + \|\nabla^{k-1}\psi\|_q),$$

$$\|\nabla^{k-2}\psi\|_{p_{k-1}} \leq A(\|\nabla^{k-1}\psi\|_q + \|\nabla^{k-2}\psi\|_q),$$

$$\|\psi\|_{p_{k-1}} \leq A(\|\nabla\psi\|_q + \|\psi\|_q);$$

thus

$$\|\psi\|_{H_{k-1}^{p_{k-1}}} \leq 2A\|\psi\|_{H_k^q}.$$

Therefore a Cauchy sequence in H_k^q of C^∞ functions is a Cauchy sequence in $H_{k-1}^{p_{k-1}}$, and the preceding inequality holds for all $\psi \in H_k^q$.

Similarly, one proves the following imbeddings: $H_k^q \subset H_{k-1}^{p_{k-1}} \subset H_{k-2}^{p_{k-2}} \subset \cdots \subset H_\ell^{p_\ell}$.

§4. Sobolev's Proof

2.12 Sobolev's lemma. *Let $p' > 1$ and $q' > 1$ two real numbers. Define λ by $1/p' + 1/q' + \lambda/n = 2$. If λ satisfies $0 < \lambda < n$, there exists a constant $K(p', q', n)$, such that for all $f \in L_{q'}(\mathbb{R}^n)$ and $g \in L_{p'}(\mathbb{R}^n)$:*

(3)
$$\int_{\mathbb{R}^n}\int_{\mathbb{R}^n} \frac{f(x)g(y)}{\|x-y\|^\lambda}\,dx\,dy \leq K(p', q', n)\|f\|_{q'}\|g\|_{p'}$$

$\|x\|$ being the Euclidean norm.

The proof of this lemma is difficult (Sobolev [255]), we assume it.

Corollary. *Let λ be a real number, $0 < \lambda < n$, and $q' > 1$. If r, defined by $1/r = \lambda/n + 1/q' - 1$, satisfies $r > 1$, then*

$$h(y) = \int_{\mathbb{R}^n} \frac{f(x)}{\|x-y\|^\lambda}\,dx \quad \textit{belongs to } L_r, \textit{ when } f \in L_{q'}(\mathbb{R}^n).$$

Moreover, there exists a constant $C(\lambda, q', n)$ such that for all $f \in L_{q'}(\mathbb{R}^n)$

$$\|h\|_r \leq C(\lambda, q', n)\|f\|_{q'}.$$

Proof. For all $g \in L_{p'}(\mathbb{R}^n)$, with $1/r + 1/p' = 1$:

$$\int_{\mathbb{R}^n} h(y)g(y)\,dy \leq K(p', q', n)\|f\|_{q'}\|g\|_{p'};$$

therefore

$$h \in L_r \simeq L_{p'}^* \quad \text{and} \quad \|h\|_r \leq K(p', q', n)\|f\|_{q'}. \qquad \blacksquare$$

Now we will prove the existence of a constant $C(n, q)$ such that all $\varphi \in \mathscr{D}(\mathbb{R}^n)$ satisfy:

(4)
$$\|\varphi\|_p \leq C(n, q)\|\nabla\varphi\|_q,$$

with $1/p = 1/q - 1/n$ and $1 < q < n$.

Since $\mathscr{D}(\mathbb{R}^n)$ is dense in $H_1^q(\mathbb{R}^n)$ Theorem 2.4 the first part of the Sobolev imbedding theorem will be proved, according to Proposition 2.11.

Let x and y be points in $\mathbb{R}^n$, and write $r = \|x - y\|$. Let $\theta \in \mathbb{S}_{n-1}(1)$, the sphere of dimension $n - 1$ and radius 1. Introduce spherical polar coordinates (r, θ), with origin at x. Obviously, because $\varphi \in \mathscr{D}(\mathbb{R}^n)$:

$$\varphi(x) = -\int_0^\infty \frac{\partial\varphi(r,\theta)}{\partial r}\,dr = -\int_0^\infty \|x - y\|^{1-n}\frac{\partial\varphi(r,\theta)}{\partial r}\,r^{n-1}\,dr$$

and

$$|\varphi(x)| \le \int_0^\infty \|x - y\|^{1-n}|\nabla\varphi(r,\theta)|\,r^{n-1}\,dr.$$

Integrating over $\mathbb{S}_{n-1}(1)$, we obtain:

$$|\varphi(x)| \le \frac{1}{\omega_{n-1}}\int_{\mathbb{R}^n}\frac{|\nabla\varphi(y)|}{\|x - y\|^{n-1}}\,dy,$$

where ω_{n-1} is the volume of $\mathbb{S}_{n-1}(1)$.

According to Corollary 2.12 with $\lambda = n - 1$, inequality (4) holds. ■

§5. Proof by Gagliardo and Nirenberg (1958)

2.13 Gagliardo [118] and Nirenberg [220] proved that for all $\varphi \in \mathscr{D}(\mathbb{R}^n)$:

$$\|\varphi\|_{n/(n-1)} \le \frac{1}{2}\prod_{i=1}^n \left\|\frac{\partial\varphi}{\partial x^i}\right\|_1^{1/n}. \tag{5}$$

It is easy to see that the Sobolev imbedding theorem follows from this inequality. First $|\partial\varphi/\partial x^i| \le |\nabla\varphi|$; therefore $\|\varphi\|_{n/(n-1)} \le \frac{1}{2}\|\nabla\varphi\|_1$. Then setting $|\varphi| = u^{p(n-1)/n}$ and applying Hölder's inequality, we obtain:

$$\|u\|_p^{p(n-1)/n} = \|\varphi\|_{n/(n-1)} \le \tfrac{1}{2}p\,\frac{n-1}{n}\,\|u^{p'}|\nabla u|\,\|_1$$

$$\le p\,\frac{n-1}{2n}\,\|\nabla u\|_q\|u^{p'}\|_{q'},$$

where $1/q + 1/q' = 1$ and $p' = p(n-1)/n - 1$. But $p'q' = p$ since $1/p = 1/q - 1/n$; hence:

$$\|u\|_p \le p\,\frac{n-1}{2n}\,\|\nabla u\|_q.$$

We now prove inequality (5). For simplicity we treat only the case $n = 3$; but the proof for $n \neq 3$ is similar.
Let P be a point of $\mathbb{R}^3$, (x, y, z) the coordinates in $\mathbb{R}^3$, (x_0, y_0, z_0) those of P, and D_x (respectively, D_y, D_z) the straight line through P parallel to the x-axis, (respectively, y-, z-axis). Since $\varphi \in \mathscr{D}(\mathbb{R}^n)$,

$$\varphi(P) = \int_{-\infty}^{x_0} \frac{\partial \varphi}{\partial x}(x, y_0, z_0)\, dx = -\int_{x_0}^{+\infty} \frac{\partial \varphi}{\partial x}(x, y_0, z_0)\, dx.$$

Thus $|\varphi(P)| \leq \frac{1}{2} \int_{D_x} |\partial_x \varphi|\, dx$. Likewise for D_y and D_z:

$$|\varphi(P)|^{3/2} \leq \left(\frac{1}{2}\right)^{3/2} \left[\int_{D_x} |\partial_x \varphi|\, dx \int_{D_y} |\partial_y \varphi|\, dy \int_{D_z} |\partial_z \varphi|\, dz\right]^{1/2}.$$

Integration of x_0 over R yields, by Holder's inequality,

$$\int_{D_x} |\varphi(x, y_0, z_0)|^{3/2}\, dx$$
$$\leq \left(\frac{1}{2}\right)^{3/2} \left[\int_{D_x} |\partial_x \varphi(x, y_0, z_0)|\, dx \int_{D_{xy}} |\partial_y \varphi(x, y, z_0)|\, dx\, dy\right.$$
$$\left.\times \int_{D_{xz}} |\partial_z \varphi(x, y_0, z)|\, dx\, dz\right]^{1/2},$$

where D_{xy} means the plane through P parallel to the x- and y-axes.

Integration of y_0 over R gives, by Hölder's inequality,

$$\int_{D_{xy}} |\varphi(x, y, z_0)|^{3/2}\, dx\, dz \leq \left(\frac{1}{2}\right)^{3/2} \left[\int_{D_{xy}} |\partial_x \varphi(x, y, z_0)|\, dx\, dy\right.$$
$$\left.\times \int_{D_{xy}} |\partial_y \varphi(x, y, z_0)|\, dx\, dy \int_{R^3} |\partial_z \varphi|\, dx\, dy\, dz\right]^{1/2}.$$

Finally, integrating z_0 over R, we obtain inequality (5). ∎

§6. New Proof

2.14 Next we give a new proof of the Sobolev imbedding theorem (Aubin (1974)), which yields the explicit value of the norm of the imbedding.

Theorem (Aubin [13] or [17], see also Talenti (257)).
If $1 \leq q < n$, all $\varphi \in H_1^q(\mathbb{R}^n)$ satisfy:

(6) $$\|\varphi\|_p \leq \mathsf{K}(n, q) \|\nabla \varphi\|_q,$$

with $1/p = 1/q - 1/n$ *and*

$$\mathsf{K}(n, q) = \frac{q-1}{n-q}\left[\frac{n-q}{n(q-1)}\right]^{1/q}\left[\frac{\Gamma(n+1)}{\Gamma(n/q)\Gamma(n+1-n/q)\omega_{n-1}}\right]^{1/n}$$

for $1 < q < n$, *and*

$$\mathsf{K}(n, 1) = \frac{1}{n}\left[\frac{n}{\omega_{n-1}}\right]^{1/n}.$$

$\mathsf{K}(n, q)$ *is the norm of the imbedding* $H_1^q \subset L_p$, *and it is attained by the functions*

$$\varphi(x) = (\lambda + \|x\|^{q/(q-1)})^{1-n/q},$$

where λ *is any positive real number.*

When $q = 1$, this gives the usual isoperimetric inequality, Federer [113]; the extremum functions are then the characteristic functions of the balls of $\mathbb{R}^n$.

The proof is carried out in three steps.

First step, Proposition 2.16: Since $\mathscr{D}(\mathbb{R}^n)$ is dense in $H_1^q(\mathbb{R}^n)$, we have only to prove inequality (6) for the functions in question in Proposition (2.16).

Second step, Proposition 2.17: it is sufficient to establish inequality (6) for functions of the kind $\varphi(x) = f(\|x\|)$, f being a positive Lipschitzian function decreasing on $[0, \infty]$ and equal to zero at infinity.

Third step, Proposition 2.18: the proof of inequality (6) for these functions.

2.15 Proposition (Milnor [200] p. 37. This is actually due to Morse).
Let M be a Riemannian manifold. Any bounded smooth function $f : M \to \mathbb{R}$ can be uniformly approximated by a smooth function g which has no degenerate critical points. Furthermore, g can be chosen so that the ith *derivatives of g on the compact set K uniformly approximate the corresponding derivatives of f for $i \leq k$.*

We recall that a point $P \in M$ is called a critical point of f if $|\nabla f(P)| = 0$, the real number $f(P)$ is a critical value of f. A critical point P is called nondegenerate if and only if the matrix $\nabla_i \nabla_j f(P)$ is nonsingular. Nondegenerate critical points are isolated.

2.16 Proposition. *Let $f \not\equiv 0$ be a C^∞ function on M_n with compact support K. f can be approximated in $H_1^q(M_n)$ by a sequence of continuous functions f_p with compact support $K_p \subset K$ (the boundary of K_p being a sub-manifold of dimension $n - 1$); moreover $f_p \in C^\infty(K_p)$ and has only nondegenerate critical points on K_p.*

Since they are isolated, the number of critical points of f_p on K_p is finite.

Proof. According to Proposition 2.15 there is a sequence of C^∞ functions g_p which have no degenerate critical points and which satisfy $|f - g_p| < 1/p$ on M_n and $|\nabla(f - g_p)| < 1/p$ on K. Choose a real number α_p satisfying $1/p < \alpha_p < 2/p$, such that neither α_p nor $-\alpha_p$ is a critical value of g_p. Then $g_p^{-1}(\alpha_p)$ and $g_p^{-1}(-\alpha_p)$ are sub-manifolds of dimension $n - 1$, unless they are empty. Let $A_p = \{x \in M_n | g_p(x) \geq \alpha_p\}$ and $A_{-p} = \{x \in M_n | g_p(x) \leq -\alpha_p\}$. Define f_p by:

$$f_p(x) = [g_p(x) - \alpha_p]\chi_{A_p}(x) + [g_p(x) + \alpha_p]\chi_{A_{-p}}(x),$$

where χ_E is the characteristic function of the set E.
The support $K_p = A_p \cup A_{-p}$ of f_p is included in K because for $x \in K_p$, $|g_p(x)| > 1/p$; thus $|f(x)| > 0$. $f_p \in C^\infty(K_p)$ and f_p is Lipschitzian, hence $f_p \in H_1^q(M_n)$.
Since $|f(x) - f_p(x)| \leq (3/p)\chi_K(x)$, $\|f - f_p\|_q \to 0$ when $p \to \infty$. Moreover, at a point x where $f(x) \neq 0$, we have $|\nabla[f(x) - f_p(x)]| \to 0$, because $x \in \bigcup_{p=1}^\infty K_p$. But the set of the points where, simultaneously, $f(x) = 0$ and $|\nabla f(x)| \neq 0$, has zero measure; consequently $|\nabla[f(x) - f_p(x)]| \to 0$ almost everywhere. Therefore $\|\nabla(f - f_p)\|_q \to 0$, according to Lebesgue's theorem, since $|\nabla(f - f_p)| \leq (\sup|\nabla f| + 1/p)\chi_K$. ∎

2.17 Proposition. *Let $f \geq 0$ be a continuous function on Σ (Σ denoting the sphere $\mathbb{S}_n$, the euclidean or hyperbolic space), which is C^∞ on its compact support K, whose boundary (if it is nonempty) is a submanifold of dimension $n - 1$, and assume f has only nondegenerate critical points. Pick P a point of Σ, and define $g(r)$, a decreasing function on $[0, \infty[$, by*

$$\mu\{Q | g[d(P, Q)] \geq a\} = \mu\{Q | f(Q) \geq a\} = \psi(a).$$

Then

$$\|\nabla g\|_q \leq \|\nabla f\|_q \quad \textit{for} \quad 1 \leq q < \infty.$$

Proof. Let $d(P, Q)$ be the distance between P and Q on Σ, and let $a > 0$ be a real number; μ denote the measure defined by the metric, and write $g(Q) = g[d(P, Q)]$.
Let $Q_i(i = 1, \ldots, k)$ be the critical points of f in K. Consider the set $\Sigma_a = f^{-1}(a)$ and note that, if $Q \in \Sigma_a$ is not one of the points Q_i, then $|\nabla f(Q)| \neq 0$. If $d\sigma(Q)$ denotes the area element on Σ_a, then we may write

$$\int_\Sigma |\nabla f|^q \, dV = \int_0^\infty \left(\int_{\Sigma_a} |\nabla f|^{q-1} \, d\sigma \right) da$$

Furthermore, when a is not a critical value of f, $\varphi(a) = \int_{\Sigma_a} |\nabla f|^{-1} \, d\sigma$ exists. $\varphi(a)$ is continuous and locally admits $-\psi(a)$ as primitive.

We consider $\varphi(a) = -\psi'(a)$ as given. Therefore $\int_{\Sigma_a} |\nabla f|^{q-1} \, d\sigma$ has a minimum, in the case $q > 1$, when $|\nabla f|$ is constant on Σ_a, according to Hölder's inequality:

$$\int_{\Sigma_a} d\sigma \leq \left(\int_{\Sigma_a} |\nabla f|^{-1} \, d\sigma \right)^{(q-1)/q} \left(\int_{\Sigma_a} |\nabla f|^{q-1} \, d\sigma \right)^{1/q}.$$

But Σ_a is the boundary of a set, whose measure $\psi(a)$ is given. Hence $\int_{\Sigma_a} d\sigma$ is greater than or equal to the area of the boundary of the ball of volume $\psi(a)$ (by A. Dinghas [110]). This completes the proof. ∎

Furthermore, one verifies that $g(r)$ is absolutely continuous and even Lipschitzian on $[0, \infty[$.

2.18 Proposition. *Let $g(r)$ be a decreasing function absolutely continuous on $[0, \infty[$, and equal to zero at infinity. Then:*

$$(7) \qquad (\omega_{n-1})^{-1/n} \left(\int_0^\infty |g(r)|^p r^{n-1} \, dr \right)^{1/p} \leq \mathsf{K}(n, q) \left(\int_0^\infty |g'(r)|^q r^{n-1} \, dr \right)^{1/q},$$

where $\mathsf{K}(n, q)$ is from Theorem 2.14.

Proof. Let us consider the following variational problem, when $q > 1$:

Maximize $I(g) = \int_0^\infty |g(r)|^p r^{n-1} \, dr$, when $J(g) = \int_0^\infty |g'(r)|^q r^{n-1} \, dr$ is a given positive constant.

The Euler equation is

$$(8) \qquad (|g'|^{q-1} r^{n-1})' = k g^{p-1} r^{n-1},$$

where k is a constant.

It is obvious that we have only to consider decreasing functions. One verifies that the functions $y = (\lambda + r^{q/(q-1)})^{1-n/q}$ are solutions of (8), $\lambda > 0$ being a real number:

$$|y'|^{q-1} r^{n-1} = \left(\frac{n-q}{q-1} \right)^{q-1} r^n (\lambda + r^{q/(q-1)})^{-n(q-1)/q},$$

$$(|y'|^{q-1} r^{n-1})' = n\lambda \left(\frac{n-q}{q-1} \right)^{q-1} r^{n-1} y^{p-1}$$

According to Bliss, Lemma 2.19, the corresponding value of the integral $I(y)$ is an absolute maximum.

The value of $\mathsf{K}(n, q)$, the best constant, is

$$\mathsf{K}(n, q) = (\omega_{n-1})^{-1/n}[I(y)]^{1/p}[J(y)]^{-1/q}.$$

Letting $q \to 1$, we establish the inequality (7) for $q = 1$:

$$\mathsf{K}(n, 1) = \lim_{q \to 1} \mathsf{K}(n, q).$$

Let us compute $\mathsf{K}(n, q)$.

$$\int_0^\infty |y'|^q r^{n-1}\, dr = \left(\frac{n-q}{q-1}\right)^q \int_0^\infty [\lambda + r^{q/(q-1)}]^{-n} r^{n+1/(q-1)}\, dr.$$

Setting $\lambda = 1$ and $r = t^{(q-1)/q}$, we obtain:

$$\int_0^\infty |y'|^q r^{n-1}\, dr = \left(\frac{n-q}{q-1}\right)^q \frac{q-1}{q} \int_0^\infty (1+t)^{-n} t^{n-n/q}\, dt = \left(\frac{n-q}{q-1}\right)^q \frac{q-1}{q} A,$$

$$\int_0^\infty y^p r^{n-1}\, dr = \frac{q-1}{q} \int_0^\infty (1+t)^{-n} t^{n-1-n/q}\, dt = \frac{q-1}{q} B.$$

Furthermore, $B/A = (n-q)/n(q-1)$, because

$$A = \int_0^\infty (1+t)^{-n} t^{n-n/q}\, dt = \frac{n}{n-1} \frac{q-1}{q} \int_0^\infty (1+t)^{1-n} t^{n-1-n/q}\, dt$$
$$= \frac{n}{n-1} \frac{q-1}{q} (A+B).$$

Hence:

$$\mathsf{K}(n, q) = (\omega_{n-1})^{-1/n} \left(\frac{q-1}{q} B\right)^{1/q - 1/n} A^{-1/q} \frac{q-1}{n-q} \left(\frac{q}{q-1}\right)^{1/q},$$

$$\mathsf{K}(n, q) = \frac{q-1}{n-q} \left(\frac{B}{A}\right)^{1/q} \left(\frac{q-1}{q} B\omega_{n-1}\right)^{-1/n}, \quad \text{with} \quad B = \frac{\Gamma(n/q)\Gamma(n - n/q)}{\Gamma(n)}.$$

■

2.19 Lemma. *Let $h(x) \geq 0$ a measurable, real-valued function defined on $\mathbb{R}$, such that $J = \int_0^\infty h^q(x)\, dx$ is finite and given. Set $g(x) = \int_0^x h(t)\, dt$. Then $I = \int_0^\infty g^p(x) x^{\alpha - p}\, dx$ attains its maximum value for the functions $h(x) = (\lambda x^\alpha + 1)^{-(\alpha+1)/\alpha}$, with p and q two constants satisfying $p > q > 1$, $\alpha = (p/q) - 1$ and $\lambda > 0$ a real number.*

This is proved in Bliss [55]. The change of variable $x = r^{(q-n)/(q-1)}$ now yields the result used in the Proposition 2.18, above. Recall that here $1/p = (1/q) - (1/n)$ and so we have $\alpha = p/n$, $(\partial x/\partial r)^{1-q} = r^{n-1}$ and $x^{1+\alpha-p} = r^n$.

§7. Sobolev Imbedding Theorem for Riemannian Manifolds

2.20 Theorem. *For compact manifolds the Sobolev imbedding theorem holds. Moreover H_k^q does not depend on the Riemannian metric.*

Proof. We are going to give the usual proof of the first part of the theorem, because it is easy for compact manifolds. But for a more precise result and a more complete proof see Theorem 2.21. Let $\{\Omega_i\}$ be a finite covering of M, $(i = 1, 2, \ldots, N)$, and (Ω_i, φ_i) the corresponding charts. Consider $\{\alpha_i\}$ a C^∞ partition of unity subordinate to the covering $\{\Omega_i\}$. We have only to prove there exist constants C_i such that every C^∞ function f on M satisfies:

$$\|\alpha_i f\|_p \leq C_i \|\alpha_i f\|_{H_1^q}. \tag{9}$$

Indeed, since $|\nabla(\alpha_i f)| \leq |\nabla f| + |f||\nabla \alpha_i|$,

$$\|f\|_p \leq \sum_{i=1}^N \|\alpha_i f\|_p \leq \sup_{1 \leq i \leq N} C_i N\left[\|\nabla f\|_q + \left(1 + \sup_{1 \leq i \leq N} |\nabla \alpha_i|\right)\|f\|_q\right],$$

and by density the theorem holds for $k = 1$.
In view of Proposition 2.11, this establishes the first part of the theorem.

On the compact set $K_i = \operatorname{supp} \alpha_i \subset \Omega_i$, the metric tensor and its derivatives of all orders are bounded in the system of coordinates corresponding to the chart (Ω_i, φ_i). Hence:

$$\begin{aligned} f \in H_k^q(M_n, g) &\Leftrightarrow [\alpha_i f \in H_k^q(M_n, g), \text{ for all } i] \\ &\Leftrightarrow [\alpha_i f \circ \varphi_i^{-1} \in H_k^q(\mathbb{R}^n), \text{ for all } i]. \end{aligned}$$

We define the functions $\alpha_i f \circ \varphi_i^{-1}$ to be zero outside $\varphi_i(K_i)$.
In particular, there exist two real numbers $\mu \geq \lambda > 0$, such that for all vectors $\xi \in \mathbb{R}^n$ and every $x \in K_i$, g_x being the metric tensor at x:

$$\lambda\|\xi\|^2 \leq g_x[(\varphi_i^{-1})_*(\xi), (\varphi_i^{-1})_*(\xi)] \leq \mu\|\xi\|^2.$$

And now according to Theorem 2.14, for any $f \in C^\infty$:

$$\left(\int_{\mathbb{R}^n} |\alpha_i f \circ \varphi_i^{-1}|^p \, dE\right)^{1/p} \leq \mathrm{K}(n, q)\left(\int_{\mathbb{R}^n} |\nabla(\alpha_i f \circ \varphi_i^{-1})|^q \, dE\right)^{1/q}.$$

Thus we obtain inequality (9):

$$\|\alpha_i f\|_p \le \mu^{n/2p}\left(\int_{\mathbb{R}^n} |\alpha_i f \circ \varphi_i^{-1}|^p \, dE\right)^{1/p}$$
$$\le \mu^{n/2p}\mathsf{K}(n, q)\mu^{1/2}\lambda^{-n/2q}\|\nabla(\alpha_i f)\|_q. \qquad \blacksquare$$

2.21 Theorem. *The Sobolev imbedding theorem holds for M_n a complete manifold with bounded curvature and injectivity radius $\delta > 0$.*
Moreover, for any $\varepsilon > 0$, there exists a constant $A_q(\varepsilon)$ such that every $\varphi \in H_1^q(M_n)$ satisfies:

(10) $\|\varphi\|_p \le [\mathsf{K}(n, q) + \varepsilon]\|\nabla\varphi\|_q + A_q(\varepsilon)\|\varphi\|_q$, with $1/p = 1/q - 1/n > 0$,

where $\mathsf{K}(n, q)$ is the smallest constant having this property.

According to Proposition 2.11 and Theorem 2.6, to prove the first part of the Sobolev imbedding theorem, it is sufficient to establish inequality (10) for the functions of $\mathscr{D}(M_n)$. The proof will be given at the end of 2.27 using Lemmas 2.24 and 2.25. First we will establish the second part of the Sobolev imbedding theorem.

2.22 Lemma. *Let M_n be a complete Riemannian manifold with injectivity radius $\delta_0 > 0$ and sectional curvature K, satisfying the bound $K \le b^2$. There exists a constant $C(q)$ such that for all $\varphi \in \mathscr{D}(M_n)$:*

(11) $$\sup|\varphi| \le C(q)\|\varphi\|_{H_1^q} \quad \text{if } q > n.$$

Proof. Let $f(t)$ be a C^∞ decreasing function on $\mathbb{R}$, which is equal to 1 in a neighbourhood of zero, and to zero for $t \ge \delta$ ($\delta \le \delta_0$ satisfying $2b\delta \le \pi$). Let P be a given point of M_n, then

$$\varphi(P) = -\int_0^\delta \partial_r[\varphi(r, \theta)f(r)]\, dr,$$

where (r, θ) is a system of geodesic polar coordinates with center P. Thus we have the estimate:

$$|\varphi(P)| \le \int_0^\delta |\nabla[\varphi(r, \theta)f(r)]|(r^{1-n})r^{n-1}\, dr.$$

Integration with respect to θ over $\mathbb{S}_{n-1}(1)$ leads, by Hölder's inequality, to

$$|\varphi(P)| \le (\omega_{n-1})^{-1}\left(\int_{B_P(\delta)} |\nabla[\varphi(r, \theta)f(r)]|^q r^{n-1}\, dr\, d\theta\right)^{1/q}$$
$$\times \left(\omega_{n-1}\int_0^\delta r^{(n-1)(1-q')}\, dr\right)^{1/q'},$$

with $1/q' = 1 - 1/q$. According to Theorem 1.53, for $r \leq \delta$, $r^{n-1}\,dr\,d\theta \leq (\pi/2)^{n-1}\,dV$. Thus:

$$|\varphi(P)| \leq \left(\frac{\pi}{2}\right)^{(n-1)/q} (\omega_{n-1})^{-1/q}\left(\|\nabla\varphi\|_q + \sup_{0<r<\delta}|f'(r)|\,\|\varphi\|_q\right) \times \left[\frac{q-1}{q-n}\delta^{(q-n)/(q-1)}\right]^{1-1/q}$$

■

2.23 *Proof of the Sobolev imbedding theorem* 2.21 (Second part). $\mathscr{D}(M_n)$ is dense in $H_1^q(M_n)$ by Theorem 2.6. So let $f \in H_1^q$ and $\{\varphi_i\}$ be a sequence of functions of $\mathscr{D}(M_n)$ such that $\|f - \varphi_i\|_{H_1^q} \to 0$ when $i \to \infty$. Clearly $\{\varphi_i\}$ is a Cauchy sequence in H_1^q. By (11), $\sup|\varphi_i - \varphi_j| \leq C(q)\|\varphi_i - \varphi_j\|_{H_1^q}$, so φ_i is a Cauchy sequence in C_B^0. Therefore $\varphi_i \to f$ in C_B^0 and $f \in C_B^0$.
Letting $i \to \infty$ in $\sup|\varphi_i| \leq C(q)\|\varphi_i\|_{H_1^q}$, we establish, for all $f \in H_1^q$ when $q > n$, the inequality:

$$\|f\|_{C^0} \leq C(q)\|f\|_{H_1^q}.$$

Let $f \in H_k^q \cap C^\infty$. Then (1) implies that $|\nabla^r f| \in H_{k-r}^q$. If $(k-r)/n > 1/q$, according to the first part of Theorem 2.21, that we are going to prove below,

$$H_{k-r}^q \subset H_1^{\tilde{q}} \quad \text{with} \quad \frac{1}{q} - \frac{k-r-1}{n} \leq \frac{1}{\tilde{q}} < \frac{1}{n}.$$

Therefore

$$\|\nabla^r f\|_{C^0} \leq C(\tilde{q})\|\nabla^r f\|_{H_1^{\tilde{q}}} \leq \text{Const} \times \|f\|_{H_k^q} \text{ and } \|f\|_{C^r} \leq \text{Const} \times \|f\|_{H_k^q}.$$

This inequality holds for all $f \in H_k^q$ since $H_k^q \cap C^\infty$ is dense in H_k^q. Let us now prove that if $q = n/(1-\alpha)$, where α satisfies $0 < \alpha < 1$, then $H_1^q \subset C^\alpha$ and the identity operator is continuous. First of all, $f \in H_1^q$ implies f is continuous.
Choose δ as in Lemma 2.22 ($\delta \leq \delta_0$ and $2b\delta \leq \pi$). When $d(P, Q) \geq \delta$, we may write:

$$|f(P) - f(Q)|[d(P, Q)]^{-\alpha} \leq 2\delta^{-\alpha}C(q)\|f\|_{H_1^q}.$$

When $d = d(P, Q) < \delta$, consider a ball $\mathfrak{B}$ of radius $d/2$ and center O, with P and Q in $\mathfrak{B}$; note $y = \exp_O^{-1} P$ and $z = \exp_O^{-1} Q$; y and z belong to a ball $\tilde{B} \subset \mathbb{R}^n$ of radius $d/2$. Consider the function $h(x) = f(\exp_O x)$ defined in $\tilde{B}$ and let (r, θ) be polar coordinates with center y.
For $\tilde{B} \ni x = (r, \theta), 0 \leq r \leq \rho(\theta)$, $(\rho(\theta), \theta)$ belonging to $\partial\tilde{B}$, the boundary of $\tilde{B}$:

$$h(x) - h(y) = \int_0^r \partial_\rho h(\rho, \theta)\,d\rho = r\int_0^1 \partial_\rho h(rt, \theta)\,dt.$$

Integration with respect to x over $\tilde{B}$ leads to

$$\int_{\tilde{B}} h(x)\,dx - \frac{1}{n}\,\omega_{n-1}(d/2)^n h(y) = \int_{\mathbb{S}_{n-1}} \int_0^{\rho(\theta)} r^n\,dr\,d\theta \int_0^1 \partial_\rho h(rt, \theta)\,dt.$$

Hence, putting $u = rt$ and using the inequality $r \le \rho(\theta) < d$, we obtain:

$$\left| \int_{\tilde{B}} h(x)\,dx - \frac{1}{n}\,\omega_{n-1}\left(\frac{d}{2}\right)^n h(y) \right|$$
$$\le d \int_0^1 t^{-n}\,dt \int_{\mathbb{S}_{n-1}} \int_0^{t\rho(\theta)} |\partial_\rho h(u, \theta)|\, u^{n-1}\,du\,d\theta$$
$$\le d\left(\int_{\tilde{B}} |\nabla_E h(x)|^q\,dE \right)^{1/q} \int_0^1 (\operatorname{vol} \tilde{B}_t)^{1-1/q} t^{-n}\,dt,$$

where we have applied Hölder's inequality. Let $\tilde{B}_t$ be the ball homothetic to $\tilde{B}$, with ratio t. Then vol $\tilde{B}_t = (1/n)\omega_{n-1}(d/2)^n t^n$. Thus, since $q > n$, the second integral converges:

$$\left| h(y) - \frac{n}{\omega_{n-1}} \left(\frac{2}{d}\right)^n \int_{\tilde{B}} h(x)\,dx \right|$$
$$\le \left(\frac{2^n n}{\omega_{n-1}} \right)^{1/q} \frac{q}{q-n}\, d^{1-n/q} \left(\int_{\tilde{B}} |\nabla_E h(x)|^q\,dE \right)^{1/q}.$$

A similar inequality holds with z in place of x, and so:

$$|h(y) - h(z)| \le \frac{2q}{q-n} \left(\frac{2^n n}{\omega_{n-1}} \right)^{1/q} d^\alpha \left(\int_{\tilde{B}} |\nabla_E h(x)|^q\,dE \right)^{1/q}.$$

According to Theorem 1.53, since $d < \delta$:

$$|f(P) - f(Q)|[d(P, Q)]^{-\alpha} \le \frac{2q}{q-n} \left(\frac{2^n n}{\omega_{n-1}} \right)^{1/q} \frac{\sinh(a\delta)}{a\delta} \left(\frac{\pi}{2}\right)^{(n-1)/q} \|\nabla f\|_q.$$

The other results are local and do not differ from the case of $\mathbb{R}^n$. ■

Proof of the Sobolev imbedding theorem 2.21 (First part). According to Proposition 2.11, if we can prove that inequality (10) holds then Theorem 2.21 will be proved. First two lemmas.

2.24 Lemma. *Let $\delta \in \mathbb{R}$ satisfy $0 < \delta < \delta_0$ and $2b\delta \le \pi$. If $B_P(\delta)$ is a ball of M_n with center P and radius δ, where the sectional curvature σ satisfies*

$-a^2 \le \sigma \le b^2$, there exists a constant $\mathsf{K}_\delta(n, q)$, such that for all functions $f \in H_1^q(M_n)$ with compact support included in $B_P(\delta)$:

$$\|f\|_p \le \mathsf{K}_\delta(n, q)\|\nabla f\|_q \tag{12}$$

For δ sufficiently small we can make $\mathsf{K}_\delta(n, q)$ is as close as we like to $\mathsf{K}(n, q)$. $\mathsf{K}_\delta(n, q)$ depends on a and b, but does not depend on P.

Proof. Let (r, θ_i), $(i = 1, 2, \ldots, n-1)$ be a system of geodesic polar coordinates with center P. According to Theorem 1.53, the components g_{ii} of the metric tensor satisfy:

$$\frac{\sin br}{br} \le \sqrt{g_{ii}(r, \theta)} \le \frac{\sinh(ar)}{ar}, \; g_{rr} = 1.$$

Let $\varepsilon > 0$ be given; one may choose δ small enough so that, when $r \le \delta$:

$$\sinh(ar)/ar \le 1 + \varepsilon \quad \text{and} \quad \sin(br)/br \ge 1 - \varepsilon.$$

Setting $\tilde{f}(x) = f(\exp_p x)$, when $\|x\| \le \delta$, then according to Theorem 2.14 we have

$$\left(\int |\tilde{f}|^p \, dE\right)^{1/p} \le \mathsf{K}(n, q)\left(\int |\nabla_E \tilde{f}|^q \, dE\right)^{1/q}.$$

Since $(1-\varepsilon)^{n-1}\, dE \le dV \le (1+\varepsilon)^{n-1}\, dE$ and $|\nabla_E \tilde{f}| \le (1+\varepsilon)|\nabla f|$; this establishes inequality (12) with

$$\mathsf{K}_\delta(n, q) = (1-\varepsilon)^{(1-n)/q}(1+\varepsilon)^{1+(n-1)/p}\mathsf{K}(n, q). \qquad \blacksquare$$

2.25 Lemma (Calabi). *Let M_n be a Riemannian manifold with injectivity radius $\delta_0 > 0$; then for all $\delta > 0$, there exist two real numbers γ and β $(0 < \gamma < \beta < \delta)$, a sequence of points $P_i \in M_n$, and $\{\Omega_i\}$ a partition of M_n, by sets, satisfying $B_{P_i}(\gamma) \subset \Omega_i \subset B_{P_i}(\beta)$ for all i, $B_P(\rho)$ being the ball with center P and radius ρ.*

According to this lemma, we are able to prove:

2.26 Lemma. *Let M_n be a Riemannian manifold with injectivity radius $\delta_0 > 0$ and bounded curvature; then there exists, if δ is small enough, a uniformly locally finite covering of M_n by a family of open balls $B_{P_i}(\delta)$.*

Uniformly locally finite means: there exists a constant k, which may depend on δ, such that each point $P \in M_n$ has a neighborhood whose intersection with each $B_{P_i}(\delta)$, at most except k, is empty.

Let us prove that the balls $B_{P_i}(\beta)$ form a uniformly locally finite covering of M_n. Let $B_{P_j}(\beta)$ be given, and suppose that k balls $B_{P_i}(\beta)$ have a nonempty intersection with $B_{P_j}(\beta)$, $i \neq j$. Since the curvature is bounded, the set of the points Q satisfying $d(P_j, Q) < 2\beta + \gamma$ has a measure less than a constant w independent of j. By theorem (1.53), if the Ricci curvature is greater than $-(n-1)\alpha^2$; therefore

$$w \leq \omega_{n-1} \int_0^{2\beta+\gamma} [\sinh(\alpha r)/\alpha]^{n-1}\, dr.$$

Also, if the sectional curvatures are less than b^2, then the measure of $B_{P_i}(\gamma)$ is greater than y, with $y = \omega_{n-1} \int_0^\gamma [\sin(br)/b]^{n-1}\, dr$.
Therefore $(k+1)\, y \leq w$, since the balls $B_{P_i}(\gamma)$ are disjoint. ∎

2.27 *Proof of Theorem* 2.21 (Continued). Consider a partition of unity $h_i \in C^\infty$ subordinate to the covering $\{B_{P_i}(\delta)\}$, such that $|\nabla(h_i^{1/q})|$ is uniformly bounded, ($|\nabla(h_i^{1/q})| \leq H$, for all i, where H is a constant). Such functions h_i exist, because the covering is uniformly locally finite.
Let $\{\tilde{h}_i\} \in C^\infty$ be a partition of unity subordinate to the covering $\{B_{P_i}(\delta)\}$ such that $|\nabla \tilde{h}_i| <$ Const. We may set $h_i = \tilde{h}_i^m / (\Sigma \tilde{h}_i^m)$, with m an integer greater than q. Let I be a finite subset of $\mathbb{N}$. By Lemma 2.24,

$$\begin{aligned}
\sum_{i\in I} \|\varphi^q h_i\|_{p/q} &= \sum_{i\in I} \|\varphi h_i^{1/q}\|_p^q \leq \mathsf{K}_\delta^q(n,q) \sum_{i\in I} \|\nabla(\varphi h_i^{1/q})\|_q^q \\
&\leq \mathsf{K}_\delta^q(n,q) \sum_{i\in I} \int (|\nabla\varphi| h_i^{1/q} + \varphi|\nabla h_i^{1/q}|)^q\, dV \\
&\leq \mathsf{K}_\delta^q(n,q) \int \sum_{i\in I} \Big[|\nabla\varphi|^q h_i + \mu|\nabla\varphi|^{q-1} h_i^{(q-1)/q} |\varphi|\,|\nabla h_i^{1/q}| \\
&\qquad + \nu|\varphi|^q |\nabla h_i^{1/q}|^q \Big], \\
&\leq \mathsf{K}_\delta^q(n,q)[\|\nabla\varphi\|_q^q + \mu k H \|\nabla\varphi\|_q^{q-1}\|\varphi\|_q + \nu k H^q \|\varphi\|_q^q], \qquad (13)
\end{aligned}$$

using Hölder's inequality. μ and ν are two constants such that for $t \geq 0$

$$(1+t)^q \leq 1 + \mu t + \nu t^q;$$

for instance, $\mu = q(q-1)$ and $\nu = (q-1)^{q-1}$, if $q > 1$. Recall that $\beta \leq \delta$. If we choose δ small enough, then $\mathsf{K}_\delta(n,q) < \mathsf{K}(n,q) + \varepsilon/2$. Since the last expression of (13) is independent of I, we establish the inequality

$$\begin{aligned}
\|\varphi\|_p^q = \|\varphi^q\|_{p/q} &= \Big\| \sum_{i\in N} \varphi^q h_i \Big\|_{p/q} \leq \sum_{i\in N} \|\varphi^q h_i\|_{p/q} \\
&\leq [\mathsf{K}(n,q) + \varepsilon/2]^q [(1+\varepsilon_0)\|\nabla\varphi\|_q^q + A(\varepsilon_0)\|\varphi\|_q^q]
\end{aligned}$$

by virtue of the inequality:

(14) $$qx^{q-1}y \leq \lambda(q-1)x^q + \lambda^{1-q}y^q,$$

valid with any x, y, and λ, three positive real numbers. To complete the proof of inequality (10), we have only to set $\lambda = q\varepsilon_0/\mu k H(q-1)$, $x = \|\nabla\varphi\|_q$, and $y = \|\varphi\|_q$, where ε_0 is small enough so that $[\mathsf{K}(n,q) + \varepsilon/2][1+\varepsilon_0]^{1/q} \leq \mathsf{K}(n,q) + \varepsilon$, and $A_q(\varepsilon) = [\mathsf{K}(n,q) + \varepsilon/2][A(\varepsilon_0)]^{1/q}$ ∎

§8. Optimal Inequalities

2.28 Theorem. *Let M_n be a C^∞ Riemannian manifold with injectivity radius $\delta_0 > 0$. If the curvature is constant or if the dimension is two and the curvature bounded, then $A_q(0)$ exists and every $\varphi \in H_1^q(M_n)$ satisfies*

$$\|\varphi\|_p \leq \mathsf{K}(n,q)\|\nabla\varphi\|_q + A_q(0)\|\varphi\|_q.$$

For $\mathbb{R}^n$ and $\mathbb{H}_n$ the hyperbolic space, the inequality holds with $A_q(0) = 0$.

For the proof, see Aubin [13] pp. 595 and 597.

2.29 Theorem. *There exists a constant $A(q)$ such that every $\varphi \in H_1^q(\mathbb{S}_n)$ satisfies:*

$$\|\varphi\|_p^q \leq \mathsf{K}^q(n,q)\|\nabla\varphi\|_q^q + A(q)\|\varphi\|_q^q \quad \text{if} \quad 1 \leq q \leq 2,$$

$$\|\varphi\|_p^{q/(q-1)} \leq \mathsf{K}^{q/(q-1)}(n,q)\|\nabla\varphi\|_q^{q/(q-1)} + A(q)\|\varphi\|_q^{q/(q-1)} \quad \text{if} \quad 2 \leq q < n.$$

Let $M_n (n \geq 3)$ be a Riemannian manifold, with constant curvature and injectivity radius $\delta_0 > 0$. There exists a constant A, such that every $\varphi \in H_1^2(M_n)$ satisfies:

$$\|\varphi\|_{2n/(n-2)}^2 \leq \mathsf{K}^2(n,2)\|\nabla\varphi\|_2^2 + A\|\varphi\|_2^2.$$

For the sphere of volume 1, the inequality holds with $A = 1$.

See Aubin [13] pp. 588 and 598. For the proof of the last part of the theorem see Aubin [14] p. 293.

§9. Sobolev's Theorem for Compact Riemannian Manifolds with Boundary

2.30 Theorem. *For the compact manifolds $\overline{W}_n$ with C^r-boundary, $(r \geq 1)$, the Sobolev imbedding theorem holds. More precisely:*
First part. The imbedding $H_k^q(W) \subset H_\ell^p(W)$ is continuous with $1/p = 1/q - (k-\ell)/n > 0$. Moreover, for any $\varepsilon > 0$, there exists a constant $A_q(\varepsilon)$ such

that every $\varphi \in \overset{\circ}{H}{}^q_1(W_n)$ *satisfies inequality* (10) *and such that every* $\varphi \in H^q_1(W_n)$ *satisfies*:

$$\|\varphi\|_p \leq [2^{1/n}\mathrm{K}(n, q) + \varepsilon]\|\nabla\varphi\|_q + A_q(\varepsilon)\|\varphi\|_q. \tag{15}$$

Second part. The following imbeddings are continuous:

(a) $H^q_k(W) \subset C^s_B(W)$, *if* $k - n/q > s \geq 0$, *s being an integer*,
(b) $H^q_k(W) \subset C^s(\overline{W})$, *if in addition* $s < r$;
(c) $H^q_k(W) \subset C^\alpha(\overline{W})$, *if* α *satisfies* $0 < \alpha < 1$ *and* $\alpha \leq k - n/q$.

Proof of the first part. Let (Ω_i, φ_i) be a finite C^r-atlas of $\overline{W}_n$, each Ω_i being homeomorphic either to a ball of $\mathbb{R}^n$ or to a half ball $D \subset \overline{E}$. As in the proof of Theorem 2.20, we have only to prove inequality (9) for all $f \in H^q_1(W) \cap C^\infty(W)$, α_i being a C^r partition of unity subordinate to the covering Ω_i. When Ω_i is homeomorphic to a ball, the proof is that of theorem (2.20). When Ω_i is homeomorphic to a half ball, the proof is similar. But one applies the following lemma:

2.31 Lemma. *Let* ψ *be* C^1*-function on* $\overline{E}$, *whose support belongs to D, then* ψ *satisfies*:

$$\|\psi\|_p \leq 2^{1/n}\mathrm{K}(n, q)\|\nabla\psi\|_q, \quad \text{with } 1/p = 1/q - 1/n > 0.$$

Proof. Recall that E is the half-space of $\mathbb{R}^n$ and $D = B \cap \overline{E}$, where B is the open ball with center 0 and radius 1.
Consider $\tilde{\psi}$ defined, for $x \in \overline{E}$, by $\tilde{\psi}(x) = \psi(x)$ and $\tilde{\psi}(\tilde{x}) = \psi(x)$, when $\tilde{x} = (-x_1, x_2, \ldots, x_n)$, $(x_1, x_2, \ldots, x_n)$ being the coordinates of x. $\tilde{\psi}$ is a Lipschitzian function with compact support, thus $\tilde{\psi} \in H^q_1(\mathbb{R}^n)$ and according to Theorem 2.14:

$$\|\tilde{\psi}\|_p \leq \mathrm{K}(n, q)\|\nabla\tilde{\psi}\|_q.$$

The lemma follows, since

$$2\int_{\overline{E}} |\psi|^p\, dE = \int_{\mathbb{R}^n} |\tilde{\psi}|^p\, dE \quad \text{and} \quad 2\int_{\overline{E}} |\nabla\psi|^q\, dE = \int_{\mathbb{R}^n} |\nabla\tilde{\psi}|^q\, dE. \qquad \blacksquare$$

2.32 The proof that every $\varphi \in \overset{\circ}{H}{}^q_1(W_n)$ satisfies inequality (10) is similar to that in 2.27. But here it is easier because the covering is finite.

Using Lemma 2.31, we can prove that all $\varphi \in H^q_1(W_n)$ satisfy (15); for a complete proof see Cherrier [97].

Proof of the second part of Theorem 2.30. a) There exist constants $C_i(\tilde{q})$ such that for all $f \in H^q_1(W) \cap C^\infty(W)$

$$\sup|\alpha_i f| \leq C_i(\tilde{q})\|f\|_{H^{\tilde{q}}_1}, \text{ if } \tilde{q} > n. \tag{16}$$

Set $h(x) = 0$ for $x \notin D$, and $h(x) = (\alpha_i f) \circ \varphi_i^{-1}(x)$ for $x \in D$.
Consider a half straight line through x, defined by $\theta \in \mathbb{S}_{n-1}(1)$, entirely included in E. We have

$$|h(x)| \leq \int_0^1 |\nabla h(r, \theta)|\, dr.$$

Now we proceed as in the proof of Lemma 2.22, but integration with respect to θ is only over half of $\mathbb{S}_{n-1}(1)$:

$$|h(x)| \leq (\omega_{n-1}/2)^{-1}\left(\int_E |\nabla h|^{\tilde{q}}\, dx\right)^{1/\tilde{q}} \left(\frac{\omega_{n-1}}{2}\int_0^1 r^{(n-1)(1-q')}\, dr\right)^{1/q'},$$

with $1/q' = 1 - 1/\tilde{q}$.
Since the metric tensor is bounded on Ω_i (by proper choice of the (Ω_i, φ_i), without loss of generality), for some constant $\tilde{C}_i(\tilde{q})$ we obtain:

$$\sup |\alpha_i f| \leq \tilde{C}_i(\tilde{q}) \|\nabla(\alpha_i f)\|_{\tilde{q}}.$$

Inequality (16) follows; thus, recalling that I is finite, we have

$$\sup |f| \leq C(\tilde{q}) \|f\|_{H_1^{\tilde{q}}}, \quad \text{with } C(\tilde{q}) = \sum_{i \in I} C_i(\tilde{q}).$$

Since $k > n/q$, there exists $\tilde{q} > n$, such that the imbedding $H_k^q(W) \subset H_1^{\tilde{q}}(W)$ is continuous; we have only to choose $1/\tilde{q} \geq 1/q - (k-1)/n$. So there exists a constant C, such that every $f \in H_k^q(W) \cap C^\infty(W)$ satisfies:

$$\sup |f| \leq C \|f\|_{H_k^q}.$$

Thus a Cauchy sequence of C^∞ functions in $H_k^q(W)$ is a Cauchy sequence in $C^0(W)$ and the preceding inequality holds for all $f \in H_k^q(W)$.
For $s > 0$, apply the preceding result to $|\nabla^\ell f|, 0 \leq \ell \leq s$, and the continuous imbedding $H_k^q(W) \subset C_B^s(W)$ is established (the proof is similar to that of 2.23).

b) Instead of taking $f \in C^\infty(W)$, we may establish an inequality of the type:

$$\|f\|_{C^s} \leq A \|f\|_{H_{s+1}^{\tilde{q}}}, \quad \text{when } 0 \leq s < r, \tag{17}$$

for the functions $f \in C^r(\overline{W})$, with A a constant and $\tilde{q} > n$.
According to Theorem 2.9, $C^r(\overline{W})$ is dense in $H_{s+1}^{\tilde{q}}(W)$. Thus $H_{s+1}^{\tilde{q}}(W) \subset C^s(\overline{W})$ and inequality (17) holds for all $f \in H_{s+1}^{\tilde{q}}(W)$. When $k - n/q > s$, we may choose $\tilde{q} > n$, such that $1/\tilde{q} \geq 1/q - (k - s - 1)/n$. In this case the imbedding $H_k^q(W) \subset H_{s+1}^{\tilde{q}}(W)$ is continuous and so $H_k^q(W) \subset C^s(\overline{W})$.

c) And now for the last part of Theorem 2.30:
Let $f \in H_k^q(W)$; according to the preceding result $f \in C^0(\overline{W})$ because $k - n/q > 0$. Consider the function on D, defined by $h(x) = (\alpha_i f) \circ \varphi_i^{-1}(x)$, for a given $i \in I$. By a proof similar to that in 2.23, we establish the existence of a constant B such that every $f \in H_1^{\tilde{q}}(W)$ satisfies

$$|h(x) - h(y)| \|x - y\|^{-\alpha} \leq B\left(\int_D |\nabla h|^{\tilde{q}}\, dx\right)^{1/\tilde{q}},$$

where $\tilde{q} = n/(1 - \alpha)$. Instead of considering a ball of radius $\|x - y\|/2$, we must integrate over a cube K with edge $\|x - y\|$, included in $\bar{E}$, with x and y belonging to K (see Adams [1] p. 109).
Then, since the metric tensor is bounded on Ω_i, there exists a constant B_i such that, for every pair (P, Q) of points of $\overline{W}$, any $f \in H_1^{\tilde{q}}$ satisfies:

$$|\alpha_i(P) f(P) - \alpha_i(Q) f(Q)| [d(P, Q)]^{-\alpha} \leq B_i \|f\|_{H_1^{\tilde{q}}}.$$

Thus we establish the desired inequality:

$$|f(P) - f(Q)| |d(P, Q)|^{-\alpha} \leq \left(\sum_{i \in I} B_i\right) \|f\|_{H_1^{\tilde{q}}} \leq \text{Const} \times \|f\|_{H_k^q},$$

where the last inequality follows from the first part of the Sobolev imbedding theorem, since $\tilde{q} = n/(1 - \alpha)$ satisfies $1/\tilde{q} \geq 1/q - (k - 1)/n$. ■

§10. The Kondrakov Theorem

2.33 *Let $k \geq 0$ be an integer, p and q two real numbers satisfying $1 \geq 1/p > 1/q - k/n > 0$. The Kondrakov Theorem asserts that, if Ω, a bounded open set of $\mathbb{R}^n$, has a sufficiently regular boundary $\partial\Omega$ ($\partial\Omega$ of class C^1, or only Lipschitzian):*

(a) *the imbedding $H_k^q(\Omega) \subset L_p(\Omega)$ is compact.*
(b) *With the same assumptions for Ω, the imbedding $H_k^q(\Omega) \subset C^\alpha(\overline{\Omega})$ is compact, if $k - \alpha > n/q$, with $0 \leq \alpha < 1$.*
(c) *For Ω a bounded open set of $\mathbb{R}^n$, the following imbeddings are compact:*

$$\mathring{H}_k^q(\Omega) \subset L_p(\Omega), \quad \mathring{H}_k^q(\Omega) \subset C^\alpha(\overline{\Omega}).$$

Proof. Roughly, the proof consists in proving that if the Sobolev imbedding theorem holds for a bounded domain Ω, then the Kondrakov theorem is true for Ω.

a) According to the Sobolev imbedding theorem 2.30, the imbedding $H_k^q \subset H_1^{\tilde{q}}$ is continuous with $1/\tilde{q} = 1/q - (k - 1)/n$. Thus we have only to prove that the imbedding of $H_1^{\tilde q} \subset L_p$ is compact when $1 \geq 1/p > 1/\tilde q - 1/n > 0$, since the composition of two continuous imbeddings is compact if one of them is compact.

Let $\mathscr{A}$ be a bounded subset of $H_1^{\tilde q}(\Omega)$, so if $f \in \mathscr{A}$,

$$\|f\|_{H_1^{\tilde q}} \leq C, \text{ a constant.}$$

By hypothesis $H_1^{\tilde q}(\Omega) \subset L_r$, with $1/r = 1/\tilde q - 1/n$, and there exists a constant A such that for $f \in H_1^{\tilde q}(\Omega)$,

$$\|f\|_r \leq A\|f\|_{H_1^{\tilde q}}.$$

Set $K_j = \{x \in \Omega/\text{dist}(x, \partial\Omega) \geq 2/j\}, j \in \mathbb{N}$. For $f \in \mathscr{A}$, by Hölder's inequality:

$$\int_{\Omega - K_j} |f|\, dx \leq \left(\int_{\Omega - K_j} |f|^r\, dx\right)^{1/r}\left(\int_{\Omega - K_j} dx\right)^{1-1/r} \leq AC\left(\int_{\Omega - K_j} dx\right)^{1-1/r},$$

which goes to zero, when $j \to \infty$. Thus, given $\varepsilon > 0$, there exists $j_0 \in \mathbb{N}$, such that $[\text{vol}(\Omega - K_{j_0})]^{1-1/r} \leq \varepsilon/AC$. Now, by Fubini's theorem:

$$\int_{K_{j_0}} |f(x+y) - f(x)|\, dx \leq \int_{K_{j_0}} dx \int_0^1 \left|\frac{d}{dt} f(x+ty)\right| dt$$

$$\leq \|y\| \int_{K_{2j_0}} |\nabla f|\, dx \leq \|y\|\,\|\nabla f\|_1$$

for $\|y\| < 1/j_0$ since $x + y \in K_{2j_0}$, if $x \in K_{j_0}$. Since $C^\infty(\Omega)$ is dense in $H_1^{\tilde q}(\Omega)$, the preceding inequality holds for any $f \in H_1^{\tilde q}(\Omega)$. Moreover, by Hölder's inequality, $\|\nabla f\|_1 \leq \|\nabla f\|_r(\text{vol}\,\Omega)^{1-1/r} \leq B$, a constant.
Theorem 3.44 with $\delta = \varepsilon/B$ then implies that $\mathscr{A}$ is precompact in $L_1(\Omega)$. Hence $\mathscr{A}$ is precompact in $L_p(\Omega)$, because if $f_m \in \mathscr{A}$ is a Cauchy sequence in L_1, it is a Cauchy sequence in L_p:

$$\|f_m - f_\ell\|_p \leq \|f_m - f_\ell\|_1^\mu \|f_m - f_\ell\|_r^{1-\mu} \leq (2AC)^{1-\mu}\|f_m - f_\ell\|_1^\mu,$$

by Hölder's inequality, with $\mu = [(r/p) - 1]/(r - 1)$.

b) *Proof of the second part of Theorem* 2.33. Let λ satisfy $\alpha < \lambda < \inf(1, k - n/q)$. Then by the Sobolev imbedding theorem 2.30, $H_k^q(\Omega)$ is included in $C^\lambda(\bar\Omega)$, and there exists a constant A such that $\|f\|_{C^\lambda} \leq A\|f\|_{H_k^q}$.

Let $\mathscr{A}$ be a bounded subset of $H_k^q(\Omega)$; if $f \in \mathscr{A}$, $\|f\|_{H_k^q} \leq C$, a constant, and $\|f\|_{C^\lambda} \leq AC$.
Thus we can apply Ascoli's Theorem, 3.15. $\mathscr{A}$ is a bounded subset of equicontinuous functions of $C^0(\bar\Omega)$, and $\bar\Omega$ is compact. So $\mathscr{A}$ is precompact in $C^0(\bar\Omega)$.

Then, since

$$|f(x) - f(y)|\,\|x - y\|^{-\alpha} = (|f(x) - f(y)|\,\|x - y\|^{-\lambda})^{\alpha/\lambda}|f(x) - f(y)|^{1-\alpha/\lambda}$$

if a sequence $f_m \in \mathscr{A}$ converges to f in $C^0(\bar{\Omega})$, $\|f\|_{C^\lambda} \leq AC$ and

$$\|f - f_m\|_{C^\alpha} \leq (2AC)^{\alpha/\lambda}(\|f - f_m\|_{C^0})^{1-\alpha/\lambda} + \|f - f_m\|_{C^0}.$$

Thus $\mathscr{A}$ is precompact in $C^\alpha(\bar{\Omega})$.

c) $\mathscr{D}(\Omega)$ is included in $\mathscr{D}(\mathbb{R}^n)$, so we can apply the theorem of Sobolev, 2.10, to the space $\mathring{H}^q_k(\Omega)$. A proof similar to those of a) and b) gives the desired result.

§11. Kondrakov's Theorem for Riemannian Manifolds

2.34 Theorem. *The Kondrakov theorem, 2.33, holds for the compact Riemannian manifolds M_n, and the compact Riemannian manifolds $\overline{W}_n$ with C^1-boundary. Namely, the following imbeddings are compact:*

(a) $H^q_k(M_n) \subset L_p(M_n)$ *and* $H^q_k(W_n) \subset L_p(W_n)$, *with* $1 \geq 1/p > 1/q - k/n > 0$.

(b) $H^q_k(M_n) \subset C^\alpha(M_n)$ *and* $H^q_k(W_n) \subset C^\alpha(\overline{W}_n)$, *if* $k - \alpha > n/q$, *with* $0 \leq \alpha < 1$.

Proof. Let (Ω_i, φ_i), $(i = 1, 2, \ldots, N)$ be a finite atlas of M_n (respectively, C^1-atlas of $\overline{W}_n$), each Ω_i being homeomorphic either to a ball of $\mathbb{R}^n$ or to a half ball $D \subset \bar{E}$. We choose the atlas so that in each chart the metric tensor is bounded. Consider a C^∞ partition of unity $\{\alpha_i\}$ subordinate to the covering $\{\Omega_i\}$. It is sufficient to prove the theorem in the special case $k = 1$ for the same reason as in the preceding proof, 2.33.

a) Let $\{f_m\}$ be a bounded sequence in H^q_1. Consider the functions defined on B (or on D), i being given:

$$h_m(x) = (\alpha_i f_m) \circ \varphi_i^{-1}(x).$$

Since the metric tensor is bounded on Ω_i, the set $\mathscr{A}_i$ of these functions is bounded in $H^q_1(\Omega)$ with $\Omega = B$ or D. (The boundary of D is only Lipschitzian, but, since $\text{supp}(\alpha_i \circ \varphi_i^{-1})$ is included in B, we may consider a bounded open set Ω with smooth boundary which satisfies $D \subset \Omega \subset E$).
According to Theorem 2.33, $\mathscr{A}_i$ is precompact. Thus there exists a subsequence which is a Cauchy sequence in L_p. Repeating this operation successively for $i = 1, 2, \ldots, N$, we may select a subsequence $\{\tilde{f}_m\}$ of the sequence $\{f_m\}$, such that $\alpha_i \tilde{f}_m$ is a Cauchy sequence in L_p for each i.

Thus $\{\tilde{f}_m\}$ is a Cauchy sequence in L_p, since

$$|\tilde{f}_m - \tilde{f}_\ell| \le \sum_{i=1}^{N} |\alpha_i \tilde{f}_m - \alpha_i \tilde{f}_\ell|.$$

b) Let λ satisfy $\alpha < \lambda < \inf(1, k - n/q)$. According to Theorems 2.21 and 2.30, the imbeddings $H_k^q(M_n) \subset C^\lambda(M_n)$ and $H_k^q(W_n) \subset C^\lambda(\overline{W}_n)$ are continuous. Thus the same proof used to show 2.33, b) establishes the result.

2.35 Remark. We have given only the main results concerning the theorems of Sobolev and Kondrakov. These theorems are proved for the compact manifolds with Lipschitzian boundary in Aubin [17]. To obtain complete results for domains of $\mathbb{R}^n$ see Adams [1].

2.36 Remark. Instead of the spaces $H_k^p(M_n)$, it is possible to introduce the spaces $H_k'^p(M_n)$, which are the completion of $\mathscr{E}_k'^p$ with respect to the norm

$$\|\varphi\|_{H_k'^p} = \sum_{0=\ell\le k/2} \|\Delta^\ell \varphi\|_p + \sum_{0=\ell\le(k-1)/2} \|\nabla\Delta^\ell\varphi\|_p,$$

with $\mathscr{E}_k'^p$ the vector space of the functions $\varphi \in C^\infty(M_n)$, such that $\Delta^\ell\varphi \in L_p(M_n)$ for $0 \le \ell \le k/2$ and such that $|\nabla\Delta^\ell\varphi| \in L_p(M_n)$ for $0 \le \ell \le (k-1)/2$. For these spaces, the Kondrakov theorem holds, as well as the Sobolev imbedding theorem when $p > 1$ (see Aubin [17]).

For $k \le 1$, $H_k^p = H_k'^p$. But for $k > 1$, the spaces $H_k'^p$ may be more convenient for the study of differential equations.

§12. Examples

2.37 The exceptional case of the Sobolev imbedding theorem (i.e., $H_k^{n/k}$ on n-dimensional manifolds).

Consider the function f on $\mathbb{R}^2$ defined by:

$$\mathbb{R}^2 \ni x \to f(x) = \log|\log\|x\||, \quad \text{for } 0 < \|x\| < 1/e, \text{ and } f(x) = 0,$$

for the other points x of $\mathbb{R}^2$.

$$\|\nabla f\|_2^2 = 2\pi \int_0^{1/e} \frac{dr}{r|\log r|^2} = 2\pi$$

and f^2 is integrable; thus $f \in H_1^2(\mathbb{R}^2)$.

As $1/q - 1/n = 0$ here, we could hope that the function f would be bounded, but it is not: $H_k^{n/k} \not\subset L_\infty$. On the other hand, e^f is integrable:

$$\|e^f\|_1 = 2\pi \int_0^{1/e} r|\log r|\,dr, \quad \text{see (2.46)}.$$

2.38 The Sobolev imbedding theorem $H_k^q \subset L_p$ holds when $1/p = 1/q - k/n > 0$, but the Kondrakov theorem does not.

Consider the sequence of functions f_k defined on $\mathbb{R}^n$ $(n > 4)$ by:

$$f_k(x) = \left(\frac{1}{k}\right)^{(n-2)/4} \left(\frac{1}{k} + \|x\|^2\right)^{1-n/2}.$$

Let us verify $f_k \in H_1^2(\mathbb{R}^n)$. Now

$$\|f_k\|_2^2 = \omega_{n-1} k^{1-n/2} \int_0^\infty \left(\frac{1}{k} + r^2\right)^{2-n} r^{n-1}\, dr = \frac{\omega_{n-1}}{k} \int_0^\infty (1 + t^2)^{2-n} t^{n-1}\, dt$$

is finite and so is

$$\|\nabla f_k\|_2^2 = \omega_{n-1} k^{1-n/2} \int_0^\infty (n-2)^2 \left(\frac{1}{k} + r^2\right)^{-n} r^{n+1}\, dr$$

$$= \omega_{n-1}(n-2)^2 \int_0^\infty (1 + t^2)^{-n} t^{n+1}\, dt = A.$$

Also, f_k belongs to L_N, with $N = 2n/(n-2)$, because

$$\int_0^\infty f_k^N r^{n-1}\, dr = \left(\frac{1}{k}\right)^{n/2} \int_0^\infty \left(\frac{1}{k} + r^2\right)^{-n} r^{n-1}\, dr$$

$$= \int_0^\infty (1 + t^2)^{-n} t^{n-1}\, dt = C < \infty.$$

Let $h_k(x) = f_k(x) - (\sqrt{k} + 1/\sqrt{k})^{1-n/2}$ for $\|x\| \leq 1$.
Then $h_k \in \overset{\circ}{H}{}_1^2(B)$, where B is the unit ball of $\mathbb{R}^n$ with center 0. Clearly, $\|h_k\|_{H_1^2(B)} \to \sqrt{A}$ when $k \to \infty$, the sequence h_k is bounded in $\overset{\circ}{H}{}_1^2(B)$, and $\|h_k\|_{L_N(B)} \to C^{1/N} \neq 0$.
Now $h_k(x) \to 0$ when $\|x\| \neq 0$. Thus a subsequence of $\{h_k\}$ cannot converge in L_N, without the limit being zero in L_N. But this contradicts the above result $(C \neq 0)$. The imbedding $H_1^2(B)$ in $L_N(B)$ is not compact.

§13. Improvement of the Best Constants

2.39 Let M_n be a complete Riemannian manifold with bounded curvature and injectivity radius $\delta > 0$. According to Theorem 2.21, if $1/p = 1/q - 1/n > 0$ then every $\varphi \in H_1^q(M_n)$ satisfies:

$$\|\varphi\|_p \leq B\|\nabla\varphi\|_q + A\|\varphi\|_q, \tag{18}$$

where A and B are constants. It is proved in 3.78 and 3.79 that $\mathsf{K} = \{\inf B$ such that all $\varphi \in H_1^q(M_n)$ satisfy inequality (18) for a certain value, $A(B)\}$ is strictly positive. By Theorem 2.21, K depends only on the dimension n and q: $\mathsf{K} = \mathsf{K}(n, q)$. For the value of $\mathsf{K}(n, q)$, see Theorem 2.14. We are going to show that the best constants $\mathsf{K}(n, q)$ can be lowered if the functions φ satisfy some additional natural orthogonality conditions.

2.40 Theorem. *Let M_n be a complete Riemannian manifold with bounded curvature and injectivity radius $\delta > 0$, and $f_i (i \in I)$ functions of class C^1, with the following properties: they change sign, their gradients are uniformly bounded, and the family $\{|f_i|^q\}$ forms a partition of unity (subordinate to a uniformly locally finite cover by bounded open sets).*
Then the functions $\varphi \in H_1^q(M_n)$, which satisfy the conditions

$$(19) \qquad \int_{M_n} |\varphi|^p f_i |f_i|^{p-1} \, dV = 0 \quad \textit{for all} \quad i \in I,$$

satisfy inequality (18) *with pairs $(B, A(B))$, with B as close as one wants to $2^{-1/n}\mathsf{K}(n, q)$. Thus the best constant B of inequality* (18) *is $2^{1/n}$ times smaller for functions φ satisfying* (19).

Proof. Set $\check{f} = \sup(f, 0)$ and $\tilde{f} = \sup(-f, 0)$; thus $f = \check{f} - \tilde{f}$. Unless otherwise stated, integration is over M_n. Since $|\nabla|\varphi|| = |\nabla\varphi|$ almost everywhere, Proposition 3.49, we can suppose, without loss of generality, $\varphi \geq 0$. On the other hand, the functions f_i may be chosen more generally, for instance, uniformly Lipschitzian.

By hypothesis, $\int \varphi^p \check{f}_i^p \, dV = \int \varphi^p \tilde{f}_i^p \, dV$, and $\varphi \tilde{f}_i$, as well as $\varphi \check{f}_i$, belong to $H_1^q(M_n)$.

Let $K > \mathsf{K}(n, q)$ and $A_0 = A(K)$, the corresponding constant in (18).

$$\|\varphi \check{f}_i\|_p^q \leq K^q \|\nabla(\varphi \check{f}_i)\|_q^q + A_0 \|\varphi \check{f}_i\|_q^q,$$

$$\|\varphi \tilde{f}_i\|_p^q \leq K^q \|\nabla(\varphi \tilde{f}_i)\|_q^q + A_0 \|\varphi \tilde{f}_i\|_q^q.$$

Suppose, for instance, that $\|\nabla(\varphi \check{f}_i)\|_q \geq \|\nabla(\varphi \tilde{f}_i)\|_q$. Then write:

$$\begin{aligned} \|\varphi f_i\|_p^q = 2^{q/p} \|\varphi \tilde{f}_i\|^q &\leq 2^{q/p}(K^q \|\nabla(\varphi \tilde{f}_i)\|_q^q + A_0 \|\varphi \tilde{f}_i\|_q^q) \\ &\leq 2^{-q/n} K^q (\|\nabla(\varphi \tilde{f}_i)\|_q^q + \|\nabla(\varphi \check{f}_i)\|_q^q) + 2^{q/p} A_0 \|\varphi \tilde{f}_i\|_q^q. \end{aligned}$$

If $\|\nabla(\varphi \check{f}_i)\|_q < \|\nabla(\varphi \tilde{f}_i)\|_q$, we obtain a similar inequality, using $\varphi \check{f}_i$ instead of $\varphi \tilde{f}_i$. Thus in all cases:

$$\|\varphi f_i\|_p^q \leq 2^{-q/n} K^q \|\nabla(\varphi f_i)\|_q^q + 2^{q/p} A_0 \|\varphi f_i\|_q^q.$$

Consider again the computation in (2.27). Set $H = \sup_{i \in I} \sup_M |\nabla f_i|$ and pick k an integer such that, at every point of M, at most k of the functions f_i are nonzero. Let J be a finite subset of I. Then there exist constants μ and v such that

$$\sum_{i \in J} \|\varphi f_i\|_p^q \leq 2^{-q/n} K^q [\|\nabla\varphi\|_q^q + \mu k H \|\nabla\varphi\|_q^{q-1} \|\varphi\|_q + v k H^q \|\varphi\|_q^q] + 2^{q/p} A_0 \|\varphi\|_q^q.$$

The second member does not depend on J, hence the corresponding series is convergent. There exist two constants β and γ, such that

$$\|\varphi\|_p^q = \|\varphi^q\|_{p/q} = \left\|\sum_{i \in I} \varphi^q |f_i|^q\right\|_{p/q} \leq \sum_{i \in I} \|\varphi^q |f_i|^q\|_{p/q}$$
$$\leq 2^{-q/n} K^q \|\nabla\varphi\|_q^q + \beta \|\nabla\varphi\|_q^{q-1} \|\varphi\|_q + \gamma \|\varphi\|_q^q.$$

Using inequality (14), for any $\varepsilon_1 > 0$, there exists an M_0 such that

$$\|\nabla\varphi\|_q^{q-1} \|\varphi\|_q \leq \varepsilon_1 \|\nabla\varphi\|_q^q + M_0 \|\varphi\|_q^q. \tag{20}$$

Thus the stated result is proved:

$$\|\varphi\|_p^q \leq (2^{-q/n} K^q + \beta\varepsilon_1) \|\nabla\varphi\|_q^q + (\gamma + M_0 \beta) \|\varphi\|_q^q.$$

To establish that $2^{-1/n} \mathsf{K}(n, q)$ is the best constant, when the functions φ satisfy (19), we have only to use the functions which are defined in Theorem 2.14. Indeed, for convenience, suppose $\tilde{f}_1$ is equal to 1 over a ball $\tilde{B}$ with center $\tilde{P}$ and radius $\rho < \delta$, and $\check{f}_1$ is equal to 1 over a ball $\check{B}$ with center $\check{P}$ and radius ρ.

Consider the sequence of functions $\psi_m (m \in N)$, which vanish outside $\tilde{B} \cup \check{B}$ and which are defined by

$$\psi_m(Q) = (r^{q/(q-1)} + 1/m)^{1-n/q} - (\rho^{q/(q-1)} + 1/m)^{1-n/q}$$

for $Q \in \tilde{B}$ with $r = d(\tilde{P}, Q)$, and for $Q \in \check{B}$ with $r = d(\check{P}, Q)$. This sequence satisfies:

$$\lim_{m \to \infty} \|\psi_m\|_p \|\nabla\psi_m\|_q^{-1} = 2^{-1/n} \mathsf{K}(n, q),$$

while

$$\lim_{m \to \infty} \|\psi_m\|_q \|\psi_m\|_p^{-1} = 0. \qquad \blacksquare$$

2.41 For applications to differential equations, the useful result is that concerning H_1. Since the best constant $\mathsf{K}(n, 2)$ is attained for the manifolds with constant curvature, Theorem 2.29, $2^{-1/n}\mathsf{K}(n, 2)$ is attained in that case.

Corollary. *Let $M_n (n \geq 3)$ be a Riemannian manifold with constant curvature and injectivity radius $\delta > 0$, and $f_i (i \in I)$ C^2-functions, having the properties required in Theorem (2.40), and satisfying $\Delta f_i^2 \leq$ Const., for all $i \in I$. Then there exists a constant A_2, such that all functions $\varphi \in \mathsf{H}_1$ satisfy:*

$$\|\varphi\|_N^2 \leq 2^{-2/n}\mathsf{K}^2(n, 2)\|\nabla\varphi\|_2^2 + A_2\|\varphi\|_2^2, \tag{21}$$

when they satisfy the conditions (19) *with* $p = N = 2n/(n-2)$.

Proof. This is similar to the preceding one. Use (14) instead of (18) and write:

$$\int |\nabla(\varphi f_i)^2|\, dV = \int f_i^2 |\nabla\varphi|^2\, dV + \tfrac{1}{2}\int \nabla^\nu \varphi^2 \nabla_\nu f_i^2\, dV + \int \varphi^2 |\nabla f_i|^2\, dV$$
$$= \int f_i^2 |\nabla\varphi|^2\, dV + \int (|\nabla f_i|^2 + \tfrac{1}{2}\Delta f_i^2)\varphi^2\, dV.$$

Since $|\nabla f_i|^2 + \frac{1}{2}\Delta f_i^2 \leq$ Const, (21) follows. ∎

2.42 The preceding theorem can be generalized. A similar proof establishes the following

Theorem. *Let M_n, $(n \geq 3)$, be a Riemannian manifold with bounded curvature and injectivity radius $\delta > 0$, and let $f_{i,j}$ $(i \in I, j = 1, 2, \ldots, m, m \geq 2$ an integer$)$ be uniformly Lipschitzian and non-negative functions, with compact support $\Omega_{i,j}$, having the following properties: $\mathring{\Omega}_{i,j} \cap \mathring{\Omega}_{i,\ell} = \varnothing$ (for $1 \leq j < \ell \leq m$ and all $i \in I$), at each point $P \in M$, only k of all the functions f_{ij} can be nonzero, and*

$$\sum_{i \in I} \sum_{j=1}^{m} |f_{i,j}|^q = 1.$$

Then the functions $\varphi \in \mathsf{H}_1$, which satisfy the conditions

$$\int |\varphi|^p f_{i,j}^p\, dV = \int |\varphi|^p f_{i,1}^p\, dV \quad (\text{for all } i \in I \text{ and each } j = 2, \ldots, m),$$

satisfy inequality (18), *where B can be chosen equal to $m^{-1/n}\mathsf{K}(n, q) + \varepsilon$, with $\varepsilon > 0$, as small as one wants. (The best constant B of inequality* (18) *is now $m^{1/n}$ times smaller than $\mathsf{K}(n, q)$.*

2.43 Remark. On a compact manifold, if we consider the Rayleigh quotient $\inf \|\nabla\varphi\|_2\|\varphi\|_2^{-1}$, when φ satisfies some well-known orthogonality conditions, we obtain, successively, the eigenvalues of the Laplacian $\lambda_0 = 0, \lambda_1, \lambda_2, \ldots$. Even if we know some properties of the sequence λ_i, we cannot compute λ_i from λ_1. It is therefore somewhat surprising that in the nonlinear case, the sequence is entirely known, the mth term being $m^{1/n}\mathsf{K}^{-1}(n, q)$.

§14. The Case of the Sphere

2.44 Definition. On the sphere $\mathbb{S}_n$, Λ will denote the vector space of functions ψ, which satisfy $\Delta\psi = \lambda_1\psi$, where λ_1 is the first nonzero eigenvalue of the Laplacian.
Recall that Λ is of dimension $n + 1$. One verifies that the eigenfunctions are $\psi(Q) = \gamma\cos[\alpha d(P, Q)]$, for any constant γ and any point $P \in \mathbb{S}_n$, with $\alpha^2 = R/n(n-1)$, R being the scalar curvature of the sphere.
There exists a family ξ_i ($i = 1, 2, \ldots, n + 1$) of functions in Λ, orthogonal in L_2 and satisfying $\sum_{i=1}^{n+1}\xi_i^2 = 1$ (see Berger (37)). In fact, if $x = (x_1, x_2, \ldots, x_{n+1})$ are the standard coordinates on $\mathbb{R}^{n+1}$, ξ_i is the restriction of x_i to $\mathbb{S}_n$.

Thus, we can apply Theorem 2.40 with $f_i = \xi_i$ when $n > 2$, $q = 2$, and $p = N = 2n/(n-2)$. But to solve the problem 5.11, we need the somewhat different conditions:

$$\int_{\mathbb{S}_n}\xi_i|\varphi|^N\,dV = 0, \quad \text{instead of} \int_{\mathbb{S}_n}\xi_i|\xi_i|^{N-1}|\varphi|^N\,dV = 0.$$

If we want to use Theorem 2.40, we must choose as functions f_i, the functions $\xi_i|\xi_i|^{1/N-1}$. But this is impossible for two good reasons. On the one hand, $\{|\xi_i|^{2/N}\}$ does not form a partition of unity; on the other, the functions $|\xi_i|^{1/N}$ do not belong to $\mathsf{H}_1(\mathbb{S}_n)$. Nevertheless, these difficulties can be overcome. We are going to establish the following.

2.45 Theorem. *The functions ξ_i are a basis of Λ; then all $\varphi \in H_1^q(\mathbb{S}_n)$, $1 \le q < n$, satisfying $\int \xi_i|\varphi|^p\,dV = 0$ (for $i = 1, 2, \ldots, n + 1$) satisfy:*

$$\|\varphi\|_p^q \le [2^{-1/n}\mathsf{K}(n, q) + \varepsilon]^q\|\nabla\varphi\|_q^q + A(\varepsilon)\|\varphi\|_q^q, \tag{22}$$

where $A(\varepsilon)$ is a constant which depends on $\varepsilon > 0$, ε as small as one wants.

Proof. Let $0 < \eta < 1/2$ be a real number, which we are going to choose very small. There exists a finite family of functions $\xi_i \in \Lambda$ ($i = 1, 2, \ldots, k$), such that:

$$1 + \eta < \sum_{i=1}^{k}|\xi_i|^{q/p} < 1 + 2\eta \quad \text{with} \quad |\xi_i| < 2^{-p}.$$

Indeed, let $Q \in \mathbb{S}_n$ and $\xi_Q(P)$ the eigenfunction of Λ, such that ξ_Q attains its maximum 1 at $P = Q$; since $\xi_Q(P) = \xi_P(Q)$, we have

$$\int |\xi_Q(P)|^{q/p}\, dV(Q) = \int |\xi_P(Q)|^{q/p}\, dV(Q) = \text{Const.}$$

From this property of the family $\{\xi_Q\}_{Q \in \mathbb{S}_n}$, clearly the above family $\{\xi_i\}$ exists.

Consider h_i, C^1 functions, such that everywhere $h_i \xi_i \geq 0$ and such that $||h_i|^q - |\xi_i|^{q/p}| < (\eta/k)^p$. Then

(23) $$1 < \sum_{i=1}^{k} |h_i|^q < 1 + 3\eta$$

and

$$\left| |h_i|^p - |\xi_i| \right| \leq \frac{p}{q}\left[|\xi_i|^{q/p} + \left(\frac{\eta}{k}\right)^p \right]^{p/q - 1} \left| |h_i|^q - |\xi_i|^{q/p} \right| \leq \frac{p}{q}\left(\frac{\eta}{k}\right)^p = \varepsilon_0^p,$$

according to Theorem 3.6, and since $|\xi_i|^{q/p} + (\eta/k)^p < 1$. As in the proof of Theorem 2.40, we suppose that $\varphi \geq 0$:

$$\|\varphi\|_p^q = \|\varphi^q\|_{p/q} \leq \left\| \sum_{i=1}^{k} \varphi^q |h_i|^q \right\|_{p/q} \leq \sum_{i=1}^{k} \|\varphi h_i\|_p^q.$$

If $\|\check{h}_i |\nabla\varphi|\|_q \geq \|\tilde{h}_i |\nabla\varphi|\|_q$, using (23) and the hypothesis, from Theorem 2.29 we obtain

$$\begin{aligned}
\|\varphi h_i\|_p^q &\leq \left[\int (\check{\xi}_i + \varepsilon_0^p)\varphi^p\, dV + \int \varphi^p \tilde{h}_i^p\, dV \right]^{q/p} \\
&= \left[\int (\tilde{\xi}_i + \varepsilon_0^p)\varphi^p\, dV + \int \varphi^p \tilde{h}_i^p\, dV \right]^{q/p} \\
&\leq 2^{q/p}\left[\int \varphi^p(\tilde{h}_i^p + \varepsilon_0^p)\, dV \right]^{q/p} \leq 2^{q/p} \|\varphi(\tilde{h}_i + \varepsilon_0)\|_p^q \\
&\leq 2^{q/p}[\mathrm{K}^q(n, q)\|\nabla[\varphi(\tilde{h}_i + \varepsilon_0)]\|_q^q + A(q)\|\varphi(\tilde{h}_i + \varepsilon_0)\|_q^q].
\end{aligned}$$

Set $H = \sup_{1 \leq i \leq k} \sup_{\mathbb{S}_n} |\nabla h_i|$. Then there exist constants μ and ν such that:

$$\|\nabla[\varphi(\tilde{h}_i + \varepsilon_0)]\|_q^q \leq \int (\tilde{h}_i + \varepsilon_0)^q |\nabla\varphi|^q\, dV + \mu H \|\nabla\varphi\|_q^{q-1} \|\varphi\|_q + \nu H^q \|\varphi\|_q^q.$$

Since $(\tilde{h}_i + \varepsilon_0)^q \leq \tilde{h}_i^q + q(\tilde{h}_i + \varepsilon_0)^{q-1}\varepsilon_0 < \tilde{h}_i^q + q\varepsilon_0$, then

$$\|\varphi\|_p^q \leq 2^{q/p}\mathrm{K}^q(n, q)\left[\frac{1}{2}\sum_{i=1}^{k} \int |h_i|^q |\nabla\varphi|^q \, dV + qk\varepsilon_0 \|\nabla\varphi\|_q^q \right.$$
$$\left. + k\mu H \|\nabla\varphi\|_q^{q-1}\|\varphi\|_q + \nu H^q k \|\varphi\|_q^q \right] + 2^{q/p}kA(q)\|\varphi\|_q^q.$$

Using (20) and (23), this leads to:

$$\|\varphi\|_p^q \leq \mathrm{K}^q(n, q)[2^{-q/n}(1 + 3\eta) + 2^{q/p}qk\varepsilon_0 + k\mu H\varepsilon_1]\|\nabla\varphi\|_q^q$$
$$+ 2^{q/p}k[\mathrm{K}^q(n, q)(\mu H M_0 + \nu H^q) + A(q)]\|\varphi\|_q^q.$$

Since $\varepsilon_0 k = (p/q)^{1/p}\eta$, setting $\varepsilon_1 = \eta/Hk$ and taking η small enough we obtain the stated result.

Thus B in (18) is as close as desired to $2^{-1/n}\mathrm{K}(n, q)$.
We applied Theorem 2.29 for $q \leq 2$. In the case $q > 2$ the proof is not different, because ε is nonzero in (22). ∎

§15. The Exceptional Case of the Sobolev Imbedding Theorem

2.46 We will expound the topic chronologically. The exceptional case of the Sobolev imbedding theorem concerns the Sobolev space $H_1^n(M)$, where n is the dimension of the manifold M, or more generally, the spaces $H_k^{n/k}(M)$.

When $\varphi \in H_1^n$ we might hope that $\varphi \in L_\infty$. Unfortunately, this is not the case. Recall Example 2.37; the function $x \to \log|\log \|x\||$ defined on the ball $B_{1/e} \subset \mathbb{R}^2$ is not bounded but belongs, however, to $\mathring{H}_1^2(B_{1/e})$. But when $\varphi \in H_1^n$ it is possible to show that e^φ, and even $\exp[\alpha|\varphi|^{n/(n-1)}]$, are locally integrable, if α is small ehough (Trüdinger [261], Aubin [10]). More precisely,

Theorem 2.46. *Let M_n be a compact Riemannian manifold with or without boundary. If $\varphi \in H_1^n(M_n)$, then e^φ and $\exp[\alpha(|\varphi| \|\varphi\|_{H_1^n}^{-1})^{n/(n-1)}]$ are integrable for α a sufficiently small real number which does not depend on φ. Moreover, there exist constants C, μ and ν such that all $\varphi \in H_1^n$ satisfy:*

$$\int_M e^\varphi \, dV \leq C \exp[\mu\|\nabla\varphi\|_n^n + \nu\|\varphi\|_n^n] \tag{24}$$

and the mapping $H_1^n \ni \varphi \to e^\varphi \in L_1$ is compact.

Proof. α) Using a finite partition of unity we see that we have only to prove Theorem 2.46 for functions belonging to $\mathring{H}_1^n(B)$, where B is the unit ball of $\mathbb{R}^n$. Indeed, if the ball carries a Riemannian metric we use the inequalities of Theorem 1.53, and if the function obtained has its support included in the half ball we consider by reflection the x_1-even function which belongs to $\overset{\circ}{H}{}^n_1(B)$, as in 2.31.

β) Now $\varphi \in \overset{\circ}{H}{}^n_1(B)$. For almost all $P \in B$

$$|\varphi(P)| \leq \omega_{n-1}^{-1} \int_B |\nabla\varphi(Q)|[d(P, Q)]^{1-n}\, dV(Q), \tag{25}$$

(see the end of 2.12). Then by Proposition 3.64, we obtain for all real $p \geq \eta$

$$\|\varphi\|_p \leq \omega_{n-1}^{-1}\|\nabla\varphi\|_n \sup_{P\in B}\left[\int_B [d(P, Q)]^{k(1-n)}\, dV(Q)\right]^{1/k},$$

with $1/k = 1/p - 1/n + 1$. This yields:

$$\|\varphi\|_p \leq \omega_{n-1}^{-1+1/k}\|\nabla\varphi\|_n\left[\int_0^1 r^{(k-1)(1-n)}\, dr\right]^{1/k}$$
$$= \omega_{n-1}^{-1+1/k}\|\nabla\varphi\|_n\left[\frac{p+1-p/n}{n}\right]^{1/k}.$$

Thus there is a constant K such that for any $p \geq 1$:

$$\|\varphi\|_p \leq K\|\nabla\varphi\|_n p^{(n-1)/n} \tag{26}$$

and we obtain

$$\int_B e^\varphi\, dV \leq \sum_{p=0}^{\infty} K^p\|\nabla\varphi\|_n^p(p!)^{-1}p^{(n-1)p/n}.$$

But according to Stirling's formula, when $p \to \infty$

$$(p/n)!(p!)^{-1}p^{(n-1)p/n} \sim (p/en)^{p/n}n^{-1/2}(e/p)^p p^{(n-1)p/n} \sim e^{(n-1)p/n}n^{-p/n-1/2},$$

so that

$$\int_B e^\varphi\, dV \leq C\exp(\mu\|\nabla\varphi\|_n^n),$$

where C and μ are two constants.

Remark. When M is compact, using Equation (15) of 4.13 after integrating by parts, and the properties of the Green's function yield immediately (without step α) an inequality of the kind (25). Thereby we can obtain a similar result when $\Delta^k\varphi \in L_{n/2k}$ instead of $\varphi \in H_{2k}^{n/2k}$ (see Aubin [10]). When

$\varphi \in H_m^s$ with $ms = n$, $\exp[\alpha|\varphi|^{s/(s-1)}]$ is locally integrable for $\alpha > 0$ small enough (see Cherrier [96]). Brezis and Wainger [67] obtained fine results in this field by using Lorentz spaces.

γ) Using (26) leads to

$$(27) \quad \int_B \exp[\nu|\varphi|^{n/(n-1)}]\, dV \leq \sum_{p=0}^{\infty} \nu^p (p!)^{-1} [K\|\nabla\varphi\|_n]^{pn/(n-1)} [pn/(n-1)]^p.$$

According to Stirling's formula $(p!)(p/e)^p p^{1/2} \leq \text{Const}$, thus the series in (27) converges for $(K\|\nabla\varphi\|_n)^{n/(n-1)}\nu\, en/(n-1) < 1$.

δ) To prove the last statement of Theorem 2.46 we will use the Kondrakov theorem 2.34: the imbedding $H_1^1(M) \subset L_1(M)$ is compact. Let $\mathscr{A}$ be a bounded set in $H_1^n(M)$. Rewriting (24) with $q\varphi$ instead of φ ($q \geq 1$) implies that the set $\{e^\varphi\}_{\varphi \in \mathscr{A}}$ is bounded in L_q for all q. Then $\|\nabla e^\varphi\|_1 \leq \|\nabla\varphi\|_n \|e^\varphi\|_{n/(n-1)}$ shows that the set $\{e^\varphi\}_{\varphi \in \mathscr{A}}$ is bounded in H_1^1. Thus the result follows. ■

§16. Moser's Results

2.47 For applications, the best values of α and μ in Theorem (2.46) are essential. On this question the following result of Moser [209] was the first.

Theorem 2.47. *Let Ω be a bounded open set in $\mathbb{R}^n$ and set $\alpha_n = n\omega_{n-1}^{1/(n-1)}$. Then all $\varphi \in \mathring{H}_1^n(\Omega)$ such that $\|\nabla\varphi\|_n \leq 1$ satisfy for $\alpha \leq \alpha_n$:*

$$(28) \qquad \int_\Omega \exp[\alpha|\varphi|^{n/(n-1)}]\, dV \leq C \int_\Omega dV.$$

Here the constant C depends only on n.
α_n is the best constant: if $\alpha > \alpha_n$ the integral on the left in (28) is finite but it can be made arbitrarily large by an appropriate choice of φ.

Proof. Making use of symmetrization as in 2.17 reduces the problem to a one-dimensional one. We have only to consider radially symmetric functions in B_ρ, the ball in $\mathbb{R}^n$ of radius ρ which has the same volume as Ω. Set $g(\|x\|) = \varphi(x)$. On the other hand, we can suppose $\varphi \in C^\infty(\bar{B}_\rho)$. Indeed, if $\varphi \in \mathring{H}_1^n(B_\rho)$ there exists $\{\psi_i\}_{i \in N}$, a sequence of C^∞ functions on $\bar{B}_\rho$ vanishing on the boundary, such that $\psi_i \to \varphi$ in H_1^n when $i \to \infty$ and such that $\psi_i \to \varphi$ almost everywhere (Proposition 3.43). Thus if we prove (28) for smooth functions, (28) holds for all $\varphi \in \mathring{H}_1^n(B_\rho)$ since

$$\int_{B(\rho)} \exp(\alpha|\varphi|^{n/(n-1)})\, dV \leq \liminf_{i\to\infty} \int_{B(\rho)} \exp(\alpha|\psi_i|^{n/(n-1)})\, dV.$$

So the problem is now:
For which $\alpha \in \mathbb{R}$, do all functions $g \in C^\infty([0, \rho])$ which vanish at ρ satisfy the inequality $\int_0^\rho \exp(\alpha |g|^{n/(n-1)})\, dV \leq C\rho^n/n$ when $\int_0^\rho |g'|^n r^{n-1}\, dr \leq \omega_{n-1}^{-1}$? Applying Proposition 2.48 below with $q = n$ and $k = \omega_{n-1}^{-1}$, C exists if and only if $\alpha \leq \alpha_n = n\omega_{n-1}^{1/(n-1)}$. ∎

The following proposition discusses the existence of the integral for all α.

2.48 Proposition. *Let g be a Lipschitzian function on $[0, \rho]$ which vanishes at ρ. If $\int_0^\rho |g'|^q r^{q-1}\, dr \leq k$ for some $q > 1$ and some $k > 0$, then $\int_0^\rho \exp(\beta |g|^{q/(q-1)}) r^{n-1}\, dr \leq C\rho^n/n$, with C a constant which depends only on q and n, if and only if $\beta \leq \beta_q = nk^{1/(1-q)}$. Moreover, the integral exists for any β, although the inequality holds only for $\beta \leq \beta_q$.*

Proof. Moser stated the result only for $q = n$, but he gave the proof for arbitrary $q > 1$. We will follow his proof.
Set $e^{-t} = (r/\rho)^n$ and $f(t) = g(\rho e^{-t/n})$. Then $f(0) = 0$ and we have $(1/q)\, d(r^q) = -(1/n)\rho^q e^{-qt/n}\, dt$ and $f'(t) = -(\rho/n) g'(r) e^{-t/n}$.
Thus $n^{q-1} \int_0^\infty |f'|^q\, dt \leq k$ and we want to have:

$$\int_0^\infty \exp(\beta |f|^{q/(q-1)} - t)\, dt \leq C. \tag{29}$$

By Hölder's inequality:

$$|f(t)| \leq \int_0^t |f'|\, dt \leq \left(\int_0^t |f'|^q\, dt \right)^{1/q} t^{(q-1)/q} \leq k^{1/q} (t/n)^{(q-1)/q} = (t/\beta_q)^{(q-1)/q}.$$

Hence for $\beta < \beta_q$ the result follows at once:

$$\int_0^\infty \exp(\beta |f|^{q/(q-1)} - t)\, dt \leq \int_0^\infty \exp[(\beta/\beta_q - 1)t]\, dt = (1 - \beta/\beta_q)^{-1}.$$

If $\beta = \beta_q$ it is not easy to establish the result (see Moser [209] for the proof).
When $\beta > \beta_q$ the integral we are studying exists, but it can be made arbitrarily large. Consider for $\tau > 0$ the function f_τ defined as follows: $f_\tau(t) = (\tau/\beta_q)^{(q-1)/q} t/\tau$ for $t \leq \tau$ and $f_\tau(t) = (\tau/\beta_q)^{(q-1)/q}$ for $t \geq \tau$. Clearly these functions satisfy the hypotheses and

$$\int_0^\infty \exp(\beta |f_\tau|^{q/(q-1)} - t)\, dt \geq \int_\tau^\infty \exp(\beta\tau/\beta_q - t)\, dt = \exp[(\beta/\beta_q - 1)\tau]$$

tends to infinity as $\tau \to \infty$.

It remains to prove the convergence of the integral. Applying Hölder's inequality we find that for $t > \tau$

$$|f(t) - f(\tau)| \leq \int_\tau^t |f'|\,dt \leq \left(\int_\tau^\infty |f'|^q\,dt\right)^{1/q} (t - \tau)^{1-1/q}.$$

Since we can choose τ so that $\int_\tau^\infty |f'|^q\,dt$ is as small as one wants, $t^{1/q-1}f(t) \to 0$ as $t \to \infty$. Hence $\beta|f(t)|^{q/(q-1)}t^{-1} \to 0$ as $t \to \infty$ and the integral in (29) exists. ■

2.49 Proposition. *Let g be as in Proposition 2.48. Then there are constants C and λ such that*

$$\int_0^\rho e^g r^{n-1}\,dr \leq C \exp\left[\lambda \int_0^\rho |g'|^q r^{q-1}\,dr\right]; \tag{30}$$

the inf *of λ such that C exists is equal to $\lambda_q = ((q-1)/n)^{q-1}q^{-q}$.*

Proof. It is easy to verify that all real numbers u satisfy

$$u \leq k\lambda_q + \beta_q|u|^{q/(q-1)};$$

thus according to Proposition 2.48,

$$\int_0^\rho e^g r^{n-1}\,dr \leq \frac{C}{n}\rho^n e^{k\lambda_q}, \quad \text{where we pick } k = \int_0^\rho |g'|^q r^{q-1}\,dr. \quad ■$$

Corollary 2.49. *Let Ω be a bounded open set of $\mathbb{R}^n$ and set $\mu_n = (n-1)^{n-1} n^{1-2n}\omega_{n-1}^{-1}$. Then all $\varphi \in \overset{\circ}{H}{}_1^n(\Omega)$ satisfy*

$$\int_\Omega e^\varphi\,dV \leq C \int_\Omega dV \exp(\mu_n \|\nabla\varphi\|_n^n), \tag{31}$$

where C depends only on n.

Proof. After symmetrization we use Proposition 2.49 with $q = n$ and we get $\mu_n = \lambda_n \omega_{n-1}^{-1}$. This result may also be obtained from (28) by using the inequality: $uv \leq \alpha_n |u|^{n/(n-1)} + \mu_n |v|^n$ with $v = \|\nabla\varphi\|_n$ and $u = \varphi\|\nabla\varphi\|_n^{-1}$. ■

§17. The Case of the Riemannian Manifolds

2.50 Return to Theorem 2.46. Set $\tilde{\alpha}_n$, the sup of α, such that $\exp[\alpha(|\varphi|\,\|\varphi\|_{H_1^n}^{-1})^{n/(n-1)}]$ is integrable and $\tilde{\mu}_n$, the inf of μ, such that C and ν exist in inequality (24). Two questions arise. Does $\tilde{\mu}_n$ depend on the manifold?

Is $\tilde{\mu}_n$ attained? (I.e., is $\mu = \tilde{\mu}_n$ allowed.) The answers were first found in Cherrier [95]. In Cherrier [96] there are similar results when $\varphi \in H_k^{n/k}$ or $\Delta^k \varphi \in L_{n/2k}$. He proved the following:

Theorem 2.50. *For Riemannian manifolds M_n with bounded curvature and global injectivity radius (in particular this is true if M is compact), the best constants $\tilde{\mu}_n$ and $\tilde{\alpha}_n$ in Theorem (2.46) depend only on n. They are equal to*

$$\mu_n = (n-1)^{n-1} n^{1-2n} \omega_{n-1}^{-1} \quad \textit{and} \quad \alpha_n = n\omega_{n-1}^{1/(n-1)}.$$

For compact Riemannian manifolds with C^1 boundary the best constants are equal, respectively, to $2\mu_n$ and $2^{-1/(n-1)}\alpha_n$.
μ_n is attained for the sphere $\mathbb{S}_n$ and μ_2 is attained for compact Riemannian manifolds of dimension 2.

2.51 The case of the sphere. We have seen that we have the best possible inequality (31) for $\varphi \in \overset{\circ}{H}{}_1^n(\Omega)$ when Ω is a bounded open set of $\mathbb{R}^n$: $\mu = \mu_n$ and $\nu = 0$ in (24).
This is also the case for the sphere $\mathbb{S}_n$. The following was proved by Moser [209] when $n = 2$, and by Aubin [21], p. 156.

Theorem 2.51. *All $\varphi \in H_1^n(\mathbb{S}_n)$ with integral equal zero ($\int_{\mathbb{S}_n} \varphi \, dV = 0$) satisfy*

$$\text{(32)} \qquad \int_{\mathbb{S}_n} e^{\varphi} \, dV \leq C \exp(\mu_n \|\nabla\varphi\|_n^n),$$

where C depends only on n and $\mu_n = (n-1)^{n-1} n^{1-2n} \omega_{n-1}^{-1}$; in particular $\mu_2 = 1/16\pi$.

2.52 As in other inequalities concerning Sobolev spaces, the best constants can be lowered when the functions φ also satisfy some natural orthogonality conditions. Theorems similar to those in 2.40 and 2.42 are proved in Aubin [21], p. 157. The sequence of best constants is $\{\mu_n/m\}_{m \in N}$. For the sphere $\mathbb{S}_n$ the following is proved.

Theorem 2.52. *Let Λ be the eigenspace corresponding to the first nonzero eigenvalue. The functions $\varphi \in H_1^n(\mathbb{S}_n)$ satisfying $\int_{\mathbb{S}_n} \xi e^{\varphi} \, dV = 0$ for all $\xi \in \Lambda$ and $\int_{\mathbb{S}_n} \varphi \, dV = 0$, satisfy the inequality*

$$\text{(33)} \qquad \int_{\mathbb{S}_n} e^{\varphi} \, dV \leq C(\mu) \exp(\mu \|\nabla\varphi\|_n^n),$$

where it is possible to choose $\mu > \mu_n/2$ as close to $\mu_n/2$ as one wants. $C(\mu)$ is a constant which depends on μ and n.

2.53 The case of the real projective space $\mathbb{P}_n$.

Theorem. *For any $\varepsilon > 0$ there is a constant $C(\varepsilon)$ which depends only on n such that all $\psi \in H_1^n(\mathbb{P}_n)$ with integral zero ($\int_{\mathbb{P}_n} \psi \, dV = 0$) satisfy*

$$\int_{\mathbb{P}_n} e^{\psi} \, dV \leq C(\varepsilon) \exp[(\mu_n + \varepsilon) \|\nabla\psi\|_n^n]. \tag{34}$$

Proof. $\mathfrak{p}: \mathbb{S}_n \to \mathbb{P}_n$, the universal covering of $\mathbb{P}_n$ has two sheets. We associate to $\psi \in H_1^n(\mathbb{P}_n)$ the function φ on $\mathbb{S}_n$ defined by $\varphi(Q) = \psi(\mathfrak{p}(Q))$ for $Q \in \mathbb{S}_n$.

The function φ so obtained satisfies the hypotheses of Theorem 2.52. $\int_{\mathbb{S}_n} \varphi \, dV = 2 \int_{\mathbb{P}_n} \psi \, dV = 0$ and e^{φ} is orthogonal to Λ. Indeed, if Q and $\tilde{Q}$ are antipodally symmetric on $\mathbb{S}_n$, $\xi(Q)e^{\varphi(Q)} = -\xi(\tilde{Q})e^{\varphi(\tilde{Q})}$ for $\xi \in \Lambda$. Thus $\int_{\mathbb{S}_n} \xi(Q)e^{\varphi(Q)} \, dV(Q) = -\int_{\mathbb{S}_n} \xi(\tilde{Q})e^{\varphi(\tilde{Q})} \, dV(\tilde{Q})$, and so vanishes.
By Theorem 2.52, for any $\varepsilon > 0$ there is $\tilde{C}(\varepsilon)$ which depends only on n, such that all $\psi \in H_1^n(\mathbb{P}_n)$ satisfy

$$2 \int_{\mathbb{P}_n} e^{\psi} \, dV = \int_{\mathbb{S}_n} e^{\varphi} \, dV \leq \tilde{C}(\varepsilon) \exp\left[(\mu_n/2 + \varepsilon) \int_{\mathbb{S}n} |\nabla\varphi|^n \, dV\right]$$

$$= \tilde{C}(\varepsilon) \exp\left[(\mu_n + 2\varepsilon) \int_{\mathbb{P}_n} |\nabla\psi|^n \, dV\right].$$

Thus we get (34) with $C(\varepsilon) = \tilde{C}(\varepsilon/2)/2$. ∎

§18. Problems of Traces

2.54 Let M be a Riemannian manifold and let $V \subset M$ be a Riemannian sub-manifold. If f is a C^k function on M, we can consider $\tilde{f}$ the restriction of f to $V, \tilde{f} \in C^k(V)$.

Now if $f \in H_k^p(M)$, it is often possible to define the trace $\tilde{f}$ of f on V by a density argument and there are imbedding theorems similar to those of Sobolev. Adams [1] discusses the case of Euclidean space. In Cherrier [97] the problem of traces is studied for Riemannian manifolds; he also considers the exceptional case.

The same problems arise for a Riemannian manifold W with boundary ∂W. We can try to define the trace on ∂W of a function belonging to $H_k^p(W)$. The results are useful for problems with prescribed boundary conditions.

Chapter 3

Background Material

§1. Differential Calculus

3.1 Definition. A *normed space* is a vector space $\mathfrak{F}$(over $\mathbb{C}$ or $\mathbb{R}$), which is provided with a norm. A *norm*, denoted by $\| \ \|$, is a real-valued functional on $\mathfrak{F}$, which satisfies:

(a) $\mathfrak{F} \ni x \to \|x\| \geq 0$, with equality if and only if $x = 0$,
(b) $\|\lambda x\| = |\lambda| \, \|x\|$ for every $x \in \mathfrak{F}$ and $\lambda \in \mathbb{C}$,
(c) $\|x + y\| \leq \|x\| + \|y\|$ for every $x, y \in \mathfrak{F}$.

A *Banach space* $\mathfrak{B}$ is a complete normed space: every Cauchy sequence in $\mathfrak{B}$ converges to a limit in $\mathfrak{B}$.

A *Hilbert space* $\mathfrak{H}$ is a Banach space where the norm comes from an *inner product*:

$$\mathfrak{H}^2 \ni (x, y) \to \langle x, y \rangle \in \mathbb{C}, \quad \text{so} \quad \|x\|^2 = \langle x, x \rangle.$$

$\mathfrak{P}$ is an inner product provided that $\mathfrak{P}$ is linear in x, that it satisfies $\langle x, y \rangle = \langle y, x \rangle$, and that $\langle x, x \rangle = 0$ if and only if $x = 0$.

3.2 Definition. Let $\mathfrak{F}$ and $\mathfrak{G}$ be two normed spaces. We denote by $\mathscr{L}(\mathfrak{F}, \mathfrak{G})$ the space of the continuous linear mappings $\mathfrak{u}$ from $\mathfrak{F}$ to $\mathfrak{G}$. $\mathscr{L}(\mathfrak{F}, \mathfrak{G})$ has the natural structure of a normed space. Its norm is

$$\|\mathfrak{u}\| = \sup\|\mathfrak{u}(x)\| \quad \text{for all } x \in \mathfrak{F} \quad \text{with } \|x\| \leq 1.$$

$\mathfrak{F}^*$, the *dual space* of $\mathfrak{F}$, is $\mathscr{L}(\mathfrak{F}, \mathbb{C})$ or $\mathscr{L}(\mathfrak{F}, \mathbb{R})$, according to whether $\mathfrak{F}$ is a vector space on $\mathbb{C}$ or $\mathbb{R}$. $\mathfrak{F}$ is said to be reflexive if the natural imbedding $\mathfrak{F} \ni x \to \tilde{x} \in \mathfrak{F}^{**}$, defined by $\tilde{x}(\mathfrak{u}) = \mathfrak{u}(x)$ for $\mathfrak{u} \in F^*$, is surjective.

3.3 Proposition. *A linear mapping* $\mathfrak{u}$ *from* $\mathfrak{F}$ *to* $\mathfrak{G}$ *(where* $\mathfrak{F}$ *and* $\mathfrak{G}$ *are two normed spaces) is continuous if and only if there exists a real number* M *such that* $\|\mathfrak{u}(x)\| \leq M\|x\|$ *for all* $x \in \mathfrak{F}$.

If $\mathfrak{B}$ *is a Banach space, then* $\mathscr{L}(\mathfrak{F}, \mathfrak{B})$ *is a Banach space.*

3.4 Definition. If Ω is an open subset of $\mathfrak{F}$, ($\mathfrak{F}$ and $\mathfrak{G}$ being two normed spaces), then $f : \Omega \to \mathfrak{G}$ is called *differentiable* at $x \in \Omega$ if there exists a $\mathfrak{u} \in \mathscr{L}(\mathfrak{F}, \mathfrak{G})$ such that:

$$f(x + y) - f(x) = \mathfrak{u}(y) + \|y\|\omega(x, y), \tag{1}$$

where $\omega(x, y) \to 0$ when $y \to 0$, for all y such that $x + y \in \Omega$. $\mathfrak{u}$, called the differential of f at x, is denoted by $f'(x)$ or $Df(x)$.

3.5 Definition. Let f be as above. f is called differentiable on Ω if f is differentiable at each point $x \in \Omega$.
f is continuously differentiable on Ω, written $f \in C^1(\Omega, \mathfrak{G})$, if the map $\psi: \Omega \ni x \to f'(x) \in \mathscr{L}(\mathfrak{F}, \mathfrak{G})$ is continuous. $f \in C^1(\Omega, \mathfrak{G})$ is twice differentiable at x if ψ is differentiable at x. We write $D^2 f(x) = \psi'(x) \in \mathscr{L}(\mathfrak{F}, \mathscr{L}(\mathfrak{F}, \mathfrak{G})) = \mathscr{L}(\mathfrak{F} \times \mathfrak{F}, \mathfrak{G})$.
f is C^2, written $f \in C^2(\Omega, \mathfrak{G})$, if $\psi \in C^1(\Omega, \mathscr{L}(\mathfrak{F}, \mathfrak{G}))$. In this case $D^2 f(x)$ is symmetric. $D^2 f(x) \in \mathscr{L}_2(\mathfrak{F}, \mathfrak{G})$, the space of continuous bilinear maps from $\mathfrak{F} \times \mathfrak{F}$ to $\mathfrak{G}$. Continuing by induction, we can define the pth differential of f at x (if it exists): $D^p f(x) = D[D^{p-1} f(x)]$.
If $x \to D^p f(x)$ is continuous on Ω, f is said to be C^p, $f \in C^p(\Omega, \mathfrak{G})$, $D^p f(x) \in \mathscr{L}_p(\mathfrak{F}, \mathfrak{G}) = \mathscr{L}(\mathfrak{F}^p, \mathfrak{G})$.

1.1. The Mean Value Theorem

3.6 Theorem. *Let $\mathfrak{F}$ and $\mathfrak{G}$ be two normed vector spaces and $f \in C^1(\Omega, \mathfrak{G})$, with $\Omega \subset \mathfrak{F}$. If a and b are two points of Ω, set*

$$[a, b] = \{x \in \mathfrak{F} \quad \text{such that } x = a + t(b - a) \quad \text{for some } t \in [0, 1]\}.$$

If $[a, b] \subset \Omega$ and if $\|f'(x)\| \leq M$ for all $x \in [a, b]$, then

$$\|f(b) - f(a)\| \leq M\|b - a\|. \tag{2}$$

3.7 Definitions. When $\mathfrak{F}$ and $\mathfrak{G}$ have finite dimension: $\mathfrak{F} = \mathbb{R}^n$, $\mathfrak{G} = \mathbb{R}^p$, a mapping f is defined on $\Omega \subset \mathfrak{F}$ by p real-valued functions $f^\alpha(x^1, x^2, \ldots, x^n)$, ($\alpha = 1, 2, \ldots, p$). Then $f \in C^1(\Omega, \mathfrak{G})$ if and only if each function f^α has continuous partial derivatives.
The matrix ($n \times p$), whose general entry is $\partial f^\alpha(x)/\partial x^i$, is called the *Jacobian matrix* of f at $x \in \Omega$.
The *rank* of f at $x \in \Omega$, is the rank of $f'(x)$, that is to say the dimension of the range of $f'(x)$.

3.8 Taylor's formula. *Let $f \in C^n(\Omega, \mathfrak{G})$, $\Omega \subset \mathfrak{F}$, ($\mathfrak{F}$ and $\mathfrak{G}$ two normed vector spaces), and $[x, x+h] \subset \Omega$. Then*

$$f(x+h) = f(x) + f'(x)h + \tfrac{1}{2}D^2f(x)h^2 + \cdots$$
$$+ \frac{1}{n!}D^nf(x)h^n + \|h\|^n\omega_n(x,h) \tag{3}$$

where $\omega_n(x, h) \to 0$, when $h \to 0$.
$D^kf(x)h^k$ means $D^kf(x)(h, h, \ldots, h)$, the h repeated k times. If $f \in C^{n+1}(\Omega, \mathfrak{G})$ and if $\mathfrak{G}$ is complete, then

$$\|h\|^n\omega_n(x,h) = \frac{1}{n!}\int_0^1 (1-t)^n D^{n+1}f(x+th)h^{n+1}\,dt.$$

3.9 Definition. A *homeomorphism* of a topological space into another is a continuous one to one map, such that the inverse function is also continuous. A C^k-diffeomorphism of an open set $\Omega \subset \mathfrak{F}$ onto an open set in $\mathfrak{G}$ is a C^k-differentiable homeomorphism, whose inverse map is C^k, where $\mathfrak{F}$ and $\mathfrak{G}$ are two normed spaces.

1.2. Inverse Function Theorem

3.10 Theorem. *Let $\mathfrak{B}$ and $\mathfrak{G}$ be two Banach spaces and $f \in C^1(\Omega, \mathfrak{G})$, $\Omega \subset \mathfrak{B}$. If at $x_0 \in \Omega$, $f'(x_0)$ is a homeomorphism of $\mathfrak{B}$ onto $\mathfrak{G}$, then there exists a neighborhood Θ of x_0, such that Φ, the restriction of f to Θ, is a homeomorphism of Θ onto $f(\Theta)$.*
If f is of class C^k, Φ is a C^k-diffeomorphism.

Implicit function theorem. *Let $\mathfrak{E}$, $\mathfrak{F}$ and $\mathfrak{B}$ be Banach spaces and let $\mathfrak{U}$ be an open set of $\mathfrak{E} \times \mathfrak{F}$. Suppose $f \in C^p(\mathfrak{U}, \mathfrak{B})$ and let $D_y f(x_0, y_0) \in \mathscr{L}(\mathfrak{F}, \mathfrak{B})$ be the differential at y_0 of the mapping $y \to f(x_0, y)$. If at $(x_0, y_0) \in \mathfrak{U}$, $D_y f(x_0, y_0)$ is invertible, then the map $(x, y) \to (x, f(x, y))$ is a C^p diffeomorphism of a neighborhood $\Omega \subset \mathfrak{U}$ of (x_0, y_0) onto an open set of $\mathfrak{E} \times \mathfrak{B}$.*

1.3. Cauchy's Theorem

3.11 *Let $\mathfrak{B}$ be a Banach space and $f(t, x)$ a continuous function on an open subset $\mathfrak{U} \subset \mathbb{R} \times \mathfrak{B}$ with range in $\mathfrak{B}$.*
Consider the initial value problem, for functions: $t \to x(t) \in \mathfrak{B}$:

$$\frac{dx}{dt} = f(t,x),\ x(t_0) = x_0 \quad \text{with } (t_0, x_0) \in \mathfrak{U}. \tag{4}$$

If, on a neighborhood of (t_0, x_0), $f(t, x)$ *is a uniformly Lipschitzian in* x, *then there exists one and only one continuous solution of* (4), *which is defined on a neighborhood of* t_0.
If f *is* C^p, *the solution is* C^{p+1}. *Moreover the solution depends on the initial conditions* (t_0, x_0); *set* $x(t, t_0, x_0)$ *the unique solution of* (4). *The map* ψ: $(t, t_0, x_0) \to x(t, t_0, x_0) \in \mathfrak{B}$ *is continuous on an open subset of* $\mathbb{R} \times \mathbb{R} \times \mathfrak{B}$. *If* f *is* C^p, ψ *is* C^p.

Recall that we say $f(t, x)$ is uniformly Lipschitzian in x on $\Theta \subset \mathbb{R} \times \mathfrak{B}$ if there exists a k such that for any $(t, x_1) \in \Theta$ and $(t, x_2) \in \Theta$,

$$\|f(t, x_1) - f(t, x_2)\| \leq k\|x_1 - x_2\|. \tag{5}$$

It is possible to have a more precise result on the interval of existence of the solution. By continuity of f, there exist M, α, and ρ, three positive numbers, such that $\|f(t, x)\| \leq M$, for any $(t, x) \in I \times \bar{B}_{x_0}(\rho) \subset \mathfrak{U}$, with $I = [t_0 - \alpha, t_0 + \alpha]$ and $B_{x_0}(\rho) = \{x \in \mathfrak{B} \mid \|x - x_0\| < \rho\}$. If $M\alpha \leq \rho$, the solution of (4) exists on I.

§2. Four Basic Theorems of Functional Analysis

2.1. Hahn–Banach Theorem

3.12 *Let* $p(x)$ *be a seminorm defined on a normed space* $\mathfrak{G}$, $\mathfrak{F}$ *a linear subspace of* $\mathfrak{G}$, *and* $f(x)$ *a linear functional defined on* $\mathfrak{F}$, *with* $|f(x)| \leq p(x)$ *for* $x \in \mathfrak{F}$. *Then* f *can be extended to a continuous linear function* $\tilde{f}$ *on* $\mathfrak{G}$ *with* $|\tilde{f}(x)| \leq p(x)$ *for all* $x \subset \mathfrak{G}$.

A seminorm is a positive real-valued functional on $\mathfrak{G}$ which satisfies b) and c) of 3.1.

2.2. Open Mapping Theorem

3.13 *Under a continuous linear map* $\mathfrak{u}$ *of one Banach space onto all of another, the image of every open set is open. If* $\mathfrak{u}$ *is one-to-one,* $\mathfrak{u}$ *has a continuous linear inverse.*

2.3. The Banach–Steinhaus Theorem

3.14 *Let* $\mathfrak{B}$ *and* $\mathfrak{F}$ *be Banach spaces and a family of* $\mathfrak{u}_\alpha \in \mathscr{L}(\mathfrak{B}, \mathfrak{F})$, ($\alpha \in A$ *a given set*). *If for each* $x \in \mathfrak{B}$, *the set* $\{\mathfrak{u}_\alpha(x)\}_{\alpha \in A}$ *is bounded, then there exists* M, *such that* $\|\mathfrak{u}_\alpha\| \leq M$ *for all* $\alpha \in A$. *In particular, if* $\mathfrak{u}_i \in \mathscr{L}(\mathfrak{B}, \mathfrak{F})$ *and if* $\lim_{i \to \infty} \mathfrak{u}_i(x)$ *exists for each* $x \in \mathfrak{B}$, *then there exists an* M *such that* $\|\mathfrak{u}_i\| \leq M$ *for all* $i \in \mathbb{N}$, *and there exists a* $\mathfrak{u} \in \mathscr{L}(\mathfrak{B}, \mathfrak{F})$ *such that* $\mathfrak{u}_i(x) \to \mathfrak{u}(x)$ *for all* $x \in \mathfrak{B}$.
But $\mathfrak{u}_i$ does not necessarily converge to $\mathfrak{u}$ in $\mathscr{L}(\mathfrak{B}, \mathfrak{F})$.

2.4. Ascoli's Theorem

3.15 *Let $\mathfrak{K}$ be a compact Haussdorff space and $C(\mathfrak{K})$ the Banach space of the continuous functions on $\mathfrak{K}$ with the norm of uniform convergence.*
A subset $A \subset C(\mathfrak{K})$ is precompact ($\bar{A}$ is compact), if and only if it is bounded and equicontinuous.
(Recall A is said to be equicontinuous if, to every $\varepsilon > 0$ and every $x \in \mathfrak{K}$, there corresponds a neighborhood $\mathfrak{U}$ of x such that $|f(x) - f(y)| < \varepsilon$ for all $y \in \mathfrak{U}$ and all $f \in A$).

§3. Weak Convergence. Compact Operators

3.16 Definition. $\{x_i\}$, a sequence in $\mathfrak{F}$ a normed space, is said to *converge weakly* to $x \in \mathfrak{F}$ if $u(x_i) \to u(x)$ for every $u \in \mathfrak{F}^*$, the dual space of $\mathfrak{F}$ (see definition (3.2)). A subset A is said to be *weakly sequentially compact*, if every sequence in A contains a subsequence which converges weakly to a point in A.

3.17 Theorem. *A weakly convergent sequence $\{x_i\}$ in a normed space $\mathfrak{F}$ has a unique limit x, is bounded, and*

$$\|x\| \leq \liminf_{i \to \infty} \|x_i\| \tag{6}$$

3.1. Banach's Theorem

3.18 Theorem. *A Banach space $\mathfrak{B}$ is reflexive, if and only if its closed unit ball $\bar{B}_1(0)$ is weakly sequentially compact.*

Particular case. In a Hilbert space, a bounded subset is weakly sequentially compact.

3.19 Definition. Let $\mathfrak{F}$ and $\mathfrak{G}$ be normed spaces and $\Omega \subset \mathfrak{F}$. A map $f: \Omega \to \mathfrak{G}$ (not necessarily linear) is said to be compact if f is continuous and maps bounded subsets of $\mathfrak{U}$ into precompact subsets of $\mathfrak{G}$.

3.20 Schauder fixed point theorem. *A compact mapping, f, of a closed bounded convex set Ω in a Banach space $\mathfrak{B}$ into itself has a fixed point.*

3.2. The Leray–Schauder Theorem

3.21 *Let T be a compact mapping of a Banach space $\mathfrak{B}$ into itself, and suppose there exists a constant M such that $\|x\| \leq M$ for all $x \in \mathfrak{B}$ and $\sigma \in [0, 1]$ satisfying $x = \sigma T x$. Then T has a fixed point.*

3.22 Definition. Let $\mathfrak{F}$ be a normed space on $\mathbb{C}$ and $T \in \mathscr{L}(\mathfrak{F}, \mathfrak{F})$. A number $\lambda \in \mathbb{C}$ is called an *eigenvalue* of T if there exists a non-zero element x in $\mathfrak{F}$ (called an *eigenvector*) satisfying $Tx = \lambda x$. The dimension of the null space of the operator $\lambda I - T$ is called the multiplicity of λ.

3.23 Theorem. *The eigenvalues of a compact linear mapping T of a normed space $\mathfrak{F}$ into itself form either a finite set, or a countable sequence converging to 0. Each non-zero eigenvalue has finite multiplicity. If $\lambda \neq 0$ is not an eigenvalue, then for each $f \in \mathfrak{F}$ the equation $\lambda x - Tx = f$ has a uniquely determined solution $x \in \mathfrak{F}$ and the operator $(\lambda I - T)^{-1}$ is continuous.*

3.3. The Fredholm Theorem

3.24 *Let T be a compact linear operator in a Hilbert space $\mathfrak{H}$ and consider the equations:*

$$x - Tx = f, \tag{7}$$

$$y - T^*y = g, \tag{8}$$

where T^* is the adjoint operator of T, ($\langle Tx, y\rangle = \langle x, T^*y\rangle$ for any x and y in $\mathfrak{H}$). Then the following alternative holds:

(α) *either there exists a unique solution of (7) and (8) for any f and g in $\mathfrak{H}$, or*

(β) *the homogeneous equation $x - Tx = 0$ has non trivial solutions. In that case the dimension of the null space of $I - T$ is finite and equals the dimension of the null space $\mathcal{N}^*$ of $I - T^*$. Furthermore (7) has a solution (not unique of course) if and only if $\langle f, y\rangle = 0$ for every $y \in \mathcal{N}^*$.*

§4. The Lebesgue Integral

3.25 Definition. Let $\mathscr{H}$ be a locally compact Haussdorff space and $C_0(\mathscr{H})$ the space of real-valued continuous functions on $\mathscr{H}$ with compact support. A *positive Radon measure* μ is a linear functionnal on $C_0(\mathscr{H})$, $\mu: f \to \mu(f) \in \mathbb{R}$, such that $\mu(f) \geq 0$ for any $f \geq 0$.

3.26 Definition. Let μ be a positive Radon measure as above. We define an *upper integral* for the non-negative functions as follows. If $g \geq 0$ is lower semicontinuous:

$$\mu^*(g) = \sup \mu(f) \quad \text{for all } f \in C_0(\mathscr{H}) \quad \text{satisfying } f \leq g,$$

and for any functions $h \geq 0$: $\mu^*(h) = \inf \mu^*(g)$ for all lower semicontinuous functions g satisfying $h \leq g$.

3.27 Definition. Let μ be a positive Radon measure as above. A function f is said to be *μ-integrable*, if there exists a sequence $f_n \in C_0(\mathscr{H})$ such that $\mu^*(|f - f_n|) \to 0$. Then $\{\mu(f_n)\}$ converges, and we set $\int f\, d\mu = \lim_{n\to\infty} \mu(f_n)$.

3.28 Definition. A set $A \subset \mathscr{H}$ is *measurable* and with finite measure $\mu(A)$, if its characteristic function χ_A is integrable (by definition $\chi_A(x) = 1$ for $x \in A$, $\chi_A(x) = 0$ for $x \notin A$). We set $\mu(A) = \int \chi_A\, d\mu$. We say that a property holds almost everywhere on $\mathscr{H}$ if it holds for all $x \in \mathscr{H}$ except on a set A of measure zero. *Almost all points* means all points, except possibly those of a set with measure zero.

3.29 Remark. By Definitions 3.26 and 3.27, a non-negative lower semicontinuous function g, with $\mu^*(g)$ finite, is integrable and $\mu^*(g) = \int g\, d\mu$. One can prove that an integrable function f is equal almost everywhere to $g_1 - g_2$, with g_1 and g_2 non-negative lower semicontinuous integrable functions.
A compact subset $\mathfrak{K} \subset \mathscr{H}$ is measurable and with finite measure. An open subset Ω is measurable (χ_Ω is lower semicontinuous).

3.30 Definition. f is said *μ-measurable* (or measurable, when there is no ambiguity) if for all compact sets $\mathfrak{K}$ and all $\varepsilon > 0$, there exists a compact set $\mathfrak{K}_\varepsilon \subset \mathfrak{K}$, such that $\mu(\mathfrak{K} - \mathfrak{K}_\varepsilon) < \varepsilon$ and such that the restriction $f/\mathfrak{K}_\varepsilon$ is continuous on $\mathfrak{K}_\varepsilon$.

If a sequence of μ-measurable functions converges almost everywhere, then the limit function is μ-measurable.

3.31 Remark. For convenience, we develop the theory for real-valued functions; however, the theory of the Lebesgue integral is similar for the functions $\vec{f}$ on $\mathscr{H}$ with values in $\mathfrak{B}$, a Banach space.

Given μ, first we define $\int \vec{f}\, d\mu$ for continuous functions $\vec{f}$ with compact support. Let $\{\vec{f}_k\}$ be a sequence of simple functions ($\vec{f}_k = \sum_{j=1}^k \vec{a}_{kj}\chi_{A_j}$, with $\vec{a}_{kj} \in \mathfrak{B}$ and A_j measurable sets), which converges uniformly to $\vec{f}$. By definition $\int \vec{f}\, d\mu = \lim_{k\to\infty} \sum_{j=1}^k \mu(A_j)\vec{a}_{kj}$.
Then we define the integrable functions $\vec{g}$, as in definition 3.27. $\vec{g}$ is μ-integrable if there exists a sequence of continuous functions $\vec{f}_n$ with compact support, such that $\mu^*(\|\vec{g} - \vec{f}_n\|) \to 0$.

4.1. Dominated Convergence Theorem

3.32 The first Lebesgue theorem. *Given μ a positive Radon measure on $\mathscr{H}$ a locally compact Haussdorff space, let $\{\vec{f}_n\}$ be a sequence of μ-integrable functions, ($\vec{f}_n : \mathscr{H} \to \mathfrak{B}$ a Banach space) which converges almost everywhere to $\vec{f}$. Then $\vec{f}$ is integrable, $\int \vec{f}_n\, d\mu \to \int \vec{f}\, d\mu$ and $\int \|\vec{f} - \vec{f}_n\|\, d\mu \to 0$, if there exists h,* *a non-negative function, satisfying* $\mu^*(h) < \infty$ *and* $\|\vec{f}_n(x)\| \leq h(x)$ *for all n and* $x \in \mathscr{H}$.

Recall $\mu^*(h)$ is finite, in particular, if h is integrable.

4.2. Fatou's Theorem

3.33 *Given* μ *as above. Let* $\{h_i\}$ *be an increasing sequence of non-negative functions on* $\mathscr{H}$, $0 \leq h_1 \leq \cdots \leq h_i \leq h_{i+1} \leq \cdots$. *Then*

$$\lim_{i \to \infty} \mu^*(h_i) = \mu^*\left(\lim_{i \to \infty} h_i\right).$$

3.34 Theorem. *From now on, we suppose that* $\mathscr{H}$, *the locally compact Haussdorff space, is a denumerable union of compact sets, and that the Banach space* $\mathfrak{B}$ *is separable. A measurable function* $\vec{f}\colon \mathscr{H} \to \mathfrak{B}$ *is* μ*-integrable if and only if* $\mu^*(\|\vec{f}\|)$ *is finite.*

4.3. The Second Lebesgue Theorem

3.35 *Let f be a real-valued function defined on* $[a, b] \subset \mathbb{R}$. *If f is integrable on* $[a, b]$ *with the Lebesgue measure, then* $F(x) = \int_a^x f(t)\,dt$ *has a derivative almost everywhere, and almost everywhere* $F'(x) = f(x)$, *for* $a \leq x \leq b$.

The Lebesgue measure on $\mathbb{R}^n$ corresponds to the positive Radon measure, defined by the Jordan integral of the continuous functions with compact support.

3.36 Theorem. *If a function* $F(x)$ *is absolutely continuous on* $[a, b]$, *then there exists* $f(x)$, *an integrable function on* $[a, b]$, *such that* $F(x) - F(a) = \int_a^x f(t)\,dt$, *and conversely. Also* $F(x)$ *has a derivative almost everywhere, which is* $f(x)$.

Recall that $F(x)$ is said to be absolutely continuous on an interval I if, for each $\varepsilon > 0$, there exists a $\delta > 0$, such that $\sum_{i=1}^k |f(y_i) - f(x_i)| < \varepsilon$, whenever $]x_i, y_i[$, $i = 1, 2, \ldots, k$, are nonoverlapping subintervals of I, satisfying $\sum_{i=1}^k |y_i - x_i| < \delta$. In particular, a Lipschitzian function is absolutely continuous.

4.4. Rademacher's Theorem

3.37 *A Lipschitzian function from an open set of* $\mathbb{R}^n$ *to* $\mathbb{R}^p$ *is differentiable almost everywhere.*

4.5. Fubini's Theorem

3.38 *Let $\mathscr{H}$ and $\mathscr{F}$ be locally compact separable metric spaces. Given two positive Radon measures μ on $\mathscr{H}$ and ν on $\mathscr{F}$, if $f(x, y)$: $\mathscr{H} \times \mathscr{F} \to \mathbb{R}$ is $\mu \otimes \nu$-integrable, then for ν-almost all y, $f_y(x) = f(x, y)$ is μ-integrable and for μ-almost all x, $f_x(y) = f(x, y)$ is ν-integrable. Moreover*

$$(9)\qquad \begin{aligned}\iint f(x, y)\, d\mu(x)\, d\nu(y) &= \int\left[\int f(x, y)\, d\nu(y)\right] d\mu(x)\\ &= \int\left[\int f(x, y)\, d\mu(x)\right] d\nu(y).\end{aligned}$$

Fubini's Theorem is very useful, but for most applications, we don't know that $f(x, y)$ is $\mu \otimes \nu$-integrable.
We overcome this difficulty as follows: More often than not, it is obvious that $f(x, y)$ is $\mu \otimes \nu$-measurable. (Is it not a recognized fact, that function is measurable, when it is defined without using the axiom of the choice!)

Then by using Theorems (3.34) and (3.39), we shall know if $f(x, y)$ is $\mu \otimes \nu$-integrable or not. Recall that a locally compact metric space is separable if and only if it is a denumerable union of compact sets.

3.39 Theorem. *Let $(\mathscr{H}, \mu)$ and $(\mathscr{F}, \nu)$ be as above, and $f(x, y)$ a $\mu \otimes \nu$-measurable function. Then*

$$(\mu \otimes \nu)^*(|f|) = \mu^*[\nu^*(|f_x|)] = \nu^*[\mu^*(|f_y|)].$$

§5. The L_p Spaces

3.40 Definition. Let $\mathscr{H}$ be a locally compact separable metric space and μ a positive Radon measure. Given $p \geq 1$ a real number, we denote by $L_p(\mathscr{H})$ the class of all measurable functions f on $\mathscr{H}$ for which $\mu^*(|f|^p) < \infty$. We identify in $L_p(\mathscr{H})$ functions that are equal almost everywhere. The elements of $L_p(\mathscr{H})$ are equivalence classes under the relation: $f_1 \sim f_2$ if $f = f_2$ almost everywhere. $L_p(\mathscr{H})$, (denoted by L_p when no confusion is possible), is a separable Banach space, the norm being defined by:

$$(10)\qquad \|f\|_p = \left(\int |f|^p\, d\mu\right)^{1/p}.$$

The Banach space $L_\infty(\mathscr{H})$ consists of all μ-essentially bounded functions. The norm is:

$$(11)\qquad \|f\|_\infty = \mu\text{-ess}\sup|f(x)| = \inf_A \sup_{x \in \mathscr{H} - A} |f(x)|,$$

where A ranges over the subsets of measure zero.

3.41 Proposition. *$C_0(\mathscr{H})$ is dense in $L_p(\mathscr{H})$ for all $1 \leq p < \infty$.*

For $p = 1$, this is true by definition 3.27. The result is not true for L_∞, of course; otherwise every function belonging to L_∞ would be continuous. If $\mathscr{H}$ is an open set Ω of $\mathbb{R}^n$ and μ the Lebesgue measure, we have a more precise result, proved by regularization (see 3.46 below): $\mathscr{D}(\Omega)$ is dense in $L_p(\Omega)$. Here $\mathscr{D}(\Omega)$ is the set of C^∞-functions with compact support lying in Ω. Likewise, if (V, g) is a C^∞ Riemannian manifold and μ the Riemannian measure, $\mathscr{D}(V)$ is dense in $L_p(V)$. $\mathscr{D}(V)$ consists of the C^∞-functions on V, with compact support.

3.42 Proposition. *For $1 \leq p < \infty$, $L_p^*(\mathscr{H})$ is isometric isomorphic to $L_q(\mathscr{H})$ with $1/p + 1/q = 1$. Hence L_p is reflexive provided $1 < p < \infty$.*

The isomorphism: $L_q \ni g \to \mathfrak{u}_g \in L_p^*$ is defined as follows:

$$L_p \ni f \to \mathfrak{u}_g(f) = \int fg \, d\mu.$$

Indeed fg, which is μ-measurable, is μ-integrable according to Hölder's inequality 3.60:

$$\|fg\|_1 \leq \|f\|_p \|g\|_q. \tag{12}$$

3.43 Proposition. *Let $\{f_k\}$ be a sequence in L_p (or in L_∞) which converges in L_p to $f \in L_p$. Then there exists a subsequence converging pointwise almost everywhere to f.*

3.44 Theorem. *Let Ω be an open set of $\mathbb{R}^n$. A bounded subset $\mathscr{A} \subset L_p(\Omega)$ is precompact in $L_p(\Omega)$ if and only if for every number $\varepsilon > 0$, there exists a number $\delta > 0$ and a compact set $\mathfrak{K} \subset \Omega$, such that for every $f \in \mathscr{A}$:*

$$\int_{\Omega - \mathfrak{K}} |f(x)|^p \, dx < \varepsilon$$

and

$$\int_{\mathfrak{K}} |f(x + y) - f(x)|^p \, dx < \varepsilon \quad \text{when } \|y\| \leq \delta,$$

where without loss of generality, we suppose that δ is smaller than the distance from $\mathfrak{K}$ to $\partial\Omega$, the boundary of Ω.

3.45 Theorem. *Let $1 < p < \infty$ and $\{f_k\}$ be a bounded sequence in $L_p(\mathscr{H})$, converging pointwise almost everywhere to f. Then f belongs to L_p and f_k converges to f weakly in L_p. The result does not hold for $p = 1$, of course (see below).*

5.1. Regularization

3.46 Let $\gamma \in \mathscr{D}(\mathbb{R}^n)$ be a non-negative function, whose integral equals 1. For convenience, we suppose $\operatorname{supp} \gamma \subset \bar{B}$.

For $k \in \mathbb{N}$, consider the function $\gamma_k(x) = k^n\gamma(kx)$, $\|\gamma_k\|_1 = 1$, and $\gamma_k(x) \to 0$ almost everywhere (except for $x = 0$) when $k \to \infty$. In fact $\{\gamma_k\}$ converges vaguely to the *Dirac measure* concentrated at zero. Let f be a function locally integrable on $\mathbb{R}^n$ (this means that $f\chi_{\mathfrak{K}}$ is integrable for any compact $\mathfrak{K} \subset \mathbb{R}^n$); we define the *regularization* of f by:

$$f_k(x) = (\gamma_k * f)(x) = \int \gamma_k(x - y) f(y)\, dy. \tag{13}$$

Obviously f_k is C^∞. Moreover, if $f \in L_p(\mathbb{R}^n)$, then $\{f_k\}$ converges to f in $L_p(1 \leq p < \infty)$, and

$$\|f_k\|_p \leq \|f\|_p. \tag{14}$$

When $f \in C_0(\mathbb{R}^n)$, the result is obvious, since f_k converges uniformly to f:

$$|f_k(x) - f(x)| = \left| \int \gamma_k(x - y)[f(y) - f(x)]\, dy \right| \leq \sup_{\|y - x\| < 1/k} |f(y) - f(x)|,$$

so the assertion follows from the uniform continuity of f. Clearly $\|f_k - f\|_p \to 0$ when $k \to \infty$. If $\tilde{f} \in L_p(\mathbb{R}^n)$, for each $\varepsilon > 0$, there exists $f \in C_0(\mathbb{R}^n)$ such that $\|\tilde{f} - f\|_p < \varepsilon$ (Proposition 3.41). But

$$\|\tilde{f}_k - \tilde{f}\|_p \leq \|\tilde{f}_k - f_k\|_p + \|f_k - f\|_p + \|f - \tilde{f}\|_p \leq 2\varepsilon + \|f_k - f\|_p,$$

where we used inequality (14), which we are going to prove now. According to Hölder's inequality (21), q being defined by $1/p + 1/q = 1$:

$$\begin{aligned} |(\gamma_k * f)(x)| &\leq \int [\gamma_k(x - y)]^{1/p + 1/q} |f(y)|\, dy \\ &\leq \left[\int \gamma_k(x - y)\, dy\right]^{1/q} \left[\int \gamma_k(x - y)|f(y)|^p\, dy\right]^{1/p} \\ &= \left[\int \gamma_k(x - y)|f(y)|^p\, dy\right]^{1/p}. \end{aligned}$$

Hence by Fubini's theorem, 3.38:

$$\int |\gamma_k * f|^p\, dx \leq \int |f|^p\, dy \int \gamma_k(x - y)\, dx = \|f\|_p^p.$$

5.2. Radon's Theorem

3.47 Theorem. *Let $\mathfrak{B}$ be a uniformly convex Banach space and $\{f_k\}$ a sequence in $\mathfrak{B}$ which converges weakly to $f \in \mathfrak{B}$.*

$$\textit{If } \|f_k\| \to \|f\|, \textit{ then } f_k \to f \textit{ strongly } (\|f - f_k\| \to 0) \textit{ as } k \to \infty.$$

Recall that $\mathfrak{B}$ is said to be uniformly convex, if $\|g_k\| = \|h_k\| = 1$ and $\|g_k + h_k\| \to 2$ implies $\|h_k - g_k\| \to 0$, when $k \to \infty$, for sequences $\{g_k\}$ and $\{h_k\}$ in $\mathfrak{B}$.

A uniformly convex Banach space is reflexive; the converse is not true. The spaces $L_p(1 < p < \infty)$ are uniformly convex. This result is due to Clarkson as a consequence of his inequality 3.63 below.
It is obvious that a Hilbert space is uniformly convex.

Proof. If $f = 0$, we have nothing to prove: $\|f_k\| \to 0$.
If $f \neq 0$, we can suppose without loss of generality that $\|f\| = 1$ and that $\|f_k\| \neq 0$ for all k. Set $g_k = \|f_k\|^{-1}f_k$ and $h_k = f$ for all k.
By the Hahn–Banach theorem 3.12, there exists $u_0 \in \mathfrak{B}^*$ such that $u_0(f) = 1$ and $\|u_0\| = 1$. Since $f_k \to f$ weakly, we have as $k \to \infty$:

$$u_0(g_k) = \|f_k\|^{-1}u_0(f_k) \to u_0(f) = 1.$$

Using $\|u_0\| = 1$ we get:

$$|1 + u_0(g_k)| \leq \|g_k + f\| \leq 2.$$

Letting $k \to \infty$, we obtain $\|g_k + f\| \to 2$. Thus $\|g_k - f\| \to 0$ and $\|f - f_k\| \to 0$, since $\|f - f_k\| \leq \|f - g_k\| + |1 - \|f_k\||$ ∎

3.48 Definition. Let $\mathfrak{u}$ be a locally integrable function in Ω, an open set of $\mathbb{R}^n$. A locally integrable function $\mathfrak{h}$ is called the *weak derivative* of $\mathfrak{u}$ with respect to x^1 if it satisfies

$$\int_\Omega \varphi\mathfrak{h}\, dV = -\int_\Omega \mathfrak{u}\, \partial_1\, \varphi\, dV \quad \text{for all } \varphi \in \mathscr{D}(\Omega).$$

By induction we define the weak derivative of $\mathfrak{u}$ of any order if it exists.

Proposition 3.48. *Let f be a Lipschitzian function on a bounded open set $\Omega \subset \mathbb{R}^n$. Then $\partial_i f$ exists almost everywhere, belongs to $L_p(\Omega)$, and coincides with the weak derivative in the sense of the distributions.*

Proof. According to Rademacher's Theorem (3.37), $\partial_i f$ exists almost everywhere, since f is a Lipschitzian function. Moreover, $\partial_i f$ is bounded almost everywhere, since $|f(x) - f(y)| \leq k\|x - y\|$ implies $|\partial_i f| \leq k$, when $\partial_i f$ exists.

On any line segment in Ω, f is absolutely continuous. Thus by Theorem (3.36), $\partial_i f$ defines the weak derivative with respect to x^i. $\partial_i f$ is the limit function, almost everywhere, of a sequence of μ-measurable functions; hence it is measurable (see definition 3.30).
Since $|\partial_i f|^p \leq k^p$, $\mu^*_{\text{on}\,\Omega}(|\partial_i f|^p) \leq k^p \mu(\Omega) < \infty$. Consequently $\partial_i f \in L_p(\Omega)$, according to theorem (3.34). ■

3.49 Proposition. *Let M_n be a C^∞ Riemannian manifold and $\varphi \in H_1^p(M)$; then almost everywhere $|\nabla|\varphi|| = |\nabla\varphi|$.*

Proof. Since a Riemannian manifold is covered by a countable set of balls, it is sufficient to establish the result for a ball B of $\mathbb{R}^n$, provided with a Riemannian metric. Denote the coordinates by $\{x^i\}$.
First of all we are going to prove the statement for C^∞ functions. Let $f \in C^\infty(B)$. When $f(x) > 0$, $\partial_i f = \partial_i |f|$, and when $f(x) < 0$, $\partial_i f = -\partial_i |f|$. Thus the result will follow, once we have shown that the set $\mathscr{A}$ of the points $x \in B$, where simultaneously $f(x) = 0$ and $|\nabla f(x)| \neq 0$, has zero measure. Since, for $x \in \mathscr{A}$, $|\nabla f(x)| \neq 0$, there exists a neighborhood Θ_x of x, such that $\Theta_x \cap f^{-1}(0)$ is a submanifold of dimension $n-1$ (see for instance Choquet-Bruhat [99] p. 12). Consequently, $\mu[\Theta_x \cap f^{-1}(0)] = 0$. As there exists a countable basis of open sets for the topology of B, $\mathscr{A}$ is covered by a countable set of $\mathscr{V}_x(x \in \mathscr{A})$. Let $\{\mathscr{V}_{x_k}\}_{k \in \mathbb{N}}$ be this set. We have $\mathscr{A} \subset \bigcup_{k=1}^\infty \mathscr{V}_{x_k}$ and $\mu[\mathscr{V}_{x_k} \cap \mathscr{A}] = 0$; thus $\mu(\mathscr{A}) = 0$.

Now let $\varphi \in H_1^p(B)$. By definition, $C^\infty(B) \cap H_1^p(B)$ is dense in $H_1^p(B)$. So using Proposition 3.43, there exists $\{f_j\}$, a sequence of C^∞ functions on B, such that $\|f_j - \varphi\|_{H_1^p} \to 0$ and such that $f_j \to \varphi$ a.e. and $\partial_i f_j \to h_i$ a.e. for all $i(1 \leq i \leq n)$ as $j \to \infty$, where h_i denotes the weak derivatives of φ.
Define for a function f, $f^+ = \sup(f, 0)$ and $f^- = \sup(-f, 0)$. Obviously $f_j^+ \to \varphi^+$ a.e. and $f_j^- \to \varphi^-$ a.e.
Moreover f_j^+ is a Cauchy sequence in H_1^p, which converges to an element of H_1^p, which is φ^+, since $\|f_j^+ - \varphi^+\|_p \to 0$. Thus $\varphi^+ \in H_1^p$ likewise $\varphi^- \in H_1^p$. According to Proposition 3.43, taking a subsequence $\{f_k\}$ of $\{f_j\}$, if necessary, we have for all $i(1 \leq i \leq n)$:
$\partial_i f_k^+ \to h_i^+$ a.e. and $\partial_i f_k^- \to h_i^-$ a.e. as $k \to \infty$, where h_i^+ (respectively, h_i^-) denote the weak derivatives of φ^+ (respectively φ^-).
Since we have proved that almost everywhere $|\partial_i f_k| = |\partial_i |f_k||$, letting $k \to \infty$ yields $|h_i| = |h_i^+ + h_i^-|$ a.e. Likewise, $\partial_{i_1} f_k \, \partial_{i_2} f_k = \partial_{i_1}|f_k| \, \partial_{i_2}|f_k|$ a.e. gives $h_{i_1} h_{i_2} = (h_{i_1}^+ + h_{i_1}^-)(h_{i_2}^+ + h_{i_2}^-)$ a.e. Hence $|\nabla\varphi| = |\nabla|\varphi||$ a.e. ■

3.50 Proposition. *Let $\overline{W}_n$ be a compact Riemannian manifold with boundary of class C^r. If $f \in C^{k-1}(\overline{W}) \cap \mathring{H}_k^p(W)$ $(1 \leq k \leq r)$, then f and its derivatives up to order $k-1$ vanish on the boundary ∂W.*

Proof. Let $\{\Omega_i, \varphi_i\}$ be an atlas of class C^r such that $\varphi_i(\Omega_i)$ is either the ball $B \subset \mathbb{R}^n$ or the half ball $D = B \cap \bar{E}$. Consider a C^∞ partition of unity $\{\alpha_i\}$ subordinate to the covering $\{\Omega_i\}$.

Choose any point P of ∂W. At least one of the α_i does not vanish at P. Let $\alpha_{i_0}(P) > 0$ and set $\tilde{f} = (\alpha_{i_0} f) \circ \varphi_{i_0}^{-1}$. $\tilde{f} \in C^{k-1}(\bar{D}) \cap \overset{\circ}{H}{}_k^p(D)$. For all $\mathfrak{u}$ and $\mathfrak{v}$ belonging to $C^1(\bar{D})$

$$\int_D \mathfrak{v}\, \partial_1 \mathfrak{u}\, dx + \int_D \mathfrak{u}\, \partial_1 \mathfrak{v}\, dx = \int_{\partial D} \mathfrak{u}\mathfrak{v}(\vec{e}_1, \vec{v})\, d\sigma, \tag{15}$$

where $\vec{v}$ is the outer normal and $\vec{e}_1$ the first unit vector of the basis of $\mathbb{R}^n$.

If $\mathfrak{u} \in \mathscr{D}(D)$, the right side vanishes. Hence, by density, the left side of (15) is zero for all $\mathfrak{u} \in \overset{\circ}{H}{}_1^p(D)$. In particular, if $\tilde{\mathfrak{u}}$ is $\tilde{f}$ or one of its derivatives up to order $k-1$, for all $\mathfrak{v} \in C^1(\bar{D})$:

$$\int_{\partial D} \tilde{\mathfrak{u}}\mathfrak{v}(\vec{e}_1, \vec{v})\, d\sigma = \int_D \mathfrak{v}\, \partial_1 \tilde{\mathfrak{u}}\, dx + \int_D \tilde{\mathfrak{u}}\, \partial_1 \mathfrak{v}\, dx = 0.$$

Consequently $\tilde{\mathfrak{u}}$ vanishes on ∂D, and f, as well as its derivatives up to order $k-1$, vanishes at P.

Remark. It is easy to prove the converse when $f \in C^k(\overline{W})$: If f and its derivatives up to order $k-1$ vanish on ∂W, then $f \in \overset{\circ}{H}{}_k^p(W)$.

§6. Elliptic Differential Operators

3.51 Definition. Let M_n be a Riemannian manifold. A *linear differential operator* $A(\mathfrak{u})$ of order $2m$ on M_n, written in a local chart (Ω, φ), is an expression of the form:

$$A(\mathfrak{u}) = \sum_{\ell=0}^{2m} a_\ell^{\alpha_1\alpha_2\cdots\alpha_\ell} \nabla_{\alpha_1\alpha_2\cdots\alpha_\ell} \mathfrak{u}, \tag{16}$$

where a_ℓ are ℓ-tensors and $\mathfrak{u} \in C^{2m}(M)$. For simplicity, we can write $A(\mathfrak{u}) = a_\ell \nabla^\ell \mathfrak{u}$. The terms of highest order, $2m$, are called the *leading part* (a_{2m} is presumed to be nonzero).

The operator is said to be *elliptic* at a point $x \in \Omega$, if there exists $\lambda(x) \geq 1$ such that, for all vectors ξ:

$$\|\xi\|^{2m} \lambda^{-1}(x) \leq a_{2m}^{\alpha_1\alpha_2\cdots\alpha_{2m}}(x) \xi_{\alpha_1} \cdots \xi_{\alpha_{2m}} \leq \lambda(x) \|\xi\|^{2m}. \tag{17}$$

We say that the operator is *uniformly elliptic* in Ω, if there exist λ_0 and $\lambda(x)$, $1 \leq \lambda(x) \leq \lambda_0$, such that (17) holds for all $x \in \Omega$.

3.52 Definition. A *differential operator* A of order $2m$ defined on M_n is written $A(\mathfrak{u}) = f(x, \mathfrak{u}, \nabla \mathfrak{u}, \ldots, \nabla^{2m} \mathfrak{u})$, where f is assumed to be a differentiable function of its arguments. Then the first variation of f at $\mathfrak{u}_0 \in C^{2m}(\Omega)$ is the linear differential operator

$$(18)\quad \begin{aligned} A'_{\mathfrak{u}_0}(\mathfrak{v}) &= \sum_{\ell=0}^{2m} \frac{\partial f(x, \mathfrak{u}_0, \nabla\mathfrak{u}_0, \ldots, \nabla^{2m}\mathfrak{u}_0)}{\partial \nabla_{\alpha_1\alpha_2\cdots\alpha_\ell}\mathfrak{u}} \nabla_{\alpha_1\alpha_2\cdots\alpha_\ell}\mathfrak{v} \\ &= \frac{\partial f}{\partial \nabla^\ell \mathfrak{u}}(x, \mathfrak{u}_0, \nabla\mathfrak{u}_0, \ldots, \nabla^{2m}\mathfrak{u}_0)\nabla^\ell \mathfrak{v}. \end{aligned}$$

If $A'_{\mathfrak{u}_0}$ is an elliptic operator, then A is called elliptic at $\mathfrak{u}_0$.

6.1. Weak Solution

3.53 a) Let A be a linear differential operator of order $2m$ defined on a Riemannian manifold M, with or without boundary. Until now, by a solution of the equation $A(u) = f$ we meant a function $u \in C^{2m}(M)$ such that the equation is satisfied pointwise. There are other quite natural ways that a more general function, such as an element of $H^p_{2m}(M)$ or a distribution, can be said to be a solution of $A(u) = f$.

b) If $f \in L_p$ and if the coefficients of A are measurable and locally bounded, we say that $\mathfrak{u} \in H^p_{2m}(M)$ is a strong solution in the L_p sense of $A(\mathfrak{u}) = f$ if there is a sequence $\{\varphi_i\}$ of C^∞ functions on M such that $\varphi_i \to \mathfrak{u}$ in $H^p_{2m}(M)$ and $A(\varphi_i) \to f$ in $L^p(M)$. Indeed, in this case the weak (distribution) derivatives up to order $2m$ are functions in $L^p(M)$ and $A(\mathfrak{u}) = f$ almost everywhere.

c) Let $A(u) = a_\ell \nabla^\ell u$. If the tensors $a_\ell \in C^\ell(M)$ for $0 \le \ell \le 2m$, then we define the *formal adjoint* of A by

$$A^*(\varphi) = (-1)^\ell \nabla^\ell(\varphi a_\ell).$$

We say that $\mathfrak{u} \in L_1(M)$ satisfies $A(\mathfrak{u}) = f$, in the sense of distributions, if for all $\varphi \in \mathscr{D}(M)$:

$$\int_M \mathfrak{u} A^*(\varphi)\, dV = \int_M f\varphi\, dV.$$

If the coefficients $a_\ell \in C^\infty(M)$, then a distribution $\mathfrak{u}$ satisfies $A(\mathfrak{u}) = f$ if for all $\varphi \in \mathscr{D}(M)$:

$$\langle \mathfrak{u}, A^*(\varphi)\rangle = \langle f, \varphi\rangle.$$

Now a given distribution is some weak derivative of some order, say r, of a locally integrable function. In this case $\langle \mathfrak{u}, A^*(\varphi)\rangle$ makes sense if the coefficients satisfy $a_\ell \in C^{\ell+r}(M)$.

d) If the operator can be written in divergence form, i.e., if we can write $A(u)$ as

$$A(u) \equiv \sum_{\substack{0 \le k \le m \\ 0 \le \ell \le m}} \nabla_{\alpha_1 \cdots \alpha_k}(a_{k,\ell}^{\alpha_1 \cdots \alpha_k \beta_1 \cdots \beta_\ell} \nabla_{\beta_1 \cdots \beta_\ell} u) + \sum_{\ell=0}^{m} b_\ell \nabla^\ell u$$

where $a_{k,\ell}$ are $k + \ell$-tensors and b_ℓ ℓ-tensors, then $u \in H_m^p(M)$ is said to be a weak solution of $A(u) = f$ with $f \in L_1(M)$ if for all $\varphi \in \mathscr{D}(M)$:

$$\sum_{\substack{0 \le k \le m \\ 0 \le \ell \le m}} (-1)^k \int_M a_{k,\ell} \nabla^\ell u \nabla^k \varphi \, dV + \sum_{\ell=0}^{m} \int_M \varphi b_\ell \nabla^\ell u \, dV = \int_M f\varphi \, dV.$$

Here we need only suppose that $a_{k,\ell}$ are measurable and are locally bounded for all pairs (k, ℓ). Note these definitions of weak or generalized solutions are not equivalent; they depend on the properties of the coefficients. But the terminology is standard and does not really cause confusion.

e) For a nonlinear differential operator of the type

$$A(u) = \sum_{\ell=0}^{m} (-1)^\ell \nabla_{\alpha_1 \cdots \alpha_\ell} A_\ell^{\alpha_1 \cdots \alpha_\ell}(x, u, \nabla u, \ldots, \nabla^m u) = (-1)^\ell \nabla^\ell A_\ell,$$

where A_ℓ are ℓ-tensors on M, $u \in C^m(M)$ is said to be a weak solution of $A(u) = 0$, if for all $\varphi \in \mathscr{D}(M)$:

$$\sum_{\ell=0}^{m} \int A_\ell^{\alpha_1 \cdots \alpha_\ell}(x, u, \nabla u, \ldots, \nabla^m u) \nabla_{\alpha_1 \cdots \alpha_\ell} \varphi \, dV = 0.$$

6.2. Regularity Theorems

3.54 And now some theorems concerning the regularity of (weak) solutions in the interior, then on the boundary, of the manifold. Roughly speaking, we can hope that if we can define in some sense, for some function u, $A(u) = f$, then u will have the maximum of regularity allowed by the coefficients. This is almost the case. Precisely, we have:

Theorem. *Let Ω be an open set of $\mathbb{R}^n$ and $A = a_\ell \nabla^\ell$ a linear elliptic operator of order $2m$ with C^∞ coefficients ($a_\ell \in C^\infty(\Omega)$ for $0 \le \ell \le 2m$). Suppose u is a distribution solution of the equation $A(u) = f$ and $f \in C^{k,\alpha}(\Omega)$ (resp., $C^\infty(\Omega)$). Then $u \in C^{k+2m,\alpha}(\Omega)$, (resp., $C^\infty(\Omega)$) with $0 < \alpha < 1$.*

If f belongs to $H_k^p(\Omega)$, $1 < p < \infty$, then u belongs locally to H_{k+2m}^p.

Proof. Although it is basic, I did not find exactly this theorem in the literature. First of all, if $f \in C^\infty(\Omega)$, $u \in C^\infty(\Omega)$; this is the well-known result of Schwartz

(250). On the other hand, if $f \in H_k^2(\Omega)$ $(k \geq 0)$, then $\mathfrak{u} \in H_{k+2m}^2$ locally according to Bers, John and Schechter (50) p. 190. Now if $f \in C^{k,\alpha}(\Omega)$, obviously f belongs locally to H_k^2. Therefore $\mathfrak{u} \in H_{k+2m}^2$ locally and according to Morrey (204) p. 246, $\mathfrak{u} \in C^{2m+k,\alpha}(\Omega)$. Finally, let us establish the result when $f \in H_k^p(\Omega)$ $p \neq 2$. There exist continuous $\mathfrak{v}$ and integer $r \geq 0$ such that $\mathfrak{u} = \Delta^r \mathfrak{v}$ (Bers, John and Schechter (50) p. 195). But $A\Delta^r$ is elliptic of order $2(m + r)$ and $\mathfrak{v}$ belongs locally to L_2. Thus if $f \in H_k^p(\Omega)$ $(1 < p < \infty)$, then $\mathfrak{v}$ belongs locally to $H_{k+2(m+r)}^p$ and $\mathfrak{u} \in H_{k+2m}^p$ locally, according to Morrey (204) p. 246. ∎

3.55 Theorem (Ladyzenskaja and Uralceva [173] p. 195). *Let Ω be an open set of $\mathbb{R}^n$ and A a linear elliptic operator of order two, with $C^{k,\alpha}$ coefficients ($k \geq 0$ an integer, $0 < \alpha < 1$). If a bounded function $\mathfrak{u} \in \mathsf{H}_2(\Omega)$ satisfies $A(\mathfrak{u}) = f$ almost everywhere, where $f \in C^{k,\alpha}(\Omega)$, then $\mathfrak{u} \in C^{k+2,\alpha}(\Omega)$. The same conclusion holds if $\mathfrak{u} \in \mathsf{H}_1(\Omega)$ is a weak solution of $A(\mathfrak{u}) = f$.*

For this last statement, the operator must be written in divergence form. In general this requires $a^{ij} \in C^1(\Omega)$. Then, according to Ladyzenskaja (173), $\mathfrak{u}$ is locally bounded (p. 199), and belongs locally to H_2 (p. 188). Thus we can apply the first part of theorem 3.55, on any bounded open set Θ with $\overline{\Theta} \subset \Omega$.

Remark. If the coefficients and f belong locally to H_q, with $q > 2 + n/2$, and if $\mathfrak{u} \in \mathsf{H}_{1,\text{loc}}$, then we can prove (see Aubin (20) p. 66) that $\mathfrak{u}$ belongs locally to H_{q+2}.

3.56 Theorem (Giraud [127], Hopf [146], and Nirenberg [216] and [217]). *Let $A(u) = F(x, u, \nabla u, \nabla^2 u)$ be a differential operator of order two, defined on Ω an open set of $\mathbf{R}^n$, F being a C^∞ differentiable function of its arguments. Suppose that A is elliptic on Ω at $u_0 \in C^2(\Omega)$, and that $A(u_0) = f \in C^{r,\beta}(\Omega)$ with $0 < \beta < 1$. Then $u_0 \in C^{r+2,\beta}(\Omega)$.*

Let Θ be a bounded subset of $C^2(\Omega)$, and suppose that A is uniformly elliptic on Ω at any $u \in \Theta$, uniformly in u (the same λ_0 is valid for all $u \in \Theta$, see definition 3.51). If $A(\Theta)$ is bounded in $C^{r,\beta}(K)$, then Θ is bounded in $C^{r+2,\beta}(K)$, for any compact set $K \subset \Omega$.

The result for $n = 2$ is due to Leray, and Nirenberg [217] established the theorem in the case $n > 2$, when there exists a modulus of continuity for the second derivatives of u_0. Previously Giraud [127] and Hopf [146] proved the result assuming that $u_0 \in C^{2,\alpha}(\Omega)$ for some $\alpha > 0$.

Remark. When A is a differential operator of order two on a compact Riemannian manifold M_n, it is possible to prove similar results: If $A(\Theta)$ is bounded in $\mathsf{H}_q(M_n)$ with $q > 2 + n/2$, then Θ is bounded in $\mathsf{H}_{q+2}(M_n)$, (see Aubin [20] p. 68).

3.57 Theorem (Agmon [2] p. 444). *Let Ω be a bounded open set of $\mathbb{R}^n$ with boundary of class C^{2m} and A be an elliptic linear differential operator of order $2m$ with coefficients $a_\ell \in C^\ell(\overline{\Omega})$. Let $u \in L_q(\Omega)$ for some $q > 1$, and $f \in L_p(\Omega)$, $p > 1$. Suppose that for all functions $v \in C^{2m}(\overline{\Omega}) \cap \mathring{H}^p_m(\Omega)$,*

$$\int uA(v)\,dV = \int fv\,dV.$$

Then $u \in H^p_{2m}(\Omega) \cap \mathring{H}^p_m(\Omega)$ and

$$\|u\|_{H^p_{2m}} \leq C[\|f\|_p + \|u\|_p],$$

where C is a constant depending only on Ω, A, n, and p.

Moreover, if $p > n/(m+1)$ then $u \in C^{m-1}(\overline{\Omega})$ and u is a solution of the Dirichlet problem

$$A^*u = f \quad \text{in } \Omega, \nabla^\ell u = 0 \quad \text{on } \partial\Omega, 0 \leq \ell \leq m-1$$

in the strong L_p sense.

3.58 Theorem (Gilbarg and Trüdinger [125] p. 177). *Let Ω be a bounded open set of $\mathbb{R}^n$ with C^{k+2} boundary ($k \geq 0$) and A a linear elliptic operator of order two, such that $a_2 \in C^{k+1}(\overline{\Omega})$ and $a_1, a_0 \in C^k(\overline{\Omega})$.*
Suppose $u \in \mathring{\mathsf{H}}_1(\Omega)$ is a weak solution of $A(u) = f$, with $f \in \mathsf{H}_k(\Omega)$. Then $u \in \mathsf{H}_{k+2}(\Omega)$ and

$$\|u\|_{\mathsf{H}_{k+2}} \leq C(\|u\|_2 + \|f\|_{\mathsf{H}_k}), \tag{19}$$

where the constant C is independent of u and f.
Thus, if the coefficients and f belong to $C^\infty(\overline{\Omega})$ and if the boundary is C^∞, then $u \in C^\infty(\overline{\Omega})$.

3.59 Theorem (Gilbarg and Trüdinger [125] p. 106). *Let Ω be a bounded open set of $\mathbb{R}^n$ with $C^{k+2,\alpha}$ boundary and let A be a linear elliptic operator of order two, with coefficients belonging to $C^{k,\alpha}(\overline{\Omega})$ ($k \geq 0$ an integer and $0 < \alpha < 1$). Suppose $u \in C^0(\overline{\Omega}) \cap C^2(\Omega)$ is a solution of the Dirichlet problem $A(u) = f$ in Ω, $u = v$ on $\partial\Omega$, with $f \in C^{k,\alpha}(\overline{\Omega})$ and $v \in C^{k+2,\alpha}(\overline{\Omega})$. Then $u \in C^{k+2,\alpha}(\overline{\Omega})$.*

Now let us prove a result which will be used in Chapter 8.

Proposition 3.59. *Let Ω be a bounded open set of $\mathbb{R}^n$ with C^∞ boundary and let $A(u) = F(x, u, \nabla u, \nabla^2 u)$ be a differential operator of order two, defined on Ω, F being a C^∞ differentiable function of its arguments on $\overline{\Omega}$. Suppose that A is uniformly elliptic on $\overline{\Omega}$ at $u_0 \in C^{2,\alpha}(\overline{\Omega})$, with $0 < \alpha < 1$. If $u_{0/\partial\Omega} \in C^\infty(\partial\Omega)$ and if $A(u_0) \in C^\infty(\overline{\Omega})$, then $u_0 \in C^\infty(\overline{\Omega})$.*

Proof. By Theorem 3.56, $u_0 \in C^\infty(\Omega)$. It remains to prove the regularity up to $\partial\Omega$. Let X be a C^∞ vector field tangent to $\partial\Omega$. Differentiating $A(u_0)$ with respect to $L = X^i\,\partial_i$ yields $A'_{u_0}(Lu_0) \in C^\alpha(\overline{\Omega})$ where A'_{u_0} is a linear elliptic operator with coefficients belonging to $C^\alpha(\overline{\Omega})$. As $Lu_0 \in C^0(\overline{\Omega}) \cap C^2(\Omega)$ and as $Lu_0/\partial\Omega \in C^\infty$, by Theorem 3.59, $Lu_0 \in C^{2,\alpha}(\overline{\Omega})$. So the third derivatives of u_0 are Hölder continuous up to $\partial\Omega$, except maybe the derivatives three times normal. Now let $P \in \partial\Omega$ and ∂_ν be the normal derivative. As A is elliptic $\partial_{\partial_{\nu\nu}} A(u_0)$ is strictly positive at P. By the inverse function theorem, $\partial_{\nu\nu}u_0$ expresses itself in a neighborhood Θ of P in function of u_0 its first derivatives and its other second derivatives which belong to $C^{1,\alpha}(\overline{\Omega} \cap \Theta)$. Thus $u_0 \in C^{3,\alpha}(\overline{\Omega})$. By induction $u_0 \in C^\infty(\overline{\Omega})$.

3.60 The Neumann Problem. Until now we talked about the Dirichlet problem. But we may wish to solve an elliptic equation with other boundary conditions.

For the Neumann Problem the normal derivative of the solution at the boundary is prescribed. For this problem, and those with mixed boundary conditions, we give as references Ladyzenskaja and Uralceva [173] p. 135, Ito [152], Friedman [116], and Cherrier [97].

6.3. The Schauder Interior Estimates[1]

3.61 Let Ω be an open set of $\mathbb{R}^n$ and let $u \in C^{2,\alpha}(\Omega)(0 < \alpha < 1)$ be a bounded solution in Ω of the equation

$$a^{ij}\,\partial_{ij}u + b^i\,\partial_i u + cu = f,$$

where f and the coefficients belong to $C^\alpha(\Omega)$, a^{ij} satisfying $a^{ij}\xi_i\xi_j \geq \lambda|\xi|^2$ with $\lambda > 0$ for all $x \in \Omega$ and $\xi \in \mathbb{R}^n$. Then on any compact set $K \subset \Omega$:

$$\|u\|_{C^{2,\alpha}(K)} \leq C[\|u\|_{C^0(\Omega)} + \|f\|_{C^\alpha(\Omega)}], \tag{20}$$

where the constant C depends on K, α, λ and Λ a bound for the C^α norm of the coefficients in Ω.

§7. Inequalities

7.1. Hölder's Inequality

3.62 Let M be a Riemannian manifold. If $f \in L_p(M)$ and $h \in L_q(M)$ with $p^{-1} + q^{-1} = 1$, then $fh \in L_1(M)$ and :

$$\|fh\|_1 \leq \|f\|_p\|h\|_q. \tag{21}$$

[1] Gilbarg and Trüdinger (125) p. 85.

More generally, if $f \in L_{p_i}(M)$, $(1 \le i \le k)$, with $\sum_{i=1}^k p_i^{-1} = 1$, then $\prod_{i=1}^k f_i \in L_1(M)$ and $\|\prod_{i=1}^k f_i\|_1 \le \prod_{i=1}^k \|f_i\|_{p_i}$.

Proposition 3.62. *Let M be a Riemannian manifold. If $f \in L_r(M) \cap L_q(M)$, $1 \le r < q \le \infty$, then $f \in L_p$ for $p \in [r, q]$ and*

$$\|f\|_p \le \|f\|_r^a \|f\|_q^{1-a} \quad \text{with } a = \frac{1/p - 1/q}{1/r - 1/q}. \tag{22}$$

The proof is just an application of Hölder's inequality.

7.2. Clarkson's Inequalities

3.63 If $u, v \in L_p(M)$, when $2 \le p < \infty$,

$$\|u + v\|_p^p + \|u - v\|_p^p \le 2^{p-1}(\|u\|_p^p + \|v\|_p^p),$$

$$\|u + v\|_p^q + \|u - v\|_p^q \ge 2(\|u\|_p^p + \|v\|_p^p)^{q-1}$$

with $p^{-1} + q^{-1} = 1$. When $1 < p \le 2$, then

$$\|u + v\|_p^q + \|u - v\|_p^q \le 2(\|u\|_p^p + \|v\|_p^p)^{q-1},$$

$$\|u + v\|_p^p + \|u - v\|_p^p \ge 2^{p-1}(\|u\|_p^p + \|v\|_p^p).$$

7.3. Convolution Product

3.64 Let $\mathfrak{u} \in L_p(\mathbb{R}^n)$, $\mathfrak{v} \in L_q(\mathbb{R}^n)$ and $p, q \in [1, \infty[$ with $p^{-1} + q^{-1} \ge 1$. Then the convolution product $(\mathfrak{u} * \mathfrak{v})(x) = \int_{\mathbb{R}^n} \mathfrak{u}(x - y)\mathfrak{v}(y)\, dy$ exists a.e., belongs to L_r with $r^{-1} = p^{-1} + q^{-1} - 1$, and satisfies

$$\|\mathfrak{u} * \mathfrak{v}\|_r \le \|\mathfrak{u}\|_p \|\mathfrak{v}\|_q \tag{23}$$

Proposition 3.64. *Let M_n, $\tilde{M}_n$ be two Riemannian manifolds and let $M_n \times \tilde{M}_n \ni (P, Q) \to f(P, Q)$ be a numerical measurable function such that, for all $P \in M_n$, $Q \to f_P(Q) = f(P, Q)$ belongs to $L_p(\tilde{M})$ with $\sup_{P \in M_n} \|f_P(Q)\|_p < \infty$, and for all $Q \in \tilde{M}_n$, $P \to \tilde{f}_Q(P) = f(P, Q)$ belongs to $L_p(M)$ with $\sup_{Q \in \tilde{M}_n} \|\tilde{f}_Q(P)\|_p < \infty$.*

If $g \in L_q(\tilde{M}_n)$ with $p^{-1} + q^{-1} \ge 1$, then $h(P) = \int_{\tilde{M}_n} f(P, Q) g(Q)\, dV(Q)$ exists for almost all $P \in M_n$ and belongs to $L_r(M_n)$ with $r^{-1} = p^{-1} + q^{-1} - 1$. Moreover:

$$\|h\|_r \le \sup_{P \in M_n} \|f_P(Q)\|_p^{1-p/r} \|g\|_q \sup_{Q \in \tilde{M}_n} \|\tilde{f}_Q(P)\|_p^{p/r}. \tag{24}$$

Proof. It is sufficient to prove inequality (24) for nonnegative C^0 functions with compact support. If $p = q = 1$, it is obvious:

$$\|h\|_1 \leq \sup_{Q \in \tilde{M}} \|\tilde{f}_Q(P)\|_1 \|g\|_1.$$

In the general case we write

$$f(P, Q)g(Q) = [f^p(P, Q)g^q(Q)]^{1/r}[f^p(P, Q)]^{1/p - 1/r}[g^q(Q)]^{1/q - 1/r}.$$

Since $1/r + (1/p - 1/r) + (1/q - 1/r) = 1$, applying Hölder's inequality, we are led to

$$|h(P)| \leq \left[\int_{\tilde{M}} f^p(P, Q)g^q(Q)\, dV(Q)\right]^{1/r}\left[\int_{\tilde{M}} f^p(P, Q)\, dV(Q)\right]^{1/p - 1/r} \times \left[\int_{\tilde{M}} g^q(Q)\, dV(Q)\right]^{1/q - 1/r},$$

and the result follows. ■

7.4. The Calderon–Zygmund Inequality

3.65 Let $\omega \in L_\infty(\mathbb{R}^n)$ with compact support satisfy $\omega(tx) = \omega(x)$ for all $0 < t \leq 1$ and $\|x\| \leq \rho$ for some $\rho > 0$, and also satisfy $\int_{\mathbb{S}_{n-1}(\rho)} \omega(x)\, d\sigma = 0$. For all $\varepsilon > 0$ let

$$(K_\varepsilon * f)(x) = \int_{\|x\| > \varepsilon} \omega(y)\|y\|^{-n} f(x - y)\, dy \quad \text{with } f \in L_p(\mathbb{R}^n).$$

If $1 < p < \infty$, then $\lim_{\varepsilon \to 0}(K_\varepsilon * f)(x)$ exists almost everywhere and the limit function denoted by $K_0 * f$ belongs to L_p.

Moreover, $K_\varepsilon * f \to K_0 * f$ in L_p and there exists a constant C, which depends on ω and p, such that

$$(25) \qquad \|K_0 * f\|_p \leq C\|f\|_p \qquad (\textit{Calderon–Zygmund inequality})$$

If $\omega \in L_\infty(\mathbb{R}^n)$ satisfies $\omega(tx) = \omega(x)$ for all $t > 0$ and $\int_{\mathbb{S}_{n-1}(1)} \omega(x)\, d\sigma = 0$, then $\|K_\varepsilon * f - K_0 * f\|_p \to 0$ when $\varepsilon \to 0$ and (25) holds.
In addition, if $\omega \in C^1(\mathbb{R}^n - \{0\})$, $K_\varepsilon * f \to K_0 * f$ a.e. (see Dunford and Schwartz [111]).

7.5. Korn–Lichtenstein Theorem

3.66 Theorem. *If $\omega(x)$ is a function with the properties described in* 3.65 *and K_0 is defined as above, there exists, for any $\alpha (0 < \alpha < 1)$, a constant $A(\alpha)$ such that*

$$\|K_0 * f\|_{C^\alpha} \leq A(\alpha)\|f\|_{C^\alpha},$$

for all $f \in C^\alpha(\mathbb{R}^n)$ with compact support.

3.67 Theorem. *Let M_n be a compact Riemannian manifold and p, q, and r real numbers satisfying $1/p = 1/q - 1/n$, $1 \leq q < n$ and $r > n$. Define $\mathscr{A} = \{\varphi \in L_1 / \int \varphi \, dV = 0\}$.*
Then there exists a constant k, such that, for all $\alpha \geq 1$, any function $\varphi \in \mathscr{A}$ with $|\nabla |\varphi|^\alpha| \in L_q$, satisfies

$$\||\varphi|^\alpha\|_p \leq k^\alpha \|\nabla |\varphi|^\alpha\|_q. \tag{26}$$

If $\varphi \in \mathscr{A}$ with $|\nabla\varphi| \in L_r$, then $\sup|\varphi| \leq \text{Const} \times \|\nabla\varphi\|_r$. If $\varphi \in L_1$ *and $\Delta\varphi \in L_q$ (in the distributional sense), then*

$$|\nabla\varphi| \in L_p \quad \textit{and} \quad \|\nabla\varphi\|_p \leq \text{Const} \times \|\Delta\varphi\|_q.$$

If $\varphi \in L_1$ and $\Delta\varphi \in L_r$, then $|\nabla\varphi|$ is bounded and

$$\sup|\nabla\varphi| \leq \text{Const} \times \|\Delta\varphi\|_r.$$

The constants do not depend on φ, of course.

Let $\overline{M}_n$ be a compact Riemannian manifolds with boundary. Then the theorem holds for functions $\varphi \in \mathscr{D}(M)$.

Proof. First of all, we are going to establish (26) for $\alpha = 1$.
Let $\mathsf{G}(P, Q)$ be the Green's function of the Laplacian. As $\int \varphi \, dV = 0$, in the distributional sense (Proposition 4.14):

$$\varphi(P) = \int \mathsf{G}(P, Q)\Delta\varphi(Q) \, dV(Q), \tag{27}$$

whence:

$$|\varphi(P)| \leq \int |\nabla_Q \mathsf{G}(P, Q)| |\nabla\varphi(Q)| \, dV(Q) \tag{28}$$

and according to Proposition 3.64, we find

$$\|\varphi\|_q \leq \|\nabla\varphi\|_q \sup_{P \in M_n} \int |\nabla_Q \mathsf{G}(P, Q)| \, dV(Q)$$

for all $\varphi \in \mathscr{A}$, such that $|\nabla\varphi| \in L_q$.

Using the Sobolev imbedding theorem 2.21, we obtain

$$\|\varphi\|_p \leq K\|\nabla\varphi\|_q + A\|\varphi\|_q \leq k_0\|\nabla\varphi\|_q, \tag{29}$$

with $k_0 = K + A \sup_{P \in M} \int |\nabla_Q \mathsf{G}(P, Q)| \, dV(Q)$.

Let us now prove (26) for $\alpha > 1$. Since the set of the C^∞ functions which have no degenerate critical points is dense in the spaces H_1^s (Proposition 2.16), we need only establish (26) for these functions. Let $\varphi \neq 0$ be such a function, with $\int \varphi \, dV = 0$.
Set $\check{\varphi} = \sup(\varphi, 0)$ and $\tilde{\varphi} = \sup(-\varphi, 0)$.
If the measure of the support of φ is less than or equal to ε (the ε of Lemma 3.68 below) (34) applied to $|\varphi|^\alpha$ gives (26) with $k = B$. Otherwise, let $a > 0$ be such that the measure of

$$\Omega_a = \{x \in M_n / |\varphi(x)| \geq a\} \quad \text{is equal to } \varepsilon\colon \mu(\Omega_a) = \varepsilon.$$

We have $a\varepsilon \leq \|\varphi\|_1$.

Since $|\varphi|^\alpha \leq \widecheck{|\varphi|^\alpha - a^\alpha} + a^\alpha$, then by (34) below we have

(30)

$$\||\varphi|^\alpha\|_p \leq \|\widecheck{|\varphi|^\alpha - a^\alpha}\|_p + a^\alpha\left(\int dV\right)^{1/p} \leq B\|\nabla|\varphi|^\alpha\|_q + a^\alpha\left(\int dV\right)^{1/p}.$$

Suppose that $\|\check{\varphi}\|_\alpha \leq \|\tilde{\varphi}\|_\alpha$, (otherwise, replace φ by $-\varphi$); then we write:

$$a\varepsilon \leq \|\varphi\|_1 = 2\|\check{\varphi}\|_1 \leq 2\|\check{\varphi}\|_\alpha\left(\int dV\right)^{1-1/\alpha}, \tag{31}$$

by using Hölder's inequality.
Now consider the function $\psi = (\check{\varphi})^\alpha - \|\check{\varphi}\|_\alpha^\alpha(\tilde{\varphi})^\alpha/\|\tilde{\varphi}\|_\alpha^\alpha$. ψ satisfies $\|\psi\|_1 = 2\|\check{\varphi}\|_\alpha^\alpha$ and $\int \psi \, dV = 0$. Thus applying (29), where we choose $k_0 \geq 1$, yields:

$$\|\psi\|_p \leq k_0\|\nabla\psi\|_q \leq k_0\|\nabla|\varphi|^\alpha\|_q \tag{32}$$

As $\|\psi\|_1 \leq \|\psi\|_p(\int dV)^{1-1/p}$, using (30), (31), and (32) leads to (26), with $k = B + 2k_0 \int dV/\varepsilon$. Indeed $k^\alpha > B + k_0\varepsilon^{-\alpha}2^{\alpha-1}(\int dV)^\alpha$.

One easily obtains the other results, by applying the properties of the Green's function from 4.13 below to (28) or (27) after differentiation:

$$|\nabla\varphi(P)| \leq \int |\nabla_P G(P, Q)||\Delta\varphi(Q)| \, dV(Q). \tag{33}$$

■

The proof for the compact manifold with boundary is similar.
Finally we must prove the following lemma which was used above.

3.68 Lemma. *Let M_n be a Riemannian manifold and p, q as above. There exist B, ε, two positive constants, such that any function $\varphi \in H_1^q$ satisfies:*

$$\|\varphi\|_p \leq B\|\nabla\varphi\|_q \tag{34}$$

when $\mu(\text{supp } \varphi) = \int_{\text{supp } \varphi} dV \leq \varepsilon$.

Proof. Since $\|\varphi\|_q \le \|\varphi\|_p[\mu(\operatorname{supp}\varphi)]^{1/n} \le \varepsilon^{1/n}\|\varphi\|_p$, using (29) we obtain (34) with any $\varepsilon < A^{-n}$ by setting $B = K(1-\varepsilon^{1/n}A)^{-1}$. ■

7.6. Interpolation Inequalities

3.69 Theorem. *Let M_n be a Riemannian manifold and q, r satisfy $1 \le q, r \le \infty$. Set $2/p = 1/q + 1/r$. Then all functions $f \in \mathscr{D}(M)$ satisfy:*

$$(35) \qquad \|\nabla f\|_p^2 \le (n^{1/2} + |p-2|)\|f\|_q\|\nabla^2 f\|_r.$$

Let $\mathfrak{F}$ denote the completion of $\mathscr{D}(M)$ under the norm $\|f\|_q + \|\nabla^2 f\|_r$.
If $f \in \mathfrak{F}$, then $|\nabla f| \in L_p(M)$ and (35) holds.
In particular, when M is compact (with boundary or without), if $f \in L_q(M)$ and $|\nabla^2 f| \in L_r$, then $f \in L_r(M)$ and (35) holds for $f \in \mathring{H}_2^r(M)$.
Moreover, if $1/q + 1/r = 1$, then all $f \in \mathscr{D}(M)$ satisfy

$$\|\nabla f\|_2^2 \le \|f\|_q\|\Delta f\|_r.$$

Proof. First of all, suppose $p \ge 2$. For $f \in \mathscr{D}(M)$:

$$(36) \qquad \nabla^\nu(f|\nabla f|^{p-2}\nabla_\nu f) = |\nabla f|^p + f|\nabla f|^{p-2}\nabla^\nu\nabla_\nu f + (p-2)|\nabla f|^{p-4} f\nabla_{\nu\mu} f\nabla^\nu f\nabla^\mu f.$$

Integrating (36) over M leads to $\|\nabla f\|_p^p = \int f\,\Delta f\,dV$ if $p = 2$, and when $p > 2$ it yields:

$$(37) \quad \|\nabla f\|_p^p = \int f\Delta f|\nabla f|^{p-2}\,dV + (2-p)\int|\nabla f|^{p-4} f\nabla_{\nu\mu} f\nabla^\nu f\nabla^\mu f\,dV.$$

But $|\Delta f|^2 \le n|\nabla^2 f|^2$ and $|\nabla_{\nu\mu} f\nabla^\nu f\nabla^\mu f| \le |\nabla^2 f||\nabla f|^2$; thus:

$$\|\nabla f\|_p^p \le (n^{1/2} + |p-2|)\int|f||\nabla^2 f||\nabla f|^{p-2}\,dV.$$

Applying the Hölder inequality 3.62, since $1/q + 1/r + (p-2)/p = 1$, we find:

$$\|\nabla f\|_p^p \le (n^{1/2} + |p-2|)\|f\|_q\|\nabla^2 f\|_r\|\nabla f\|_p^{p-2},$$

and the desired result follows. When $1 \le p < 2$, the proof is similar, but a little more delicate (see Aubin [22]). When M is compact, if $f \in L_q(M)$ and $|\nabla^2 f| \in L_r$, then by the properties of the Green's function $f \in L_r(M)$. ■

3.70 Theorem (See Nirenberg [220] p. 125). *M_n will be either $\mathbb{R}^n$, or a compact Riemannian manifold with or without boundary. Let q, r be real numbers $1 \le q, r \le \infty$ and j, m integers $0 \le j < m$.*

Then there exists k, a constant depending only on n, m, j, q, r, and a, and on the manifold, such that for all $f \in \mathscr{D}(M)$ *(with* $\int f\, dV = 0$*, in the compact case without boundary):*

$$\|\nabla^j f\|_p \leq k \|\nabla^m f\|_r^a \|f\|_q^{1-a}, \tag{38}$$

where

$$\frac{1}{p} = \frac{j}{n} + a\left(\frac{1}{r} - \frac{m}{n}\right) + (1-a)\frac{1}{q}, \tag{39}$$

for all a in the interval $j/m \leq a \leq 1$*, for which p is non-negative.*

If $r = n/(m-j) \neq 1$*, then (38) is not valid for* $a = 1$.

Proof. α) The result holds also for $j = m = 0$, with $k = 1$. This is just proposition (3.62).

Once the two cases $j = 0$, $m = 1$, and $j = 1$, $m = 2$ are proved, the general case will follow by induction, by applying the inequality

$$|\nabla |\nabla^\ell f|| \leq |\nabla^{\ell+1} f| \text{(see Proposition (2.11))}. \tag{40}$$

For the proof, we are going to use Hölder's inequality, Theorem (3.69), and the Sobolev imbedding theorem. It may be written (Corollary 2.12, and Theorem 3.67):

$$\|h\|_s \leq \text{Const} \times \|\nabla h\|_t, \quad \text{where } \frac{1}{s} = \frac{1}{t} - \frac{1}{n} > 0, \text{ for all } h \in \mathscr{D}(M_n) \tag{41}$$

(with $\int h\, dV = 0$, when the manifold is compact without boundary).

β) *The case* $j = 0$, $m = 1$, $p < \infty$. By (41), with $t = r < n$ and Proposition (3.62):

$$\|f\|_p \leq \|f\|_s^a \|f\|_q^{1-a} \leq k \|\nabla f\|_r^a \|f\|_q^{1-a}, \tag{42}$$

with $1/p - 1/q = a(1/s - 1/q) = a(1/r - 1/n - 1/q)$. Thus for $j = 0$ and $m = 1$, if $r < n$, then (38) holds for $0 \leq a \leq 1$ and p runs from q to $s = rn/(n-r)$.

If $r \geq n$, use (41), with $1/ap = 1/\mu - 1/n$. Putting $h = |f|^{1/a}$, we find the desired result when $p < \infty$. Indeed, $\|h\|_{ap} \leq C \|\nabla h\|_\mu$ becomes

$$\|f\|_p^{1/a} \leq \frac{C}{a} \| |\nabla f| |f|^{(1/a)-1} \|_\mu \leq \frac{C}{a} \|\nabla f\|_r \|f\|_q^{(1/a)-1} \tag{43}$$

by using Hölder's inequality, since $1/r + (1/a - 1)/q = 1/\mu = 1/ap + 1/n$.

γ) *The case* $j = 0$, $m = 1$, $p = +\infty$. If $r > n$, let $s \in [(n + r)/2, r]$. When $M_n \neq \mathbb{R}^n$, all $f \in \mathscr{D}(M)$ (with $\int f\, dV = 0$ in the compact case) satisfy (Theorem 3.67):

$$\|f\|_p \leq \text{Const} \times \|\nabla f\|_s$$

for all p, $1 \leq p \leq \infty$, and the constant does not depend on p and s. Thus C does not depend on p in (43). Letting $p \to \infty$ in (43), we obtain the inequality for $p = +\infty$.
If $r > n$ and $M_n = \mathbb{R}^n$, a proof similar to that of the Sobolev imbedding theorem, yields:
There exists a constant $C(v)$, such that for all $f \in \mathscr{D}(\mathbb{R}^n)$:

$$\text{(44)} \qquad \sup|f| \leq C(v)(\|\nabla f\|_r + \|f\|_v), \quad \text{when } v > n.$$

Consider the function $\varphi(x) = f(tx)$, with $0 < t < \infty$. Applying (44) to φ and setting $y = tx$ lead to:

$$\sup|f| \leq C(v)(t^{1-n/r}\|\nabla f\|_r + t^{-n/v}\|f\|_v).$$

Choosing $t = (\|f\|_v \|\nabla f\|_r^{-1})^{(n/v+1-n/r)-1}$, we find:

$$\text{(45)} \qquad \sup|f| \leq 2C(v)\|\nabla f\|_r^d \|f\|_v^{1-d},$$

with $d^{-1} = 1 + v(1/n - 1/r)$.
If $q > n$, we can choose $v = q$, and the result follows for $p = +\infty$, $(d = a)$.
If $q \leq n$, since $\|f\|_v \leq \|f\|_q^{q/v}(\sup|f|)^{1-q/v}$, (45) gives:

$$(\sup|f|)^{1-(1-d)(1-q/v)} \leq 2C(v)\|\nabla f\|_r^d \|f\|_q^{(1-d)q/v},$$

which is the result for $p = +\infty$, with

$$a^{-1} = q/v\, d + 1 - q/v = 1 + q(1/n - 1/r).$$

δ) *The case* $j = 1$, $m = 2$. We have established (Theorem 3.69) inequality (38) for $a = j/m = 1/2$. If $r < n$, inequality (38) for $a = 1$ is just the Sobolev imbedding theorem (Corollary 2.12, Theorem 3.67). By interpolation (22), we find the inequality for $\frac{1}{2} < a < 1$. If $r \geq n$, according to (38) with $j = 0$, $m = 1$, applied to the function $|\nabla f|$:

$$\text{(46)} \qquad \|\nabla f\|_p \leq \text{Const} \times \|\nabla^2 f\|_r^b \|\nabla f\|_s^{1-b},$$

with $1/p = 1/s + b(1/r - 1/n - 1/s) > 0$ and $0 \leq b \leq 1$.
Using (35) in (46) yields the desired inequality. Indeed, $\|\nabla f\|_s^2 \leq \text{Const} \times \|\nabla^2 f\|_r \|f\|_q$ with $2/s = 1/r + 1/q$. Thus we find inequality (38) where $j = 1$, $m = 2$ and $a = (1 + b)/2$.

We can verify that $a \le a_1 = [1 + (1/n - 1/r)/(1/n + 1/q)]^{-1}$ implies $b \le a_0 = (1 + s/n - s/r)$. Thus (38) holds. ■

§8. Maximum Principle

8.1. Hopf's Maximum Principle[2]

3.71 *Let Ω be an open connected set of $\mathbb{R}^n$ and $L(u)$ a linear uniformly elliptic differential operator in Ω of order 2:*

$$L(u) = \sum_{i,j} a_{ij}(x) \frac{\partial^2 u}{\partial x^i \, \partial x^j} + \sum_{i=1}^{n} b_i(x) \frac{\partial u}{\partial x^i} + h(x)u$$

with bounded coefficients and $h \le 0$.

Suppose $u \in C^2(\Omega)$ satisfies $L(u) \ge 0$.

If u attains its maximum $M \ge 0$ in Ω, then u is constant equal to M on Ω. Otherwise if at $x_0 \in \partial\Omega$, u is continuous and $u(x_0) = M \ge 0$, then the outer normal derivative at x_0, if it exists, satisfies $\partial u/\partial \nu(x_0) > 0$, provided x_0 belongs to the boundary of a ball included in Ω.

Moreover, if $h \equiv 0$, the same conclusions hold for a maximum $M < 0$.

Remark 3.71. We can state a maximum principle for weak solution (see Gilbarg and Trüdinger [125] p. 168).

Let $Lu = \partial_i(a^{ij}\,\partial_j u) + b^i\,\partial_i u + hu$ be an elliptic operator in divergence form defined on an open set Ω of $\mathbb{R}^n$, where the coefficients a^{ij}, b^i and h are assumed to be measurable and locally bounded.

$u \in \mathrm{H}_1(\Omega)$ is said to satisfy $Lu \ge 0$ *weakly* if for all $\varphi \in \mathscr{D}(\Omega)$, $\varphi \ge 0$:

$$\int_\Omega [a^{ij}\,\partial_i u\,\partial_j \varphi - (b^i\,\partial_i u + hu)\varphi]\,dx \le 0.$$

In this case, if $h \le 0$ then $\sup_\Omega u \le \sup_{\partial\Omega} \max(u, 0)$.

The last term is defined in the following way: we say that $v \in \mathrm{H}_1(\Omega)$ satisfies $v/\partial\Omega \le k$ if $\max(v - k, 0) \in \mathring{\mathrm{H}}_1(\Omega)$.

8.2. Uniqueness Theorem

3.72 *Let $\overline{W}$ be a compact Riemannian manifold with boundary and $L(u)$ a linear uniformly elliptic differential operator on $\overline{W}$:*

$$L(u) = a^{ij}(x)\nabla_i\nabla_j u + b^i(x)\nabla_i u + h(x)u$$

with bounded coefficients and $h \le 0$.

[2] Protter and Weinberger [239].

Then, the Dirichlet problem $L(u) = f$, $u|\partial W = g$ (f and g given) has at most one solution.

Proof. Suppose $\tilde{\mathfrak{u}}$ and $\mathfrak{u}$ are solutions of the Dirichlet problem. Then $\mathfrak{v} = \tilde{\mathfrak{u}} - \mathfrak{u}$ satisfies $L\mathfrak{v} = 0$ in W and $\mathfrak{v}|\partial W = 0$. According to the maximum principle $\mathfrak{v} \leq 0$ on W. But the same result holds for $-\mathfrak{v}$. Thus $\mathfrak{v} = 0$ in W. ∎

3.73 Theorem. *Let $\overline{W}$ and $L(u)$ be as above. If $\mathfrak{w} \in C^2(W) \cap C^0(\overline{W})$ is a subsolution of the above Dirichlet problem, i.e. $\mathfrak{w}$ satisfies:*

$$L\mathfrak{w} \geq f \quad \text{in } W, \ \mathfrak{w}/\partial W \leq g,$$

then $\mathfrak{w} \leq \mathfrak{u}$ everywhere, if $\mathfrak{u}$ is the solution of the Dirichlet problem. Likewise, if $\mathfrak{v}$ is a supersolution, i.e. $\mathfrak{v}$ satisfies $L\mathfrak{v} \leq f$ in W and $\mathfrak{v}/\partial W \geq g$ then $\mathfrak{u} \leq \mathfrak{v}$ everywhere.

8.3. Maximum Principle for Nonlinear Elliptic Operator of Order Two

3.74 Let $\overline{W}$ be a Riemannian compact manifold with boundary, and $A(\mathfrak{u}) = f(x, \mathfrak{u}, \nabla\mathfrak{u}, \nabla^2\mathfrak{u})$ a differential operator of order two defined over W, where f is supposed to be a differentiable function of its arguments. Suppose $\mathfrak{v}$, $\mathfrak{w} \in C^2(W)$ satisfy $A(\mathfrak{v}) = 0$ and $A(\mathfrak{w}) \geq 0$. Define $\mathfrak{v}_t$ by $[\mathfrak{v} + t(\mathfrak{w} - \mathfrak{v})]$.

Theorem 3.74. *Let $A(\mathfrak{u})$ be uniformly elliptic with respect to $\mathfrak{v}_t$, for all $t \in]0, 1[$. Then $\varphi = \mathfrak{w} - \mathfrak{v}$ cannot achieve a nonnegative maximum $M \geq 0$ in W, unless it is a constant, if $\partial f(x, \mathfrak{v}_t, \nabla\mathfrak{v}_t, \nabla^2\mathfrak{v}_t)/\partial\mathfrak{u} \leq 0$ on W.*

Moreover suppose $\mathfrak{v}$, $\mathfrak{w} \in C^0(\overline{W})$ and $\mathfrak{w} \leq \mathfrak{v}$ on the boundary, then $\mathfrak{w} \leq \mathfrak{v}$ everywhere provided the derivatives of $f(x, \mathfrak{v}_t, \nabla\mathfrak{v}_t, \nabla^2\mathfrak{v}_t)$ are bounded (in the local charts of a finite atlas) for all $t \in]0, 1[$.
If in addition at $x_0 \in \partial W$, $\varphi(x_0) = 0$ and $\partial\varphi/\partial\nu(x_0)$ exists, then $\partial\varphi/\partial\nu(x_0) > 0$, unless φ is a constant, provided the boundary is C^2.

Proof. Consider $\gamma(t) = f(x, \mathfrak{v}_t)$. For some $\theta \in]0, 1[$ the mean value theorem shows that $0 \leq A(\mathfrak{w}) - A(\mathfrak{v}) = \gamma(1) - \gamma(0) = \gamma'(\theta)$ with

$$\gamma'(\theta) = \frac{\partial f(x, \mathfrak{v}_\theta)}{\partial \nabla_{ij}\mathfrak{u}} \nabla_{ij}\varphi + \frac{\partial f(x, \mathfrak{v}_\theta)}{\partial \nabla_i \mathfrak{u}} \nabla_i \varphi + \frac{\partial f(x, \mathfrak{v}_\theta)}{\partial u} \varphi = L(\varphi)$$

Thus $\varphi = \mathfrak{w} - \mathfrak{v}$ satisfies $L(\varphi) \geq 0$. Applying the above theorems yields the present statements. ∎

3.75 As an application of the maximum principle we are going to establish the following lemma, which will be useful to solve Yamabe's problem.

Proposition 3.75. *Let M_n be a compact Riemannian manifold. If a function $\psi \geq 0$, belonging to $C^2(M)$, satisfies an inequality of the type $\Delta\psi \geq \psi f(P, \psi)$, where $f(P, t)$ is a continuous numerical function on $M \times \mathbb{R}$, then either ψ is strictly positive, or ψ is identically zero.*

Proof. According to Kazdan. Since M is compact and since ψ is a fixed non-negative continuous function, there is a constant $a > 0$ such that $\Delta\psi + a\psi \geq 0$. By the maximum principle 3.71, the result follows: $u = -\psi$ cannot have a local maximum ≥ 0 unless $u \equiv 0$. Here $L = -\Delta - a$. ■

8.4. Generalized Maximum Principle

3.76 There is a generalized maximum principle on complete noncompact manifolds Cheng and Yau (90). Namely:

Theorem. *Let (M, g) be a complete Riemannian manifold. Suppose that for any $x \in M$ there is a C^2 non-negative function φ^x on M with support K^x in a compact neighborhood of x which satisfies $\varphi^x(x) = 1$, $\varphi^x \leq k$, $|\nabla\varphi^x| \leq k$, and $\varphi^x_{ii} \geq -kg_{ii}$ for all directions i, where k is a constant independent of x.*

If f is a C^2 function on M which is bounded from above, then there exists a sequence $\{x_j\}$ in M such that $\lim f(x_j) = \sup f$,

$$\lim |\nabla f(x_j)| = 0 \quad \textit{and} \quad \lim\sup \nabla_{ii} f(x_j) \leq 0$$

for all directions i.

Proof. Denote by L the sup of f, which we suppose not attained; otherwise the theorem is obvious by the usual maximum principle. Let $\{y_j\}$ be a sequence in M such that $\lim f(y_j) = L$. On $\mathring{K}^{y_j}$ consider the function $(L - f)/\varphi^{y_j}$. This is strictly positive and goes to ∞ when $x \to \partial\mathring{K}^{y_j}$.
Let $x_j \in \mathring{K}^{y_j}$ be a point where this function attains its minimum. We have

$$\left(\frac{L - f}{\varphi^{y_j}}\right)(x_j) \leq \left(\frac{L - f}{\varphi^{y_j}}\right)(y_j) = L - f(y_j)$$

$$\left(\frac{\nabla_i(L - f)}{L - f}\right)(x_j) = \left(\frac{\nabla_i\varphi^{y_j}}{\varphi^{y_j}}\right)(x_j)$$

$$\left(\frac{\nabla_{ii}(L - f)}{L - f}\right)(x_j) \geq \left(\frac{\nabla_{ii}\varphi^{y_j}}{\varphi^{y_j}}\right)(x_j) \quad \text{for all direction } i.$$

From these we get

$$0 < L - f(x_j) \leq k[L - f(y_j)]$$

$$|\nabla f(x_j)| \leq k[L - f(y_j)]$$

$$\nabla_{ii} f(x_j) \leq k[L - f(y_j)]g_{ii}.$$

Thus $\{x_j\}$ is a sequence having the required properties. ■

§9. Best Constants

3.77 Theorem (Lions [188]). *Let $\mathfrak{B}_1, \mathfrak{B}_2, \mathfrak{B}_3$ be three Banach spaces and* $\mathfrak{u}, \mathfrak{v}$, *two linear operators*: $\mathfrak{B}_1 \xrightarrow{\mathfrak{u}} \mathfrak{B}_2 \xrightarrow{\mathfrak{v}} \mathfrak{B}_3$. *Suppose* $\mathfrak{u}$ *is compact and* $\mathfrak{v}$ *continuous and one to one. Then given any* $\varepsilon > 0$, *there is* $A(\varepsilon) > 0$ *such that for all* $x \in \mathfrak{B}_1$:

$$\|\mathfrak{u}(x)\|_{\mathfrak{B}_2} \leq \varepsilon\|x\|_{\mathfrak{B}_1} + A(\varepsilon)\|\mathfrak{v} \circ \mathfrak{u}(x)\|_{\mathfrak{B}_3}$$

Proof. Suppose the contrary. Then there exists $\varepsilon_0 > 0$ and a sequence $\{x_k\}$ in $\mathfrak{B}_1$, satisfying $\|x_k\|_{\mathfrak{B}_1} = 1$ such that

(47) $$\|\mathfrak{u}(x_k)\|_{\mathfrak{B}_2} > \varepsilon_0\|x_k\|_{\mathfrak{B}_1} + k\|\mathfrak{v} \circ \mathfrak{u}(x_k)\|_{\mathfrak{B}_3},$$

Since $\mathfrak{u}$ is compact, a subsequence of $\{\mathfrak{u}(x_k)\}$ converges in $\mathfrak{B}_2$, Say $\mathfrak{u}(x_{k_i}) \to y_0 \in \mathfrak{B}_2$. Rewriting (47) for this subsequence gives:

$$\|\mathfrak{u}(x_{k_i})\|_{\mathfrak{B}_2} > \varepsilon_0 + k_i\|\mathfrak{v} \circ \mathfrak{u}(x_{k_i})\|_{\mathfrak{B}_3}.$$

Whether $y_0 = 0$ or not, letting $k_i \to \infty$ yields the desired contradiction. ∎

3.78 Theorem (Aubin [17]). *Let $\mathfrak{B}_1, \mathfrak{B}_2, \mathfrak{B}_3$ be three Banach spaces and* $\mathfrak{u}, \mathfrak{w}$ *two continuous linear operators*: $\mathfrak{B}_1 \xrightarrow{\mathfrak{u}} \mathfrak{B}_2$, $\mathfrak{B}_1 \xrightarrow{\mathfrak{w}} \mathfrak{B}_3$.
Suppose $\mathfrak{u}$ *is not compact and* $\mathfrak{w}$ *is compact. And consider all pairs of real numbers* C, A, *such that all* $x \in \mathfrak{B}_1$ *satisfy*:

(48) $$\|\mathfrak{u}(x)\|_{\mathfrak{B}_2} \leq C\|x\|_{\mathfrak{B}_1} + A\|\mathfrak{w}(x)\|_{\mathfrak{B}_3}.$$

Define $\mathsf{K} = \inf C$ *such that some* A *exists. Then* $\mathsf{K} > 0$.

Proof. Since $\mathfrak{u}$ is not compact, there exists a sequence $\{x_i\}$ in $\mathfrak{B}_1$ with $\|x_i\|_{\mathfrak{B}_1} = 1$, such that no subsequence of $\{\mathfrak{u}(x_i)\}$ converges in $\mathfrak{B}_2$. But $\mathfrak{w}$ is compact. Hence there exists $\{\mathfrak{w}(x_k)\}$, a subsequence of $\{\mathfrak{w}(x_i)\}$, which converges in $\mathfrak{B}_3$. Because $\{\mathfrak{u}(x_k)\}$ is not a Cauchy sequence in $\mathfrak{B}_2$, there exist $\eta > 0$ and $\{k_j\}$ an increasing sequence in $\mathbb{N}$ such that

$$\|\mathfrak{u}(y_j)\| > \eta, \quad \text{where } y_j = x_{k_{2j+1}} - x_{k_{2j}}.$$

Write (48) for y_j:

$$\|\mathfrak{u}(y_j)\|_{\mathfrak{B}_2} \leq C\|y_j\|_{\mathfrak{B}_1} + A\|\mathfrak{w}(y_j)\|_{\mathfrak{B}_3}.$$

Letting $j \to \infty$ leads to $\eta \leq 2C$, since $\mathfrak{w}(y_j) \to 0$ in $\mathfrak{B}_3$. Thus $\mathsf{K} \geq \eta/2 > 0$. ∎

9.1. Application to Sobolev Spaces

3.79 Let M_n be a compact Riemannian manifold with boundary or without. Consider the following Banach spaces $\mathfrak{B}_1 = H_1^q(M)$, $\mathfrak{B}_2 = L_p(M)$ and $\mathfrak{B}_3 = L_q(M)$ with $q < n$ and $1/p = 1/q - 1/n$.

Recall Sobolev's and Kondrakov's theorems, 2.21 and 2.34. The imbedding $\mathfrak{B}_1 \subset \mathfrak{B}_2$ is not compact (example 2.38) and the imbedding $\mathfrak{B}_1 \subset \mathfrak{B}_3$ is compact. Thus there exist constants A, C such that

$$\|f\|_p \leq C(\|\nabla f\|_q + \|f\|_q) + A\|f\|_q, \tag{49}$$

for instance $(C_0, 0)$, and $\mathsf{K} = \inf\{C \text{ such that some } A \text{ exists}\} > 0$.

Of course K depends on n, q and on the manifold. But we have proved (Theorem 2.21) that $\mathsf{K} = \mathsf{K}(n, q)$ is the same constant for all compact manifolds of dimension n and that K is the norm of the imbedding $H_1^q(\mathbb{R}^n) \subset L_p(\mathbb{R}^n)$.

Chapter 4

Green's Function for Riemannian Manifolds

§1. Linear Elliptic Equations

4.1 To prove the existence of Green's function, first of all we have to solve linear elliptic equations. We also need some results concerning the eigenvalues of the Laplacian. Let (M_n, g) be a C^∞ Riemannian manifold. We are going to consider equations of the type

$$(1) \qquad -\nabla^i[a_{ij}(x)\nabla^j\varphi] = f(x),$$

where $a_{ij}(x)$ are the components, in a local chart, of a C^∞ Riemannian metric (see 1.15) and where f belongs to $L_2(M)$.

1.1. First Nonzero Eigenvalue λ of Δ.

4.2 Theorem. *Let (M_n, g) be a compact C^∞ Riemannian manifold. The eigenvalues of the Laplacian $\Delta = -\nabla^\nu\nabla_\nu$ are nonnegative. The eigenfunctions, corresponding to the eigenvalue $\lambda_0 = 0$, are the constant functions. The first nonzero eigenvalue λ_1 is equal to μ, defined by: $\mu = \inf\|\nabla\varphi\|_2^2$ for all $\varphi \in \mathscr{A}$, with $\mathscr{A} = \{\varphi \in \mathsf{H}_1$ satisfying $\|\varphi\|_2 = 1$ and $\int \varphi\, dV = 0\}$.*

Proof. The first statement is proved in 1.77; the second in 1.71.
Let $\{\varphi_i\}_{i\in\mathbb{N}}$ be a sequence in $\mathscr{A}$, such that $\|\nabla\varphi_i\|_2^2 \to \mu$ when $i \to \infty$. $\{\varphi_i\}$ is called a minimizing sequence. Obviously $\{\varphi_i\}$ is bounded in H_1.
According to Kondrakov's theorem, 2.33, there exists a subsequence $\{\varphi_j\}$ of $\{\varphi_i\}$ and $\varphi_0 \in L_2$, such that $\varphi_j \to \varphi_0$ strongly in L_2 ($\|\varphi_j - \varphi_0\|_2 \to 0$). Hence $\|\varphi_j - \varphi_0\|_1 \to 0$ when $j \to \infty$, since the manifold is compact. Thus φ_0 satisfies $\|\varphi_0\|_2 = 1$ and $\int \varphi_0\, dV = 0$.
By Theorem 3.18, there exist $\tilde\varphi_0 \in \mathsf{H}_1$ and $\{\varphi_k\}$ a subsequence of $\{\varphi_j\}$ such that $\varphi_k \to \tilde\varphi_0$ weakly in H_1.
Furthermore, $\varphi_k \to \varphi_0$ weakly in L_2 (strong convergence implies weak convergence). Thus, since the imbedding $\mathsf{H}_1 \subset L_2$ is continuous, φ_0 and $\tilde\varphi_0$ are functions in L_2 which define the same distribution on $\mathscr{D}(M)$. Therefore $\varphi_0 = \tilde\varphi_0$, and $\varphi_0 \in \mathscr{A}$.

According to Theorem 3.17, $\|\varphi_0\|_{\mathsf{H}_1} \leq \liminf_{k\to\infty} \|\varphi_k\|_{\mathsf{H}_1}$, since $\varphi_k \to \varphi_0$ weakly in H_1. This implies $\|\nabla\varphi_0\|_2^2 \leq \lim_{k\to\infty} \|\nabla\varphi_k\|_2^2 = \mu$, because $\|\varphi_k\|_2 = \|\varphi_0\|_2 = 1$. But $\varphi_0 \in \mathscr{A}$, hence $\|\nabla\varphi_0\|_2^2 = \mu$, and the minimum μ is attained.

Writing Euler's equation of our variational problem (see, for instance, Berger [42], p. 123), leads to: There exist numbers α and β, such that for all $\psi \in \mathsf{H}_1$:

$$\int \nabla^\nu \varphi_0 \nabla_\nu \psi \, dV = \alpha \int \varphi_0 \psi \, dV + \beta \int \psi \, dV.$$

Picking $\psi = 1$, gives $\beta = 0$. Choosing next $\psi = \varphi_0$ leads to $\alpha = \mu$. So $\varphi_0 \in \mathsf{H}_1$ satisfies weakly

(2) $$\Delta\varphi_0 = \mu\varphi_0.$$

Thus, by the regularity theorem 3.54, $\varphi_0 \in C^\infty$. In fact, here it would be easy to prove this regularity result without using Theorem 3.54. Let us begin by observing that by (2), $\Delta\varphi_0 \in L_2$. Since

(3) $$\int \nabla^{ij}\varphi \nabla_{ij}\varphi \, dV = \int (\Delta\varphi)^2 \, dV - \int R_{ij} \nabla^i \varphi \nabla^j \varphi dV,$$

$\varphi_0 \in \mathsf{H}_2$. Taking the Laplacian of Equation (2) gives $\Delta^2\varphi_0 = \Delta\Delta\varphi_0 = \mu\Delta\varphi_0 \in L_2$. By induction $\Delta^k\varphi_0 \in L_2$ and $|\nabla\Delta^k\varphi_0| \in L_2$ for all $k \in \mathbb{N}$. A straightforward calculation gives equalities like (3) for derivatives of higher order. So $\varphi_0 \in \mathsf{H}_k$ for all $k \in \mathbb{N}$, and the Sobolev imbedding theorem implies $\varphi_0 \in C^\infty$.

Thus μ is an eigenvalue of Δ, and φ_0 an eigenfunction. On the other hand, let γ be an eigenfunction satisfying $\Delta\gamma = \lambda_1\gamma$; then $\lambda_1 = \|\nabla\gamma\|_2^2 \|\gamma\|_2^{-2} \geq \mu$, according to the definition of μ, because $\int \gamma \, dV = 0$. ∎

4.3 Corollary. *Let M be a compact C^∞ Riemannian manifold. If $\varphi \in \mathsf{H}_1$ satisfies $\int \varphi \, dV = 0$, then $\|\varphi\|_2 \leq \lambda_1^{-1/2} \|\nabla\varphi\|_2$, where λ_1 is the first nonzero eigenvalue of Δ.*

4.4 Theorem. *Let $\overline{W}_n$ be a compact Riemannian manifold with boundary of class C^∞. The eigenvalues of the Laplacian Δ are strictly positive. The eigenfunctions, corresponding to the first eigenvalue λ_1, are proportional (the eigenspace is of dimension one). They belong to $C^\infty(\overline{W})$ and they are either strictly positive on W or strictly negative. Moreover λ_1 is equal to μ defined by: $\mu = \inf \|\nabla\varphi\|_2^2$ for all $\varphi \in \mathscr{A}$, with $\mathscr{A} = \{\varphi \in \mathring{\mathsf{H}}_1(W)$ satisfying $\|\varphi\|_2 = 1\}$.*

Proof. We recall that λ is said to be an eigenvalue of the Laplacian Δ if the Dirichlet problem

(4) $$\Delta u = \lambda u \quad \text{in} \quad W, u = 0 \quad \text{on} \quad \partial W$$

has a nontrivial solution (in a given space—here in $C^\infty(\overline{W})$).

First of all, we are going to establish the existence of $\varphi_0 \geq 0$, a nontrivial weak solution of $\Delta u = \mu u$ in W, belonging to $\overset{\circ}{H}_1(W)$. By Proposition 3.49, if $\varphi \in \mathscr{D}(W)$, $|\nabla|\varphi|| = |\nabla\varphi|$ almost everywhere. And $|\varphi| \in \overset{\circ}{H}_1(W)$ by Proposition 3.48. Thus, since $\mathscr{D}(W)$ is dense in $\overset{\circ}{H}_1(W)$, by definition $\mu = \inf\|\nabla\varphi\|_2^2$ for all $\varphi \in \mathscr{A}_+ = \{\varphi \in \overset{\circ}{H}_1(W)$ satisfying $\|\varphi\|_2 = 1$ and $\varphi \geq 0\}$. We proceed now as in Theorem 4.2. Let $\{\varphi_i\}_{i\in\mathbb{N}}$ be a minimizing sequence in $\mathscr{A}_+$. There exist a subsequence $\{\varphi_k\}$ and a $\varphi_0 \in \overset{\circ}{H}_1(W)$, such that $\|\varphi_k - \varphi_0\|_2 \to 0$ and $\varphi_k \to \varphi_0$ weakly in $\overset{\circ}{H}_1$, when $k \to \infty$.
According to Proposition 3.43, there exists a subsequence of $\{\varphi_k\}$ which converges almost everywhere to φ_0. Since $\varphi_k \geq 0$ for all k, $\varphi_0 \geq 0$ and $\varphi_0 \in \mathscr{A}_+$. μ is attained by φ_0, and writing the Euler equation yields: φ_0 satisfies weakly in $\overset{\circ}{H}_1(W)$, $\Delta\varphi_0 = \mu\varphi_0$. It remains to prove the regularity of the solution. According to Theorem 3.54, we have at once $\varphi_0 \in C^\infty(W)$. But the regularity at the boundary is more difficult to establish.
Let $\{\Omega_i, \psi_i\}_{i\in\mathbb{N}}$ be a C^∞ atlas of $\overline{W}$, such that $\psi_i(\Omega_i)$ is either the ball $B = B_0(1) \subset \mathbb{R}^n$ or the half ball $D = \bar{E} \cap B$. And let $\{\alpha_i\}$ be a C^∞ partition of unity subordinate to the cover $\{\Omega_i\}$. We have only to prove that the functions $f_i = (\alpha_i\varphi_0) \circ \psi_i^{-1}$ belong either to $\mathscr{D}(B)$ or to $C^\infty(\bar{D})$. Theorem 3.54 implies the result for the former. For the latter we can prove (see Nirenberg [218], p. 665) that $f_i \in H_k(W)$ for all $k \in \mathbb{N}$. Hence, according to the Sobolev imbedding theorem, $f_i \in C^\infty(\bar{D})$.

So $\varphi_0 \in \overset{\circ}{H}_1(W) \cap C^\infty(\overline{W})$. Consequently φ_0 is zero on ∂W, by Proposition 3.50. If μ is zero, $\|\nabla\varphi_0\|_2 = 0$ and $\varphi_0 = 0$ in W. This fact contradicts $\|\varphi_0\|_2 = 1$. Hence $\mu > 0$ and $\mu = \lambda_1$. Indeed, let $u \in C^2$ be a solution of (4). Multiplying the equation by u and integrating over W lead to $\|\nabla u\|_2^2 = \lambda\|u\|_2^2$, by means of an integration by parts of the first integral. Hence $\lambda \geq \mu$. As $\varphi_0 \not\equiv 0$ satisfies $\varphi_0 \geq 0$, $\varphi_0 \in C^\infty(\overline{W})$, $\varphi_0/\partial W = 0$, and $\Delta\varphi_0 = \mu\varphi_0$ in W, according to the maximum principle 3.71, $\varphi_0 > 0$ in W.

Let ψ_0 be an eigenfunction with $\Delta\psi_0 = \lambda_1\psi_0$. We are going to prove that ψ_0 and φ_0 are proportional. Define $\beta = \sup\{\nu \in \mathbb{R}$ such that $\varphi_0 - \nu\psi_0 > 0$ in $W\}$.
Obviously $\varphi_0 - \beta\psi_0 \geq 0$ in W. But we have more: there is a point $P \in W$ where the function $\varphi_0 - \beta\psi_0$ vanishes. Indeed, suppose $\varphi_0 - \beta\psi_0 > 0$ in W. According to the maximum principle 3.71, $(\partial/\partial\nu)\,(\varphi_0 - \beta\psi_0) < 0$ on ∂W, since

$$\Delta(\varphi_0 - \beta\psi_0) = \lambda_1(\varphi_0 - \beta\psi_0) \geq 0. \tag{5}$$

But the first derivatives are continuous on $\overline{W}$, so there exists an $\varepsilon > 0$ such that $\varphi_0 - (\beta + \varepsilon)\psi_0 > 0$ in W. Hence our initial supposition is false and P does exist. Applying the maximum principle yields $\varphi_0 - \beta\psi_0 = 0$ everywhere, since (5) implies that $\varphi_0 - \beta\psi_0$ cannot achieve its minimum in W, unless it is constant. ∎

4.5 Remark. If $\overline{W}_n$ is only of class C^k, the preceding proof is valid, except for the regularity on the boundary.

When $k > n/2$, we can prove that $\varphi_0 \in C^r(\overline{W})$ with $r < k - n/2$. The proof is similar to the previous one, except that now the atlas $\{\Omega_i, \psi_i\}_{i \in \mathbb{N}}$ is of class C^k. The functions f_i satisfy elliptic equations on $\bar{D}$ with coefficients belonging to $C^{k-2}(\bar{D})$; then by Theorem 3.58, $f_i \in \mathsf{H}_k(\bar{D})$. Applying Theorem 2.30, we have $f_i \in C^r(\bar{D})$.

4.6 Corollary. *Let $\overline{W}$ be a compact Riemannian manifold with boundary of class C^k, $k \geq 1$, or at least Lipschitzian. There exists $\lambda_1 > 0$, such that $\|\varphi\|_2^2 \leq \lambda_1^{-1} \|\nabla\varphi\|_2^2$ for all $\varphi \in \mathring{\mathsf{H}}_1(W)$. Thus $\|\nabla\varphi\|_2$ is an equivalent norm for $\mathring{\mathsf{H}}_1(W)$.*

Proof. Since the Sobolev imbedding theorem 2.30 and the Kondrakov theorem 2.34 hold, the proof of Theorem (4.4) is valid, except for the regularity at the boundary. Thus μ is attained in $\mathscr{A}$ and consequently $\mu = \lambda_1$ is strictly positive. ∎

1.2. Existence Theorem for the Equation $\Delta\varphi = f$

4.7 Theorem. *Let (M_n, g) be a compact C^∞ Riemannian manifold. There exists a weak solution $\varphi \in \mathsf{H}_1$ of (1) if and only if $\int f(x)\, dV = 0$. The solution φ is unique up to a constant. If $f \in C^{r+\alpha}$, ($r \geq 0$ an integer or $r = +\infty$, $0 < \alpha < 1$), then $\varphi \in C^{r+2+\alpha}$.*

Proof. α) if φ is a weak solution of (1) in H_1, by Definition 3.53, $\int a_{ij}\nabla^i\varphi\nabla^j\psi\, dV = \int \psi f\, dV$ for all $\psi \in \mathsf{H}_1$. Choosing $\psi = 1$, we find $\int f\, dV = 0$. This condition is necessary.

β) *Uniqueness up to a constant.* Let φ_1 and φ_2 be weak solutions of (1) in H_1. Set $\tilde{\varphi} = \varphi_2 - \varphi_1$. For all $\psi \in \mathsf{H}_1$, $\int a_{ij}\nabla^i\psi\nabla^j\tilde{\varphi}\, dV = 0$. Choosing $\psi = \tilde{\varphi}$ leads to $\int a_{ij}\nabla^i\tilde{\varphi}\nabla^j\tilde{\varphi}\, dV = 0$. Thus $\tilde{\varphi} =$ Constant.

γ) *Existence of φ.* If $f \equiv 0$, the solutions of (1) are $\varphi =$ Constant. Henceforth suppose $f \not\equiv 0$. Let us consider the functional $I(\varphi) = \int a_{ij}\nabla^i\varphi\nabla^j\varphi\, dV$. Define $\mu = \inf I(\varphi)$ for all $\varphi \in \mathscr{B}$, with $\mathscr{B} = \{\varphi \in \mathsf{H}_1$ satisfying $\int \varphi\, dV = 0$ and $\int \varphi f\, dV = 1\}$.
μ is a nonnegative real number, $0 \leq \mu \leq I(f\|f\|_2^{-2})$. Let $\{\varphi_i\}_{i \in \mathbb{N}}$ be a minimizing sequence in $\mathscr{B}$: $I(\varphi_i) \to \mu$. Since $a_{ij}(x)$ are the components of a Riemannian metric, there exists $\alpha > 0$ such that $I(\varphi_i) \geq \alpha\|\nabla\varphi_i\|_2^2$. Thus the set $\{|\nabla\varphi_i|\}_{i \in \mathbb{N}}$ is bounded in L_2. Moreover, since $\int \varphi_i\, dV = 0$, it follows by Corollary (4.3), that $\|\varphi_i\|_2 \leq \lambda_1^{-1/2}\|\nabla\varphi_i\|_2$. Consequently $\{\varphi_i\}_{i \in \mathbb{N}}$ is bounded in H_1. Applying Theorems 3.17, 3.18, and 2.34 gives: There exist a subsequence $\{\varphi_k\}$ of $\{\varphi_i\}$ and $\varphi_0 \in \mathsf{H}_1$ such that $\|\varphi_k - \varphi_0\|_2 \to 0$ and such that $I(\varphi_0) \leq \mu$, (see the proof of Theorem 4.2).

Hence $\varphi_0 \in \mathscr{B}$ and $I(\varphi_0) = \mu$. Since φ_0 minimizes the variational problem, there exist two constants β and γ such that for all $\psi \in \mathrm{H}_1$:

$$\int a_{ij} \nabla^i \varphi_0 \nabla^j \psi \, dV = \beta \int f \psi \, dV + \gamma \int \psi \, dV.$$

Picking $\psi = 1$ yields $\gamma = 0$. Choosing $\psi = \varphi_0$ implies $\beta = \mu$. Since $\int \varphi_0 f \, dV = 1$, φ_0 is not constant and $\mu = I(\varphi_0) > 0$. Set $\tilde{\varphi}_0 = \varphi_0/\mu$. Then $\tilde{\varphi}$ satisfies Equation (1) weakly in H_1 and Theorem 3.54 implies the last statement. ■

4.8 Theorem. *Let $\overline{W}_n$ be a compact Riemannian manifold with boundary of class C^∞. There exists a unique weak solution of (1): $\varphi \in \mathring{\mathrm{H}}_1(W)$. If $f \in C^\infty(\overline{W})$, then $\varphi \in C^\infty(\overline{W})$ and φ vanishes on the boundary.*

Proof. Uniqueness is obvious. Let φ_1 and φ_2 be weak solutions of (1) in $\mathring{\mathrm{H}}_1(W)$. Set $\tilde{\varphi} = \varphi_2 - \varphi_1$. For all $\psi \in \mathring{\mathrm{H}}_1$:

$$\int_W a_{ij} \nabla^i \psi \nabla^j \tilde{\varphi} \, dV = 0. \quad \text{Choosing } \psi \in \tilde{\varphi} \text{ leads to } \tilde{\varphi} \equiv 0.$$

For existence, let us consider the functional

$$J(\varphi) = \int_W a_{ij} \nabla^i \varphi \nabla^j \varphi \, dV - 2 \int_W f \varphi \, dV.$$

Define $\mu = \inf J(\varphi)$ for all $\varphi \in \mathring{\mathrm{H}}_1(W)$. μ is finite. Indeed, since a_{ij} are the components of a C^∞ Riemannian metric on $\overline{W}$ which is compact, there exists an $\alpha > 0$ such that

$$\int_W a_{ij} \nabla^i \varphi \nabla^j \varphi \, dV \geq \alpha \|\nabla \varphi\|_2^2.$$

On the other hand, by Corollary 4.6, $\|\varphi\|_2^2 \leq \lambda_1^{-1} \|\nabla \varphi\|_2^2$. Thus for all $\varepsilon > 0$,

$$(6) \quad J(\varphi) \geq \alpha \|\nabla \varphi\|_2^2 - \varepsilon \|\varphi\|_2^2 - \frac{1}{\varepsilon} \|f\|_2^2 \geq (\alpha - \varepsilon \lambda_1^{-1}) \|\nabla \varphi\|_2^2 - \frac{1}{\varepsilon} \|f\|_2^2.$$

Choosing $\varepsilon = \alpha \lambda_1/2$ gives $\mu \geq -2(\alpha \lambda_1)^{-1} \|f\|_2^2$.

Let $\{\varphi_i\}_{i \in \mathbb{N}}$ be a minimizing sequence of J in $\mathring{\mathrm{H}}_1(W)$. According to (6), $\{\varphi_i\}_{i \in \mathbb{N}}$ is bounded in $\mathring{\mathrm{H}}_1$. Applying Theorems 3.17, 3.18, and 2.34 yields: There exist a subsequence $\{\varphi_k\}$ of $\{\varphi_i\}$ and $\varphi_0 \in \mathring{\mathrm{H}}_1$ such that $\|\varphi_k - \varphi_0\|_2 \to 0$ and such that $J(\varphi_0) \leq \mu$. Hence $J(\varphi_0) = \mu$, and φ_0 satisfies Equation (1) weakly in $\mathring{\mathrm{H}}_1$. Theorem 3.58 and Proposition (3.50) imply the last statement. ■

§2. Green's Function of the Laplacian

4.9 The Laplacian in a local chart can be written as follows:

$$\Delta\varphi = -\nabla_i(g^{ij}\nabla_j\varphi) = -\partial_i(g^{ij}\partial_j\varphi) - g^{kj}\partial_j\varphi\Gamma^i_{ik},$$
$$\Delta\varphi = -|g|^{-1/2}\partial_i[g^{ij}\sqrt{|g|}\partial_j\varphi], \text{ because } \Gamma^i_{ik} = \partial_k \log\sqrt{|g|}.$$

If $\varphi = f(r)$ in geodesic polar coordinates:

$$-\Delta f(r) = \frac{1}{r^{n-1}\sqrt{|g|}}\partial_r[r^{n-1}\sqrt{|g|}\partial_r f] = f'' + \frac{n-1}{r}f' + f'\partial_r \log\sqrt{|g|}.$$

By Theorem 1.53, there exists a constant A such that

$$|\partial_r \log\sqrt{|g|}| \leq Ar. \tag{7}$$

2.1. Parametrix

4.10 In $\mathbb{R}^n$, $n \geq 3$, $\Delta_{Q\,\text{distr.}}(r^{2-n}) = (n-2)\omega_{n-1}\delta_P$ and in $\mathbb{R}^2$, $\Delta_{Q\,\text{distr.}} \log r = -2\pi\delta_P$, where δ_P is the Dirac function at P, $r = d(P, Q)$ and ω_{n-1} is the volume of the unit sphere of dimension $n-1$.
On a Riemannian manifold, r is only a Lipschitzian function. For this reason we must consider $f(r)$ a positive decreasing function, which is equal to 1 in a neighborhood of zero, and to zero for $r \geq \delta(P)$ the injectivity radius at P (see Theorem 1.36). We define:

$$\mathsf{H}(P, Q) = [(n-2)\omega_{n-1}]^{-1}r^{2-n}f(r), \quad \text{for } n > 2 \quad \text{and} \tag{8}$$
$$\mathsf{H}(P, Q) = -(2\pi)^{-1}f(r)\log r, \quad \text{for } n = 2,$$

and compute when $n > 2$:

$$\Delta_Q \mathsf{H}(P, Q) = [(n-2)\omega_{n-1}]^{-1}r^{1-n}[(n-3)f' - rf'' + ((n-2)f - rf')\partial_r \log\sqrt{|g|}].$$

According to (7), there exists a constant B such that

$$|\Delta_Q \mathsf{H}(P, Q)| \leq Br^{2-n}; \tag{9}$$

B depends on P. But on a compact Riemannian manifold, we can choose $f(r)$ zero for $r \geq \delta$ the injectivity radius, and then B does not depend on P. Henceforth in the compact case, $f(r)$ will be chosen in this way.

2.2. Green's Formula

4.10

$$(10)\qquad \psi(P) = \int_M \mathsf{H}(P, Q)\Delta\psi(Q)\,dV(Q) - \int_M \Delta_Q \mathsf{H}(P, Q)\psi(Q)\,dV(Q),$$

for all $\psi \in C^2$. *Recall* (8), *the definition of* $\mathsf{H}(P, Q)$.

For the proof, we compute $\int_{M-B_P(\varepsilon)} \mathsf{H}(P, Q)\Delta\psi(Q)\,dV(Q)$, integrating by parts twice. Letting $\varepsilon \to 0$ yields Green's formula.
If $\psi \in C^\infty$, by definition $\langle \Delta_{\text{distr.}} \mathsf{H}, \psi\rangle = \langle \mathsf{H}, \Delta\psi\rangle$ in the sense of distributions, and $\langle \mathsf{H}, \Delta\psi\rangle = \int_M \mathsf{H}(P, Q)\Delta\psi(Q)\,dV(Q)$. Thus:

$$(11)\qquad \Delta_{Q\,\text{distr.}} \mathsf{H}(P, Q) = \Delta_Q \mathsf{H}(P, Q) + \delta_P(Q).$$

Picking $\psi \equiv 1$ in (10), gives $\int_M \Delta_Q \mathsf{H}(P, Q)\,dV(Q) = -1$.
From (10), after interchanging the order of integration, we establish that all $\varphi \in C^2$ satisfy:

$$(12)\qquad \varphi(Q) = \Delta_Q \int_M \mathsf{H}(P, Q)\varphi(P)\,dV(P) - \int_M \Delta_Q \mathsf{H}(P, Q)\varphi(P)\,dV(P).$$

4.11 Definition. Let $\overline{W}_n$ be a compact Riemannian manifold with boundary of class C^∞. The *Green's function* $\mathsf{G}(P, Q)$ of the Laplacian is the function which satisfies in $W \times W$:

$$(13)\qquad \Delta_{Q\,\text{distr.}} \mathsf{G}(P, Q) = \delta_P(Q),$$

and which vanishes on the boundary (for P or Q belonging to ∂W).

Let M_n be a compact C^∞ Riemannian manifold having volume V. The *Green's function* $\mathsf{G}(P, Q)$ is a function which satisfies:

$$(14)\qquad \Delta_{Q\,\text{distr.}} \mathsf{G}(P, Q) = \delta_P(Q) - V^{-1}.$$

The Green's function is defined up to a constant in this case. Recall that δ_P is the Dirac function at P.

4.12 Proposition (Giraud [126] p. 150). *Let* Ω *be a bounded open set of* $\mathbb{R}^n$ *and let* $X(P, Q)$ *and* $Y(P, Q)$ *be continuous functions defined on* $\Omega \times \Omega$ *minus the diagonal which satisfy*

$$|X(P, Q)| \le \text{Const} \times [d(P, Q)]^{\alpha-n} \quad\textit{and}\quad |Y(P, Q)| \le \text{Const} \times [d(P, Q)]^{\beta-n}$$

for some real numbers α, β *belonging to* $]0, n[$. *Then*

$$Z(P, Q) = \int_\Omega X(P, R)Y(R, Q)\, dV(R)$$

is continuous for $P \neq Q$ *and satisfies*:

$$|Z(P, Q)| \leq \text{Const} \times [d(P, Q)]^{\alpha+\beta-n} \quad \text{if } \alpha + \beta < n,$$

$$|Z(P, Q)| \leq \text{Const} \times [1 + |\log d(P, Q)|] \quad \text{if } \alpha + \beta = n,$$

$$|Z(P, Q)| \leq \text{Const} \quad \text{if } \alpha + \beta > n;$$

in the last case $Z(P, Q)$ *is continuous on* $\Omega \times \Omega$.

Proof. The integral which defines $Z(P, Q)$ is less than the sum of three integrals, an upper bound of which is easily found. The integrals are over the sets $\Omega \cap B_P(\rho)$, $[B_Q(3\rho) - B_P(\rho)] \cap \Omega$, and $\Omega - \Omega \cap B_Q(3\rho)$, with $2\rho = d(P, Q)$ small enough. ∎

2.3. Green's Function for Compact Manifolds

4.13 Theorem. *Let* M_n *be a compact* C^∞ *Riemannian manifold. There exists* $\mathsf{G}(P, Q)$, *a Green's function of the Laplacian which has the following properties*:

(a) *For all functions* $\varphi \in C^2$:

$$(15) \quad \varphi(P) = V^{-1} \int_M \varphi(Q)\, dV(Q) + \int_M \mathsf{G}(P, Q)\Delta\varphi(Q)\, dV(Q).$$

(b) $\mathsf{G}(P, Q)$ *is* C^∞ *on* $M \times M$ *minus the diagonal* (*for* $P \neq Q$).

(c) *There exists a constant* k *such that*:

$$|\mathsf{G}(P, Q)| < k(1 + |\log r|) \quad \text{for} \quad n = 2 \quad \text{and}$$

$$(16) \quad |\mathsf{G}(P, Q)| < kr^{2-n} \quad \text{for} \quad n > 2, |\nabla_Q \mathsf{G}(P, Q)| < kr^{1-n},$$

$$|\nabla_Q^2 \mathsf{G}(P, Q)| < kr^{-n} \quad \text{with} \quad r = d(P, Q).$$

(d) *There exists a constant* A *such that* $\mathsf{G}(P, Q) \geq A$. *Because the Green function is defined up to a constant, we can thus choose the Green's function everywhere positive.*

(e) $\int \mathsf{G}(P, Q)\, dV(P) = \text{Const}$. *We can choose the Green's function so that its integral equals zero.*

(f) $\mathsf{G}(P, Q) = \mathsf{G}(Q, P)$.

Proof of existence. Define $\Gamma(P, Q) = \Gamma_1(P, Q) = -\Delta_Q \mathsf{H}(P, Q)$ and $\Gamma_{i+1}(P, Q) = \int_M \Gamma_i(P, R)\Gamma(R, Q)\, dV(R)$ for $i \in \mathbb{N}$. Pick $\mathbb{N} \ni k > n/2$ and set

$$(17) \quad \mathsf{G}(P, Q) = \mathsf{H}(P, Q) + \sum_{i=1}^{k} \int_M \Gamma_i(P, R)\mathsf{H}(R, Q)\, dV(R) + F(P, Q).$$

By (11), (12), and (14), $F(P, Q)$ satisfies

$$(18) \qquad \Delta_Q F(P, Q) = \Gamma_{k+1}(P, Q) - V^{-1}.$$

According to (9), $|\Gamma(P, Q)| \leq Br^{2-n}$. Thus from Proposition 4.12, $\Gamma_k(P, Q)$ is bounded and consequently $\Gamma_{k+1}(P, Q)$ is C^1.

Now for P fixed, there exists a weak solution of (18), (Theorem 4.7), unique up to a constant. Using the theorem of regularity 3.54, the solution is C^2. $\mathsf{G}(P, Q)$, defined by (17), satisfies (14). And $Q \to \mathsf{G}(P, Q)$ is C^∞ for $P \neq Q$ (Theorem 3.54). For the present, we choose $\mathsf{G}(P, Q)$ such that:

$$\int_M \mathsf{G}(P, Q)\, dV(Q) = 0.$$

Proof of the properties. a) (14) applied to $\varphi \in C^\infty$ leads to (15) and $\mathscr{D}(M)$ is dense in C^2.

b) We are going to prove that $P \to \mathsf{G}(P, Q)$ is continuous for $P \neq Q$. Since we know that $Q \to \mathsf{G}(P, Q)$ is C^∞ for $Q \neq P$, the result is a consequence of f): $\mathsf{G}(P, Q) = \mathsf{G}(Q, P)$, using for the derivatives a proof similar to the following. Iterating k-times ($k > n/2$) Green's formula (10) leads to:

$$\psi(P) = \int_M \mathsf{H}(P, Q)\Delta\psi(Q)\, dV(Q)$$
$$+ \sum_{i=1}^{k} \int_M \left[\int_M \Gamma_i(P, R)\mathsf{H}(R, Q)\, dV(R)\right]\Delta\psi(Q)\, dV(Q)$$
$$+ \int_M \Gamma_{k+1}(P, Q)\psi(Q)\, dV(Q).$$

Using (8), (9), and Proposition 4.12 gives:

$$(19a) \qquad |\psi(P)| \leq \text{Const} \times (\sup |\Delta\psi| + \|\psi\|_2).$$

According to Corollary 4.3, if $\int \psi\, dV = 0$:

$$(19b) \qquad \|\psi\|_2^2 \leq \lambda_1^{-1} \|\nabla\psi\|_2^2 \leq \lambda_1^{-1} \|\Delta\psi\|_2 \|\psi\|_2;$$

the last inequality arises after integrating by parts and using Hölder's inequality. Hence, there exists a constant C such that the solution of $\Delta\psi = f$ with $\int \psi\, dV = 0$ and $\int f\, dV = 0$ (Theorem 4.7) satisfies:

$$\sup|\psi| \leq C \sup|f|.$$

Applying this result to (18):

$$\sup_Q \left| [F(P,Q) - F(R,Q)] - V^{-1} \int_V [F(P,Q) - F(R,Q)]\, dV(Q) \right|$$
$$\leq C \sup_Q |\Gamma_{k+1}(P,Q) - \Gamma_{k+1}(R,Q)|.$$

Since $\int_M \mathsf{G}(P,Q)\, dV(Q) = 0$, it follows from (17) that $\int_M F(P,Q)\, dV(Q)$ is a continuous function of P.
Thus $P \to F(P,Q)$ is continuous, and for $P \neq Q$, $P \to \mathsf{G}(P,Q)$ is also.

Using only this continuity of $\mathsf{G}(P,Q)$, we will shortly prove parts d)–f). Assuming this has been done, we complete the proof of part b). By f) $\mathsf{G}(P,Q) = \mathsf{G}(Q,P)$. Thus $\mathsf{G}(P,Q)$ is C^∞ in P for $P \neq Q$ and any r-derivative at P $\partial_P^r \mathsf{G}(P,Q)$ is a distribution in Q which satisfies $\Delta_Q\, \partial_P^r \mathsf{G}(P,Q) = 0$ on $M - \{P\}$. $\partial_P^r \mathsf{G}(P,Q)$ is then C^∞ in Q for $Q \neq P$ according to Theorem 3.54.

c) The inequalities follow from (17). When $Q \to P$, the leading part of $\mathsf{G}(P,Q)$ is $\mathsf{H}(P,Q)$.

d) From this fact, there exists an open neighborhood Ω of the diagonal in $M \times M$, where $\mathsf{G}(P,Q)$ is positive. On $M \times M - \Omega$, which is compact, $\mathsf{G}(P,Q)$ is continuous. Thus $\mathsf{G}(P,Q)$ has a minimum on $M \times M$.

e) Since $P \to \mathsf{G}(P,Q)$ is continuous for $P \neq Q$ and $|\mathsf{G}(P,Q)| \leq \text{Const}\, r^{2-n}$, we can consider $\int_M \mathsf{G}(P,Q)\, dV(P)$, and the transposition of (15):

$$\psi(Q) = V^{-1} \int_M \psi(P)\, dV(P) + \Delta_Q \int_M \mathsf{G}(P,Q)\psi(P)\, dV(P). \tag{20}$$

Picking $\psi \equiv 1$ gives $\int_M \mathsf{G}(P,Q)\, dV(P) = \text{Const}$.

f) Choosing $\psi = \Delta\varphi$ in (20) leads to

$$\Delta\varphi(Q) = \Delta_Q \int_M \mathsf{G}(P,Q)\Delta\varphi(P)\, dV(P).$$

Thus $\varphi(Q) = \int_M \mathsf{G}(P, Q)\Delta\varphi(P)\,dV(P) + \text{Const}$, by (15), and this equality yields

$$\int_M [\mathsf{G}(P, Q) - \mathsf{G}(Q, P)]\Delta\varphi(Q)\,dV(Q) = \text{Const} \tag{21}$$

for all $\varphi \in C^2$. Integrating (21) proves that the constant is zero, since

$$\int \mathsf{G}(Q, P)\,dV(P) = 0 \quad \text{and} \quad \int \mathsf{G}(P, Q)\,dV(P) = \text{Const}.$$

Thus $\mathsf{G}(P, Q) - \mathsf{G}(Q, P) = \text{Const}$. Interchanging P and Q implies the second member is zero. ∎

4.14 Proposition. *Equality* (15) *holds when the integrals make sense.*

Proof. Suppose that $\Delta\varphi \in L_1$. Since $\mathscr{D}(M)$ is dense in L_1 there exists a sequence $\{g_m\}$ in $\mathscr{D}(M)$ such that $\|g_m - \Delta\varphi\|_1 \to 0$. Thus $\int_M g_m\,dV \to 0$ and $g_m - V^{-1}\int_M g_m\,dV \to \Delta\varphi$ in L_1.

Therefore we can choose $\{g_m\}$ with $\int_M g_m\,dV = 0$ and, according to Theorem 4.7, there exists $\{f_m\}$ such that $\int_M f_m\,dV = \int_M \varphi\,dV$ and $\Delta f_m = g_m$. f_m belongs to C^∞ and satisfies

$$f_m(P) = V^{-1}\int_M f_m\,dV + \int_M \mathsf{G}(P, Q)g_m(Q)\,dV(Q).$$

According to Proposition 3.64, $f_m \to V^{-1}\int_M \varphi\,dV + \int_M \mathsf{G}(P, Q)\Delta\varphi(Q)\,dV(Q)$ in L_1. On the other hand, $f_m \to \varphi$ in the distributional sense, since $\int_M f_m\,dV = \int_M \varphi\,dV$ and $\Delta f_m \to \Delta\varphi$ in L_1. Thus φ satisfies (15) almost everywhere. ∎

4.15 Remark. It is possible to define the Green's function as the sum of a series (see Aubin [12]). This alternate definition allows one to obtain estimates on the Green's function in terms of the diameter D, the injectivity radius, d, the upper bound b, of the curvature and the lower bound a of the Ricci curvature. As a consequence, Aubin ([12] p. 367) proved that λ_1 the first nonzero eigenvalue is bounded away from zero:
There exist three positive constants C, k, and ξ which depend only on n, such that $\lambda_1 \geq CD^{-2}k^{D/\delta}$, δ satisfying $-a\delta^2 \leq \xi$, $2\delta\sqrt{\sup(0, b)} \leq \pi$, and $0 < \delta \leq d$.
Other positive lower bounds were found by various authors. Let us mention only Cheeger [82] and Yau [274].

4.16 Remark. The Green's function was introduced by Hilbert [140] and the Green's form on a compact Riemannian manifold by G. De Rham [105] and Bidal and De Rham [52].

2.4. Green's Function for Compact Manifolds with Boundary

4.17 Theorem. *Let $\overline{W}_n$ be an oriented compact Riemannian manifold with boundary of class C^∞. There exists $\mathsf{G}(P, Q)$, the Green's function of the Laplacian, which has the following properties:*

(a) *All functions $\varphi \in C^2(\overline{W})$ satisfy*

$$(22)\quad \varphi(P) = \int_W \mathsf{G}(P, Q)\Delta\varphi(Q)\, dV(Q) - \int_{\partial W} \nu^i \nabla_{iQ} \mathsf{G}(P, Q)\varphi(Q)\, ds(Q),$$

where ν is the unit normal vector oriented to the outside and ds is the volume element on ∂W corresponding to the Riemannian metric $\mathfrak{j}^*g$ ($\mathfrak{j}: \partial W \to \overline{W}$ the canonical imbedding).

(b) *$\mathsf{G}(P, Q)$ is C^∞ on $\overline{W} \times \overline{W}$ minus the diagonal (for $P \neq Q$).*

(c) *$|\mathsf{G}(P, Q)| < kr^{2-n}$ for $n > 2$, $|\mathsf{G}(P, Q)| < k(1 + |\log r|)$ for $n = 2$, $|\nabla_Q \mathsf{G}(P, Q)| < kr^{1-n}$, $|\nabla^2_Q \mathsf{G}(P, Q)| < kr^{-n}$, with $r = d(P, Q)$ and k a constant which depends on the distance of P to the boundary.*

(d) *$\mathsf{G}(P, Q) > 0$ for P and Q belonging to the interior of the manifold*

(e) *$\mathsf{G}(P, Q) = \mathsf{G}(Q, P)$.*

Proof of existence. Let $P \in W$ given. We define $\mathsf{H}(P, Q)$ as in (8), where $f(r)$ is a function equal to zero for $r > \delta(P)(k + 1)^{-1}$ with $\mathbb{N} \ni k > n/2$ and $\delta(P)$ the injectivity radius at P.
$F(P, Q)$ defined by (17) satisfies

$$\Delta_Q F(P, Q) = \Gamma_{k+1}(P, Q),\ F(P, Q) = 0 \quad \text{for } Q \in \partial W.$$

According to Theorem 4.8, there exists a solution in $\overset{\circ}{\mathsf{H}}_1(W)$.
$\mathsf{G}(P, Q)$ defined by (17) satisfies (13), is C^∞ on $\overline{W} - P$, and equals zero for $Q \in \delta W$. (We can apply the theorems of regularity to $\overline{W} - B_P(\varepsilon)$ with $\varepsilon > 0$ small enough.)

Proof of the properties. a) The result is obtained by using Stokes' formula (see 1.70).

b) The proof is similar to that of Theorem 4.13 b).

c) The leading part of $\mathsf{G}(P, Q)$ is $\mathsf{H}(P, Q)$

d) Let $P \in W$ given. According to the previous result $\mathsf{G}(P, Q) > 0$ for Q belonging to a ball $B_P(\varepsilon)$ with $\varepsilon > 0$ small enough. Applying the maximum principle 3.71, $\mathsf{G}(P, Q)$ achieves a minimum on the boundary of $W - B_P(\varepsilon)$, since $\Delta_Q \mathsf{G}(P, Q) = 0$. Thus $\mathsf{G}(P, Q) > 0$ for $Q \in W$.

e) Transposing (22) with φ and ψ belonging to $\mathcal{D}(W)$ yields:

$$\psi(Q) = \Delta_Q \int_W \mathsf{G}(P, Q)\psi(P)\, dV(P).$$

Choose $\psi(Q) = \Delta\varphi(Q)$. By Theorem 4.8,

$$\varphi(Q) = \int_W \mathsf{G}(P, Q)\Delta\varphi(P)\, dV(P).$$

Hence $\mathsf{G}(P, Q)$ satisfies

$$\Delta_{P\,\text{distr.}}\, \mathsf{G}(P, Q) = \delta_Q(P)$$

and $\mathsf{G}(P, Q) = \mathsf{G}(Q, P)$.
Indeed $\Delta_Q[\mathsf{G}(P, Q) - \mathsf{G}(Q, P)] = 0$ and $\mathsf{G}(P, Q) - \mathsf{G}(Q, P)$ vanishes for $Q \in \partial W$. Applying Theorem 4.8 yields the claimed result. ∎

4.18 Let us now prove a result similar to that of Theorem 4.7, a result which we will use in Chapter 7.

On a compact Riemannian manifold M, let $\mathfrak{Q}$ be a $C^{r+\alpha}$ section of $T^*(M) \otimes T^*(M)$, which defines everywhere a positive definite bilinear symmetric form ($\mathfrak{Q}$ is a $C^{r+\alpha}$ Riemannian metric) where $r \geq 1$ is an integer and α a real number $0 < \alpha < 1$. Consider the equation

$$-\nabla^i[a_{ij}(x)\nabla^j\varphi] + b(x)\varphi = f(x) \tag{23}$$

where $a_{ij}(x)$ are the components in a local chart of $\mathfrak{Q}$ and where $b(x)$ and $f(x)$ are functions belonging to $C^{r+\alpha}$. Moreover, we suppose that $-\nabla^i a_{ij}(x)$ belongs to $C^{r+\alpha}$.

Theorem 4.18. *If* $b(x) > 0$, *Equation* (23) *has a unique solution belonging to* $C^{r+2+\alpha}$.

Proof. Suppose at first that $a_{ij}(x)$, $b(x)$ and $f(x)$ belong to C^∞. In that case we consider the functional $I(\varphi) = \int a_{ij}\nabla^i\varphi\nabla^j\varphi\, dV + \int b\varphi^2\, dV$ and $\mu = \inf I(\varphi)$, for all $\varphi \in \mathsf{H}_1$ satisfying $\int \varphi f\, dV = 1$. A proof similar to that of 4.7 establishes the existence of a solution, which belongs to C^∞ by the regularity theorem 3.54 and which is unique by the maximum principle 3.71.

Now in the general case we approximate in $C^{1+\alpha}$ the coefficients of Equation (23) by coefficients belonging to C^∞. We obtain a sequence of equations

$$E_k\colon -\nabla^i[a_{kij}(x)\nabla^j\varphi] + b_k(x)\varphi = f_k(x)$$

with C^∞ coefficients ($k = 1, 2, \ldots$). And we can choose E_k so that $b_k(x) > b_0$ and $a_{kij}(x)\xi^i\xi^j \geq \lambda|\xi|^2$ for some $b_0 > 0$ and $\lambda > 0$ independent of k.

By the first part of the proof, E_k has a C^∞ solution φ_k. And these solutions ($k = 1, 2, \ldots$) are uniformly bounded. Indeed, considering the maximum and then the minimum of φ_k, we get

$$\|\varphi_k\|_{C^0} \leq b_0^{-1}\|f_k\|_{C^0}.$$

Now by the Schauder interior estimates 3.61, the sequence $\{\varphi_k\}$ is bounded in $C^{2,\alpha}$. To apply the estimates we consider a finite atlas $\{\Omega_i, \psi_i\}$ and compact sets $K_i \subset \Omega_i$ such that $M = \bigcup_i K_i$.
As $\{\varphi_k\}$ is bounded in $C^{2,\alpha}$, by Ascoli's theorem 3.15, there exist $\varphi \in C^2$ and a subsequence $\{\varphi_j\}$ of $\{\varphi_k\}$ such that $\varphi_j \to \varphi$ in C^2. Thus $\varphi \in C^{2,\alpha}$ and satisfies (23). Lastly, according to Theorem 3.55, the solution φ belongs to $C^{r+2+\alpha}$ and is unique (uniqueness does not use the smoothness of the coefficients).

Remark. For the proof of Theorem 4.18, we can also minimize over H_1 the functional

$$J(\varphi) = \int a_{ij} \nabla^i \varphi \nabla^j \varphi \, dV + \int b\varphi^2 \, dV - 2 \int f\varphi \, dV.$$

We considered a similar functional in the proof of Theorem 4.8.

Chapter 5

The Methods

5.1 Preliminaries. We do not intend to discuss the various methods from a general theoretical point of view. For that the reader may wish to look at Berger's book [42]. Instead, we are going to apply some of these methods to specific geometric problems, to see how they are used in a concrete way.

5.2 Some methods. The *variational method* will be used in this chapter for the *Yamabe Problem* 6.2 on the scalar curvature and for the same problem in two dimensions according to Melvyn Berger [5.8].

The *continuity method* will be used to prove the existence of Einstein–Kähler metric on some compact Kählerian manifold (Chapter 7).

The *method of lower and upper solutions* will be mentioned in Chapter 7 but the reader can look at Kazdan and Kramer [156], where the method is used for equations with the right-hand side a function of the gradient (a surprise). In this article also, the *fixed point method* is applied. In Inouë [150] the *steepest descent method* is used. For more details on *topological methods* and also on the other methods, the reader should look at Nirenberg's article [223].

§1. Yamabe's Equation

5.3 On a C^∞ compact Riemannian manifold M_n of dimension $n \geq 3$, let us consider the differential equation

$$(1) \qquad \Delta\varphi + h(x)\varphi = \lambda f(x)\varphi^{N-1},$$

where $h(x)$ and $f(x)$ are C^∞ functions on M_n, with $f(x)$ everywhere strictly positive and $N = 2n/(n-2)$.

The problem is to prove the existence of a real number λ and of a C^∞ function φ, everywhere strictly positive, satisfying (1).

1.1. Yamabe's Method

5.4 Yamabe considered, for $2 < q \leq N$, the functional

$$(2) \qquad I_q(\varphi) = \left[\int_M \nabla^i\varphi\nabla_i\varphi\, dV + \int_M h(x)\varphi^2\, dV\right]\left[\int_M f(x)\varphi^q\, dV\right]^{-2/q},$$

where $\varphi \not\equiv 0$ is a nonnegative function belonging to H_1, the first Sobolev space. The denominator of $I_q(\varphi)$ makes sense since, according to Theorem 2.21, $\mathsf{H}_1 \subset L_N \subset L_q$. Define

$$\text{(3)} \qquad \mu_q = \inf I_q(\varphi) \quad \text{for all } \varphi \in \mathsf{H}_1,\ \varphi \geq 0,\ \varphi \not\equiv 0.$$

It is impossible to prove directly that μ_N is attained and thus to solve Equation (1). (We shall soon see why). This is the reason why Yamabe considered the approximate equations for $q < N$:

$$\text{(4)} \qquad \Delta\varphi + h(x)\varphi = \lambda f(x)\varphi^{q-1}$$

and proved (Theorem B of Yamabe [269]):

5.5 Theorem. *For $2 < q < N$, there exists a C^∞ strictly positive function φ_q satisfying Equation (4) with $\lambda = \mu_q$ and $I_q(\varphi_q) = \mu_q$.*

Proof.

a) For $2 < q \leq N$, μ_q is finite. Indeed

$$\text{(5)} \qquad I_q(\varphi) \geq \left[\inf_{x \in M}(0, h(x))\right]\left[\sup_{x \in M} f(x)\right]^{-2/q} \|\varphi\|_2^2 \|\varphi\|_q^{-2}$$

and

$$\text{(6)} \qquad \|\varphi\|_2^2 \|\varphi\|_q^{-2} \leq V^{1-2/q} \leq \sup(1, V^{2/n}),$$

with $V = \int_M dV$. On the other hand,

$$\mu_q \leq I_q(1) = \left[\int_M h(x)\, dV\right]\left[\int_M f(x)\, dV\right]^{-2/q}.$$

b) Let $\{\varphi_i\}$ be a minimizing sequence such that $\int_M f(x)\varphi_i^q\, dV = 1$:

$$\varphi_i \in \mathsf{H}_1,\ \varphi_i \geq 0,\ \lim_{i \to \infty} I_q(\varphi_i) = \mu_q.$$

First we prove that the set of the φ_i is bounded in H_1,

$$\|\varphi_i\|_{\mathsf{H}_1}^2 = \|\nabla\varphi_i\|_2^2 + \|\varphi_i\|_2^2 = I_q(\varphi_i) - \int_M h(x)\varphi_i^2\, dV + \|\varphi_i\|_2^2.$$

Since we can suppose that $I_q(\varphi_i) < \mu_q + 1$, then

$$\|\varphi_i\|_{\mathsf{H}_1}^2 \leq \mu_q + 1 + \left[1 + \sup_{x \in M}|h(x)|\right]\|\varphi_i\|_2^2$$

and

$$\|\varphi_i\|_2^2 \leq [V]^{1-2/q}\|\varphi_i\|_q^2 \leq [V]^{1-2/q}\left[\inf_{x\in M} f(x)\right]^{-2/q}$$

c) If $2 < q < N$, there exists a nonnegative function $\varphi_q \in \mathrm{H}_1$, satisfying

$$I_q(\varphi_q) = \mu_q \quad \text{and} \quad \int_M f(x)\varphi_q^q \, dV = 1.$$

Indeed, for $2 < q < N$, the imbedding $\mathrm{H}_1 \subset L_q$ is compact by Kondrakov's theorem 2.34 and, since the bounded closed sets in H_1 are weakly compact (Theorem 3.18), there exists $\{\varphi_j\}$ a subsequence of $\{\varphi_i\}$, and a function $\varphi_q \in \mathrm{H}_1$ such that:

(α) $\varphi_j \to \varphi_q$ in L_q,
(β) $\varphi_j \to \varphi_q$ weakly in H_1.
(γ) $\varphi_j \to \varphi_q$ almost everywhere

The last assertion is true by Proposition 3.43. (α) $\Rightarrow \int_M f(x)\varphi_q^q \, dV = 1$; γ) $\Rightarrow \varphi_q \geq 0$, and β) implies

$$\|\varphi_q\|_{\mathrm{H}_1} \leq \liminf_{i\to\infty} \|\varphi_j\|_{\mathrm{H}_1} \qquad \text{(Theorem 3.17).}$$

Hence $I_q(\varphi_q) \leq \lim_{j\to\infty} I_q(\varphi_j) = \mu_q$ because $\varphi_j \to \varphi_q$ in L_2, according to (α) since $q \geq 2$. Therefore, by definition of μ_q, $I_q(\varphi_q) = \mu_q$.

d) φ_q satisfies Equation (4) weakly in H_1. We compute Euler's equation. Set $\varphi = \varphi_q + v\psi$ with $\psi \in \mathrm{H}_1$ and v a small real number. An asymptotic expansion gives:

$$I_q(\varphi) = I_q(\varphi_q)\left[1 + vq\int_M f(x)\varphi_q^{q-1}\psi \, dV\right]^{-2/q}$$
$$+ 2v\left[\int_M \nabla^\nu \varphi_q \nabla_\nu \psi \, dV + \int_M h(x)\varphi_q \psi \, dV\right] + 0(v).$$

Thus φ_q satisfies for all $\psi \in \mathrm{H}_1$:

$$\text{(7)} \qquad \int_M \nabla^\nu \varphi_q \nabla_\nu \psi \, dV + \int_M h(x)\varphi_q \psi \, dV = \mu_q \int_M f(x)\varphi_q^{q-1}\psi \, dV.$$

To check that the preceding computation is correct, we note that since $\mathscr{D}(M)$ is dense in H_1 and $\varphi \not\equiv 0$, then

$$\inf_{\varphi\in \mathrm{H}_1} I_q(\varphi) = \inf_{\varphi\in C^\infty} I_q(\varphi) = \inf_{\varphi\in C^\infty} I_q(|\varphi|) \geq \inf_{\substack{\varphi\in \mathrm{H}_1\\ \varphi\geq 0}} I_q(\varphi) \geq \inf_{\varphi\in \mathrm{H}_1} I_q(\varphi).$$

$I(\varphi) = I(|\varphi|)$ when $\varphi \in C^\infty$ because the set of the point P where simultaneously $\varphi(P) = 0$ and $|\nabla\varphi(P)| \neq 0$ has zero measure (or we can use Proposition 3.49 directly).

e) *$\varphi_q \in C^\infty$ for $2 \leq q < N$ and the functions φ_q are uniformly bounded for $2 \leq q \leq q_0 < N$.*

Let $\mathsf{G}(P, Q)$ be the Green's function (see 4.13). φ_q satisfies the integral equation (see 4.14)

$$(8) \qquad \varphi_q(P) = V^{-1}\int_M \varphi_q(Q)\, dV(Q) + \int_M \mathsf{G}(P, Q)[\mu_q f(Q)\varphi_q^{q-1} - h(Q)\varphi_q]\, dV(Q).$$

We know that $\varphi_q \in L_{r_0}$ with $r_0 = N$. Since, by Theorem 4.13c there exists a constant B such that $|\mathsf{G}(P, Q)| \leq B[d(P, Q)]^{2-n}$, then according to Sobolev's lemma 2.12 and its corollary, $\varphi_q \in L_{r_1}$, for $2 < q \leq q_0$ with

$$\frac{1}{r_1} = \frac{n-2}{n} + \frac{q_0 - 1}{r_0} - 1 = \frac{q_0 - 1}{r_0} - \frac{2}{n}$$

and there exists a constant A_1 such that $\|\varphi_q\|_{r_1} \leq A_1\|\varphi_q\|_{r_0}^{q-1}$.
By induction we see that $\varphi_q \in L_{r_k}$ with

$$\frac{1}{r_k} = \frac{q_0 - 1}{r_{k-1}} - \frac{2}{n} = \frac{(q_0 - 1)^k}{r_0} - \frac{2}{n}\,\frac{(q_0 - 1)^k - 1}{q_0 - 2}$$

and there exists a constant A_k such that $\|\varphi_q\|_{r_k} \leq A_k\|\varphi_q\|_{r_0}^{(q-1)^k}$.
If for k large enough, $1/r_k$ is negative, then $\varphi_q \in L_\infty$. Indeed, suppose $1/r_{k-1} > 0$ and $1/r_k < 0$. Then $(q_0 - 1)/r_{k-1} - 2/n < 0$, and Hölder's inequality 3.62 applied to (8) yields $\|\varphi_q\|_\infty \leq \text{Const} \times \|\varphi_q\|_{r_{k-1}}^{q-1}$.
There exists a k such that

$$\frac{1}{r_k} = (q_0 - 1)^k\left[\frac{1}{r_0} - \frac{2}{n(q_0 - 2)}\right] + \frac{2}{n(q_0 - 2)} < 0$$

because $n(q_0 - 2) < 2r_0 = 2N$, since $q_0 < N = 2n/(n - 2)$.
Moreover, there exists a constant A_k, which does not depend on $q \leq q_0$, such that:

$$\|\varphi_q\|_\infty \leq A_k\|\varphi_q\|_N^{(q-1)^k}.$$

But the set of the functions φ_q is bounded in H_1 (same proof as in b)).

Thus by the Sobolev imbedding theorem 2.21 the functions φ_q are uniformly bounded. Since $\varphi_q \in L_\infty$, differentiating (8) yields $\varphi_q \in C^1$. φ_q satisfies (4); thus $\Delta\varphi_q$ belongs to C^1 and $\varphi_q \in C^2$ according to Theorem 3.54.

f) φ_q is strictly positive. This is true because $\int_M f(x)\varphi_q^q\,dV = 1$, Proposition 3.75 establishes this result since φ_q cannot be identically zero. Lastly $\varphi_q \in C^\infty$ by induction according to Theorem 3.54. ∎

5.6 Remark. The proof of Theorem 5.5 does not work for $q = N$. The problem is that if $q = N$, we cannot apply Krondrakov's theorem in c), and therefore only have

$$\int_M f(x)\varphi_N^N\,dV \leq 1.$$

Moreover, the method in e) yields nothing when $q_0 = N$. In this case $r_k = r_0 = N$ for all k. Indeed, we shall see below that if $q = N$ then Equation (4) may not have a positive solution (see Theorem 6.13).

5.7 Remark. Using the same method, one can study equations of the type $\Delta\varphi + h(x)\varphi = \lambda f(\varphi, x)$, where $f(t, x)$, a C^∞ function on $\mathbb{R} \times M$, satisfies some conditions. In particular, $|f(t, x)| \leq \text{Const} \times (1 + |t|^{q_0})$ with $q_0 < N$ (see Berger [39]) or with $q_0 \leq N$ (see Aubin [16]). The idea is to consider the variational problem:

$$\text{inf of } \int_M \nabla^i\varphi\nabla_i\varphi\,dV + \int_M h(x)\varphi^2\,dV \quad \text{when} \quad \int_M F(\varphi, x)\,dV = \text{Const},$$

where $F(t, x) = \int_0^t f(u, x)\,du$.

§2. Berger's Problem

5.8 Let (M, g) be a compact Riemannian manifold of dimension two. It is well known that there exists a metric on M whose curvature is constant. Considering conformal metrics g' of g, Melvyn Berger ([40]) wanted to prove this result by using the variational method.

Set $g' = e^\varphi g$. Then the problem is equivalent to solving the equation

$$\Delta\varphi + R = R'e^\varphi \tag{9}$$

with R' some constant (see 6.3). Here R denotes the scalar curvature of (M, g) (in dimension two R is twice the sectional curvature) and R' is the scalar curvature of (M, g').

By Theorem 4.7 we can write $R = \tilde{R} + \Delta\gamma$ with $\tilde{R} = \text{Const}$ and $\gamma \in C^\infty(M)$ satisfying $\int \gamma\,dV = 0$. Setting $\psi = \varphi + \gamma$, Equation (9) becomes

$$\Delta\psi + \tilde{R} = R'e^{-\gamma}e^\psi. \tag{10}$$

Note that we know the sign of R'; it is that of χ, the Euler–Poincaré characteristic. Indeed, by the Gauss–Bonnet theorem $4\pi\chi = \int R\, dV$, so integrating (9) over M gives $R' \int e^{\varphi}\, dV = \int R\, dV$, which is equal to $\tilde{R} \int dV$.

5.9 *Solution for* $\chi \leq 0$. If $\chi = 0$, $R' = \tilde{R} = 0$, $\psi = 0$ is a solution of (10), so the metric $g' = e^{-\gamma} g$ solves the problem.

If $\chi < 0$, $\tilde{R} < 0$ and R' will be negative. We are going to use the variational method to solve Equation (10).

Consider the functional

$$I(\psi) = \frac{1}{2}\int \nabla^{\nu}\psi \nabla_{\nu}\psi\, dV + \tilde{R}\int \psi\, dV \tag{11}$$

and set $\nu = \inf I(\psi)$ for all $\psi \in \mathsf{H}_1$ satisfying $\int e^{\psi - \gamma}\, dV = 1$.

a) ν *is finite*. Since the exponential function is convex

$$\int (\psi - \gamma)\, dV \leq \log \int e^{\psi - \gamma}\, dV = 0.$$

Thus $\int \psi\, dV \leq 0$ and $\nu \geq 0$.

b) ν *is attained*. Let $\{\psi_i\}$ be a minimizing sequence. We can choose it such that $I(\psi_i)$ is smaller than $\nu + 1$. Then

$$\int \nabla^{\nu}\psi_i \nabla_{\nu}\psi_i\, dV < 2(1 + \nu) \quad \text{and} \quad \tilde{R}\int \psi_i\, dV < 1 + \nu.$$

Thus the set $\{\psi_i\}_{i \in \mathbb{N}}$ is bounded in H_1, since $\|\nabla\psi_i\|_2 \leq$ Const and $(1 + \nu)/\tilde{R} < \int \psi_i\, dV \leq 0$. Indeed, by Corollary 4.3 $\|\psi_i\|_2 \leq |\int \psi_i\, dV|[\int dV]^{-1/2} + \lambda_1^{-1/2}\|\nabla\psi_i\|_2$ where λ_1 is the first nonzero eigenvalue. Using Theorems 3.18, 2.34, and 2.46 we see that there exist $\tilde{\psi} \in \mathsf{H}_1$ and a subsequence $\{\psi_j\}$ such that $\psi_j \to \tilde{\psi}$ weakly in H_1, strongly in L_1, and such that $e^{\psi_j} \to e^{\tilde{\psi}}$ strongly in L_1. Therefore $\int e^{\tilde{\psi} - \gamma}\, dV = 1$ and according to Theorem 3.17, $I(\tilde{\psi}) \leq \nu$. But $\tilde{\psi}$ satisfies the constraint. Thus by definition of ν we have $I(\tilde{\psi}) = \nu$. Hence ν is attained.

c) Writing Euler's equation yields

$$\int \nabla^{\nu}\tilde{\psi}\nabla_{\nu} h\, dV + \tilde{R}\int h\, dV = K\int h e^{\tilde{\psi} - \gamma}\, dV$$

for any $h \in \mathsf{H}_1$. Picking $h = 1$ gives the value of K, the Lagrange multiplier $K = \tilde{R}\int dV = 4\pi\chi$. Thus $\tilde{\psi}$ is a weak solution of

$$\Delta\tilde{\psi} + \tilde{R} = 4\pi\chi e^{\tilde{\psi} - \gamma} \tag{12}$$

By Theorem 2.46 the right-hand side belongs to L_p for all p. Therefore $\Delta\tilde{\psi} \in L_p$ for all p. Differentiating Equation (15) in 4.13 and using the properties of the Green's function show that $\tilde{\psi} \in C^{1+\alpha}$ for $0 < \alpha < 1$. Thus $\Delta\tilde{\psi} \in C^{1+\alpha}$ and according to Theorem 3.54, $\tilde{\psi} \in C^{3+\alpha}$. By induction $\tilde{\psi} \in C^\infty$.

2.1. The Positive Case $\chi > 0$

5.10 There are only two compact manifolds which are involved $\mathbb{S}_2$ if $\chi = 2$ and the real projective space $\mathbb{P}_2$ if $\chi = 1$. We suppose M is one of these manifolds.

From now on $\tilde{R}$ is a positive real number. More generally than previously, we will consider the equation

$$\Delta\varphi + \tilde{R} = fe^\varphi, \tag{13}$$

where $f \in C^\infty$ is a function positive at least at one point.
This property of f is necessary in order that Equation (13) have a solution, since $\tilde{R}\int dV = \int fe^\varphi\, dV$.
Henceforth, without loss of generality, we suppose the volume equals 1. Set $\nu = \inf I(\varphi)$ for all $\varphi \in \mathsf{H}_1$ satisfying $\int fe^\varphi\, dV = \tilde{R}$, where $I(\varphi)$ is the functional (11).

Theorem 5.10. *Equation* (13) *has a* C^∞ *solution if* $\tilde{R} < 8\pi$.

Proof

a) ν *is finite.* First of all there are functions satisfying $\int fe^\varphi\, dV = \tilde{R}$ since f is positive somewhere.
On the other hand, according to Theorem 2.51 or 2.53

$$\tilde{R} = \int fe^\varphi\, dV \leq \sup f \int e^\varphi\, dV \leq C(\varepsilon)\sup f \exp\left[(\mu_2 + \varepsilon)\|\nabla\varphi\|_2^2 + \int \varphi\, dV\right]. \tag{14}$$

Thus

$$I(\varphi) \geq [\tfrac{1}{2} - (\mu_2 + \varepsilon)\tilde{R}]\|\nabla\varphi\|_2^2 + \tilde{R}\log(\tilde{R}/C(\varepsilon)\sup f). \tag{15}$$

$\mu_2 = 1/16\pi$, if $\tilde{R} < 8\pi$ we can choose $\varepsilon = \varepsilon_0 > 0$ small enough so that $2(\mu_2 + \varepsilon_0)\tilde{R} < 1$. Therefore ν is finite, $\nu \geq \tilde{R}\log(\tilde{R}/C(\varepsilon_0)\sup f)$.

b) ν *is attained.* Let $\{\varphi\}_{i\in\mathbb{N}}$ be a minimizing sequence; (15) implies

$$[\tfrac{1}{2} - (\mu_2 + \varepsilon_0)\tilde{R}]\|\nabla\varphi_i\|_2^2 \leq \text{Const, thus } \|\nabla\varphi_i\|_2 \leq \text{Const}.$$

Moreover, $I(\varphi_i) \leq$ Const implies $\int \varphi_i\, dV \leq$ Const, and by (14) $\int \varphi_i\, dV \geq$ Const. Therefore $\{\varphi_i\}_{i\in\mathbb{N}}$ is bounded in H_1 (Corollary 4.3).

As in 5.9b it follows that ν is attained. There exists $\tilde{\varphi} \in \mathrm{H}_1$ such that $I(\tilde{\varphi}) = \nu$ and $\int fe^{\tilde{\varphi}}\, dV = \tilde{R}$.

c) Writing Euler's equation yields

$$\int \nabla^\nu \tilde{\varphi} \nabla_\nu h\, dV + \tilde{R} \int h\, dV = K \int hfe^{\varphi}\, dV$$

for any $h \in \mathrm{H}_1$. Picking $h = 1$ gives $K = 1$. $\tilde{\varphi}$ satisfies Equation (13) weakly and by the proof in 5.9c, $\tilde{\varphi} \in C^\infty$. ∎

§3. Nirenberg's Problem

5.11 Given F, a C^∞ function on the sphere $\mathbb{S}_2$ (if necessary "close" to zero), does there exist a conformal metric g' whose scalar curvature $R' = R + F$ with R the scalar curvature of $\mathbb{S}_2$? We saw in 5.8 that this problem is equivalent to solving Equation (9). Theorem 5.10 asserts that this equation has a solution if $R < 8\pi$.

Unfortunately, since the volume is equal to one, $R = 4\pi\chi = 8\pi$ on $\mathbb{S}_2$ and Theorem 5.10 yields nothing for the sphere. However, on $\mathbb{P}_2$, $R = 4\pi\chi = 4\pi$ so we get the following:

Corollary (Moser [210]). *Any C^∞ function which is positive somewhere on $\mathbb{P}_2$ is the scalar curvature of a conformal metric to the standard metric on $\mathbb{P}_2$.*

3.1. A Nonlinear Theorem of Fredholm

5.12 For $\mathbb{S}_2$, $\mu_2 = 1/16\pi$ is attained (Theorem 2.51), so in (15) we can pick $\varepsilon = 0$; thus ν is finite. But in 5.10b it is impossible to find a convergent subsequence from the minimizing sequence. This conclusion is correct because Kazdan and Warner [157] showed that Equation (9) on $\mathbb{S}_2$ with $R' = R + F$ has no solution if F is an eigenfunction corresponding to the first nonzero eigenvalue λ_1.

They established the following result. If R' and φ satisfy (9), then for all $\xi \in \Lambda = \{\psi \in C^\infty / \Delta\psi = \lambda_1 \psi\}$,

$$\int_{\mathbb{S}_2} \nabla^\nu \xi \nabla_\nu R' e^{\varphi}\, dV = 0.$$

Thus if $R' = R + F$ with $F \in \Lambda$, $F \not\equiv 0$, we get a contradiction by choosing $\xi = F$.

Therefore the operator $\varphi \to e^{-\varphi}(\Delta\varphi + R)$ is not locally invertible at $\varphi = 0$, and it is not even globally surjection.

Theorem 5.12 (Aubin [21]). *Given $f \in C^\infty(\mathbb{S}_2)$ satisfying $\int f\, dV > 0$, there exists $h(f) \in \Lambda$ such that Equation (9) with $R' = f - h(f)$ has a solution $\varphi_0 \in C^\infty$, e^{φ_0} being orthogonal to Λ (in the L_2 sense).*

Proof. Recall that $\Lambda = \{\varphi \in C^\infty / \Delta\varphi = \lambda_1 \varphi\}$ and the volume of $\mathbb{S}_2$ is assumed equal to one. We consider again the functional

$$I(\varphi) = \frac{1}{2}\int \nabla^\nu \varphi \nabla_\nu \varphi\, dV + R\int \varphi\, dV$$

but now set $\tilde{\nu} = \inf I(\varphi)$ for all $\varphi \in \mathsf{H}_1$ which satisfy $\int f e^\varphi\, dV = R$ and $\int \xi e^\varphi\, dV = 0$ for all $\xi \in \Lambda$. We are going to see that the proof in 5.10 works.

As before, $\tilde{\nu}$ is finite with $R \log(R/C(\varepsilon) \sup f) \le \tilde{\nu} \le R \log(R/\int f\, dV)$. Now, however, although $R = 8\pi$, the minimum $\tilde{\nu}$ is attained because we can use (33) in 2.52 instead of (32) in 2.51 since the best constant is half its previous value. We get instead of (15)

$$I(\varphi) \ge [\tfrac{1}{2} - (\mu_2/2 + \varepsilon)R]\|\nabla\varphi\|_2^2 + R\log(R/C(\varepsilon)\sup f).$$

Therefore a minimizing sequence is bounded in H_1 and a subsequence converges in H_1 to φ_0 which solves Euler's equation: there exist $h(f) \in \Lambda$ and a constant β such that

$$\Delta\varphi_0 + R = [\beta f - h(f)]e^{\varphi_0}. \tag{16}$$

Moreover, φ_0 satisfies the constraints, e^{φ_0} is orthogonal to Λ ,and $\int f e^{\varphi_0}\, dV = R$. Integrating (16) yields $\beta = 1$. Lastly, $\varphi_0 \in C^\infty$, as in 5.9c. We may write Euler's equation (16) because $f \notin \Lambda$, since $\int f\, dV \neq 0$. ■

Corollary 5.12. *A necessary condition for solving* (9) *with $R' = f$ is that f is positive somewhere. This condition is also sufficient up to a space of three dimensions.*

Proof. Suppose f is positive at P, and consider a conformal transformation of the sphere with pole at P. The new metric is of the form $\tilde{g}(Q) = (\beta - \cos \alpha r)^{-2} g(Q)$ for some $\beta > 1$, $1/\alpha$ being the radius of the sphere ($R = 2\alpha^2$). The new scalar curvature is constant $\tilde{R} = R(\beta^2 - 1)$. Then on the sphere we have to solve an equation like (13) in the metric $\tilde{g}$. Since we can choose β so that $\int f\, d\tilde{V} > 0$, we can apply theorem 5.12 in the metric $\tilde{g}$.

Remark 5.12. The same problem arises on the sphere $\mathbb{S}_n (n > 2)$ (see Aubin [21]). The equation to solve is then Equation (1) of 6.3.

There exist obstructions, as Kazdan and Warner [160] pointed out (see 6.13). Lowering the best constant in the Sobolev inequality overcomes the difficulty. Here the proof is more delicate because the imbedding $\mathsf{H}_1 \subset L_{2n/(n-2)}$ is not compact.

3.2. Open Questions

5.13 We know that we can solve Nirenberg's problem up to a space of dimension three, by Corollary 5.12. But in Theorem 5.12 we have no information on $h(f)$.

What conditions must f satisfy in order for Equation (9) to have a solution with $R' = f$? What is the set of these functions? The theorem asserts there exists some $h(f) \in \Lambda$, but it is conceivable that there are many such functions. How can one compute one of them in terms of f?

All these questions are open on $\mathbb{S}_2$ as well as on $\mathbb{S}_n$.

Chapter 6

The Scalar Curvature

§1. The Yamabe Problem

6.1 The problems concerning scalar curvature turn out to be very special when the dimension is two, the scalar curvature is then twice the Gaussian curvature. When $n = 2$ in chapter 5 we studied a problem in connection with the uniformization problem Berger (42) p. 224. Now in this chapter we shall deal with some problems concerning the scalar curvature of compact Riemannian manifolds of dimension $n \geq 3$. We shall deal in particular with

6.2 The Yamabe Problem ([269]). Let (M_n, g) be a C^∞ compact Riemannian manifold of dimension $n \geq 3$, R its scalar curvature. The problem is: does there exist a metric g', conformal to g, such that the scalar curvature R' of the metric is constant.

6.3 The differential equation. Let us consider the conformal metric $g' = e^f g$ with $f \in C^\infty$. By 1.19, if Γ'^l_{ij} and Γ^l_{ij} denote the Christoffel symbols relating to g' and g respectively:

$$\begin{aligned}\Gamma'^l_{ij} - \Gamma^l_{ij} &= \tfrac{1}{2}[g_{kj}\,\partial_i f + g_{ki}\,\partial_j f - g_{ij}\,\partial_k f]g^{kl} \\ &= \tfrac{1}{2}[\delta^l_j\,\partial_i f + \delta^l_i\,\partial_j f - g_{ij}\nabla^l f].\end{aligned}$$

According to 1.13,

$$\begin{aligned}R'_{ij} = R'^k_{ikj} = R_{ij} &- \frac{n-2}{2}\nabla_{ij} f + \frac{n-2}{4}\nabla_i f \nabla_j f \\ &- \frac{1}{2}\left(\nabla^\nu_\nu f + \frac{n-2}{2}\nabla^\nu f \nabla_\nu f\right)g_{ij}\end{aligned}$$

so

$$R' = e^{-f}\left[R - (n-1)\nabla^\nu_\nu f - \frac{(n-1)(n-2)}{4}\nabla^\nu f \nabla_\nu f\right].$$

If we consider the conformal deformation in the form $g' = \varphi^{4/(n-2)}g$ (with $\varphi \in C^\infty$, $\varphi > 0$), the scalar curvature R' satisfies the equation:

(1) $4((n-1)/(n-2))\Delta\varphi + R\varphi = R'\varphi^{(n+2)/(n-2)}$, where $\Delta\varphi = -\nabla^\nu\nabla_\nu\varphi$.

So the Yamabe problem is equivalent to solving equation (1) with $R' =$ Const, and the solution φ must be smooth and strictly positive.

1.1. Yamabe's Functional

6.4 To solve Equation (1), Yamabe used the variational method. He considered the functional, for $2 \le q \le N = 2n/(n-2)$,

$$(2) \qquad J_q(\varphi) = \left[4\frac{n-1}{n-2}\int_M \nabla^i\varphi\nabla_i\varphi\, dV + \int_M R(x)\varphi^2\, dV\right]\|\varphi\|_q^{-2}$$

and defined $\mu_q = \inf J_q(\varphi)$ for all $\varphi \ge 0$, $\varphi \not\equiv 0$, belonging to H_1. Set $\mu = \mu_N$ and $J(\varphi) = J_N(\varphi)$.

Proposition. *μ is a conformal invariant.*

Proof. Consider a change of conformal metric defined by $g' = \varphi^{4/(n-2)}g$. We have $dV' = \varphi^N\, dV$ and

$$J(\varphi\psi) = \frac{4\dfrac{n-1}{n-2}\left[\displaystyle\int_M \varphi^2\nabla^\nu\psi\nabla_\nu\psi\, dV + \int_M \varphi\psi^2\Delta\varphi\, dV\right] + \displaystyle\int_M R\varphi^2\psi^2\, dV}{\left[\displaystyle\int_M \varphi^N\psi^N\, dV\right]^{2/N}}.$$

Using (1) yields $J(\varphi\psi) = J'(\psi)$ and consequently $\mu = \mu'$. J' is the functional related to g', and μ' the inf of J'. ∎

By a homothetic change of metric we can set the volume equal to one. So henceforth, without loss of generality, we suppose the volume equal to one.

1.2. Yamabe's Theorem

6.5 In his article [269], Yamabe proved Theorem 5.5 and then he claimed that the C^∞ strictly positive functions φ_q, $q \in]2, N[$ satisfying

$$(3) \qquad 4\frac{n-1}{n-2}\Delta\varphi_q + R\varphi_q = \mu_q\varphi_q^{q-1} \quad \text{and} \quad \|\varphi_q\|_q = 1$$

are uniformly bounded.

But this does not hold in general. The functions $\varphi(r)$ of Theorem 6.12 are not uniformly bounded on the sphere. This counterexample shows that Yamabe's proof is wrong. Indeed (p. 35 of Yamabe [269]), the inequality $\|v^{(q)}\|_{q_n} \leq$ Const $\times$ $\|v^{(q)}\|_{q_1}$ must be replaced by $\|v^{(q)}\|_{q_n} \leq$ Const $\times$ $\|v^{(q)}\|_{q_1}^{(q-1)^{n-1}}$ and this does not yield the result.
In the negative or zero case, ($\mu \leq 0$), it is easy to overcome the mistake. But in the positive case ($\mu > 0$), it is impossible. Yamabe did not solve his problem but he proved the following.

Theorem (See Aubin [11] p. 386). *Let M_n be a C^∞ compact Riemannian manifold; there exists a conformal metric whose scalar curvature is either a non-positive constant or is everywhere positive.*

Proof. α) *the positive case* ($\mu > 0$). If $\mu_{q_0} > 0$, μ_q is positive for $q \in]2, N[$. Indeed

$$\mu_q = J_q(\varphi_q) = J_{q_0}(\varphi_q)\|\varphi_q\|_{q_0}^2\|\varphi_q\|_q^{-2} \geq \mu_{q_0}\|\varphi_q\|_{q_0}^2.$$

Then consider the conformal metric $g' = \varphi_{q_0}^{4/(n-2)}g$; applying (1) and (3) leads to

$$R'(x) = \mu_{q_0}\varphi_{q_0}^{q_0-N}(x), \tag{4}$$

the scalar curvature R' is everywhere strictly positive. Moreover we can prove that $\mu > 0$. Indeed the functional J' corresponding to g' satisfies

$$J'(\psi) \geq \inf_{x\in M}\left[4\frac{n-1}{n-2}, R'(x)\right]\left[\int_M g'^{ij}\nabla_i\psi\nabla_j\psi\, dV' + \int_M \psi^2\, dV'\right]$$
$$\times\left[\int \psi^N\, dV'\right]^{-2/N}.$$

According to the Sobolev imbedding theorem, $J'(\psi) \geq$ Const > 0 for all $\psi \in \mathsf{H}_1$. Thus $\mu' > 0$ and we have $\mu = \mu'$ (Proposition 6.4).

β) *The null case* ($\mu = 0$). If $\mu_{q_0} = 0$, by (4) the scalar curvature R' vanishes, and $\mu_q = 0$ for all $q \in]2, N]$, because for all ψ and q, $J_q(\psi) \geq 0$.

γ) *The negative case* ($\mu < 0$). If $\mu_{q_0} < 0$ there exists a $\psi \in C^\infty$ such that $J_{q_0}(\psi) < 0$. Hence $J_q(\psi) < 0$ for all $q \in [2, N]$ and $\mu_q < 0$. In particular, $\mu < 0$. Moreover, $\mu_q \leq J_q(\psi) = J(\psi)\|\psi\|_N^2\|\psi\|_q^{-2} \leq J(\psi)$. Thus $\mu_q(q \in [2, N])$ is bounded away from zero.
Now we are able to prove very simply that the functions $\varphi_q(q \in]q_0, N[)$ are uniformly bounded with $q_0 \in]2, N[$. At a point P where φ_q is maximum

$\Delta\varphi_q \geq 0$, hence $\mu_q \varphi_q^{q-1}(P) \geq R(P)\varphi_q(P)$. We find at once that $\varphi_q^{q-2} \leq |\inf R| |J(\psi)|^{-1}$ and $\varphi_q \leq 1 + [|\inf R| |J(\psi)|^{-1}]^{1/(q_0-2)}$. By (3), φ_q satisfies:

$$(5) \quad \varphi_q(P) = \int_M \varphi_q(Q)\, dV(Q) + \int_M \mathsf{G}(P, Q) \frac{n-2}{4(n-1)} [\mu_q \varphi_q^{q-1}(Q) - R(Q)\varphi_q(Q)]\, dV(Q).$$

Differentiating (5) yields $\varphi_q \in C^1$ uniformly, and according to Ascoli's theorem 3.15, it is possible to exhibit a sequence φ_{q_i} with $q_i \to N$, such that φ_{q_i} converges uniformly to a nonnegative function φ_N.
But $0 > \mu_q \geq \inf R(x) \|\varphi_q\|_2^2 \geq \inf R(x)$. Therefore a subsequence μ_{q_i} converges to a real number ν (in fact μ_q is a continuous function of q for $q \in]2, N]$ by Proposition 6.6, so $\mu = \nu$).
Letting $q_i \to N$ in (5), shows that φ_N is a weak solution of

$$(6) \qquad 4\frac{n-1}{n-2} \Delta\varphi_N + R\varphi_N = \nu\varphi_N^{N-1}.$$

Since $\|\varphi_q\|_q = 1$, $\|\varphi_N\|_N = 1$. Multiplying (6) by φ_N and integrating yield $J(\varphi_N) = \nu$.

The second term in (6) is continuous; thus, by (5), $\varphi_N \in C^1$. Now apply the regularity theorem 3.54: the second member of (6) is C^1; thus $\varphi_N \in C^2$. Now according to Proposition 3.75 φ_N is strictly positive everywhere, since $\|\varphi_N\|_N = 1$ implies $\varphi_N \not\equiv 0$. We can use the regularity theorem again to prove by induction that $\varphi_N \in C^\infty$. Thus the C^∞ function $\varphi_N > 0$ satisfies (1) with $R' = \text{Const}$ (in fact, $R' = \mu$).
In the negative case it is therefore possible to make the scalar curvature constant and negative. ■

6.6 Proposition. *μ_q is a continuous function of q for $q \in]2, N]$, which is either everywhere positive, everywhere zero, or everywhere negative. Moreover, $|\mu_q|$ is decreasing in q if we suppose the volume equal to* 1.

Proof. If the volume is equal to 1 for $\psi \in C^\infty$, $\|\psi\|_q$ is an increasing function of q. Thus $|J_q(\psi)| \leq |J_p(\psi)|$ when $p < q$ and this implies $|\mu_p| \geq |\mu_q|$ since C^∞ functions are dense in H_1.
Moreover, $J_q(\psi)$ is a continuous function of q. It follows that μ_q is an upper semicontinuous function of q. Indeed, for all $\varepsilon > 0$, there exists $\psi \in C^\infty$ such that $J_p(\psi) < \mu_p + \varepsilon$ and since $\mu_q \leq J_q(\psi)$, $\lim_{q\to p} J_q(\psi) = J_p(\psi)$ yields $\limsup_{q\to p} \mu_q \leq \mu_p + \varepsilon$.
Let q_i be a sequence converging to $p \in]2, N]$.
In the negative case, we saw, 6.5γ, that the functions φ_q are uniformly bounded. Therefore $\|\varphi_{q_i}\|_p \to 1$ and as $\mu_p \leq J_p(\varphi_{q_i}) = \mu_{q_i}\|\varphi_{q_i}\|_p^{-2}$, $\liminf_{q\to p} \mu_q \geq \mu_p$.

This establishes the continuity of $q \to \mu_q$ in the negative case. Similarly we can prove that this function is continuous on $]2, N[$ in the positive case, because if $q_0 < N$, the functions φ_q are uniformly bounded for $q \in]2, q_0]$ by (5.5e). Finally $\mu_q \to \mu_N$ when $q \to N$ because the function $q \to \mu_q$ is upper semi-continuous and decreasing in the positive case. If the volume is not one, we consider a homothetic change of metric such that the volume in the new metric is equal to one.

§2. The Positive Case

6.7 Theorem (Aubin [14]). *$\mu \le n(n-1)\omega_n^{2/n}$. If $\mu < n(n-1)\omega_n^{2/n}$, there exists a strictly positive solution $\varphi_0 \in C^\infty$ of (1) with $R' = \mu$ and $\|\varphi_0\|_N = 1$. Here ω_n is the volume of the unit sphere of radius* 1 *and dimension n. For the definition of μ see* 6.4.

Proof. α) Recall that $\mathsf{K}(n, 2) = 2(\omega_n)^{-1/n}[n(n-2)]^{-1/2}$ is the *best constant* in the Sobolev inequality (Theorem 2.14). By theorem (2.21), the best constant is the same for all compact manifolds. Thus there exists a sequence of C^∞ functions ψ_i such that

$$\|\psi_i\|_N = 1,\ \|\psi_i\|_2 \to 0 \quad \text{and} \quad \|\nabla\psi_i\|_2 \to K^{-1}(n, 2),$$

when $i \to +\infty$. Therefore $J(\psi_i) \to n(n-1)\omega_n^{2/n}$ and $\mu \le n(n-1)\omega_n^{2/n}$.

β) Let us again consider the set of functions φ_q ($q \in]2, N[$) which are solutions of (3). This set is bounded in H_1 since we have $\|\varphi_q\|_2 \le 1$ and

$$4\frac{n-1}{n-2}\|\nabla\varphi_q\|_2^2 \le \mu_q + \sup|R| \le \int R\, dV + \sup|R|.$$

Therefore there exists $\varphi_0 \in \mathsf{H}_1$ and a sequence $q_i \to N$ such that $\varphi_{q_i} \to \varphi_0$ weakly in H_1 (the unit ball in H_1 is weakly compact), strongly in L_2 (Kondrakov's theorem) and almost everywhere (Proposition 3.43). The weak limit in H_1 is the same as that in L_2 because H_1 is continuously imbedded in L_2, and strong convergence implies weak convergence.

γ) Since φ_{q_i} satisfies (3), then for all $\psi \in \mathsf{H}_1$:

$$4\frac{n-1}{n-2}\int \nabla^\nu\psi\nabla_\nu\varphi_{q_i}\, dV + \int R\psi\varphi_{q_i}\, dV = \mu_{q_i}\int \psi\varphi_{q_i}^{q_i-1}\, dV.$$

Letting $q_i \to N$ gives us

$$(7) \qquad 4\frac{n-1}{n-2}\int \nabla^\nu\psi\nabla_\nu\varphi_0\, dV + \int R\psi\varphi_0\, dV = \mu\int \psi\varphi_0^{N-1}\, dV.$$

Indeed, according to Theorem 3.45, $\varphi_{q_i}^{q_i-1}$ converges weakly to φ_0^{N-1} in $L_{N/(N-1)}$ since $\varphi_{q_i}^{q_i-1} \to \varphi_0^{N-1}$ almost everywhere and

$$\|\varphi_{q_i}^{q_i-1}\|_{N/(N-1)} = \|\varphi_{q_i}\|_{(q_i-1)N/(N-1)}^{q_i-1} \leq \|\varphi_{q_i}\|_N^{q_i-1} \leq \text{Const} \times \|\varphi_{q_i}\|_{H_1}^{q_i-1} \leq \text{Const}.$$

Therefore φ_0 satisfies (7) for all $\psi \in \mathrm{H}_1 \subset L_N$. According to Trüdinger's theorem ([262] p. 271) $\varphi_0 \in C^\infty$ and satisfies (1) with $R' = \mu$.

δ) The problem is not solved yet because the maximum principle implies that either $\varphi_0 > 0$ everywhere or $\varphi_0 \equiv 0$, and for the moment we cannot exclude the latter case. In order to prove that φ_0 is not identically zero, we must use Theorem 2.21. We write, using (3),

$$(8)\quad 1 = \|\varphi_q\|_q^2 \leq \|\varphi_q\|_N^2 \leq [\mathsf{K}^2(n, 2) + \varepsilon]\frac{n-2}{4(n-1)}\left[\mu_q - \int R\varphi_q^2\, dV\right] + A(\varepsilon)\|\varphi_q\|_2^2,$$

where $\varepsilon > 0$ is arbitrary and $A(\varepsilon)$ is a constant which depends on ε. When $\mu < n(n-1)\omega_n^{2/n}$, if we choose ε small enough, there exist $\varepsilon_0 > 0$ and $\eta > 0$ such that for $N - q < \eta$,

$$(9)\qquad 0 < \varepsilon_0 \leq 1 - \left[\frac{\omega_n^{-2/n}}{n(n-1)} + \varepsilon\frac{n-2}{4(n-1)}\right]\mu_q$$

since $\mu_q \to \mu$ when $q \to N$.
In this case, (8) and (9) imply

$$\liminf_{q\to N} \|\varphi_q\|_2 \geq \text{Const} > 0.$$

Because φ_{q_i} converges strongly to φ_0 in L_2, $\|\varphi_0\|_2 \neq 0$. Thus $\varphi_0 \not\equiv 0$ and $\varphi_0 > 0$. Picking $\psi = \varphi_0$ in (7) gives $J(\varphi_0) = \mu\|\varphi_0\|_N^{N-2}$; thus $\|\varphi_0\|_N \geq 1$ since $J(\varphi_0) \geq \mu$. But since the sequence $\varphi_{q_i}^{q_i/N}$ of 6.7β converges weakly to φ_0 in L_N by Theorem 3.45, $\|\varphi_0\|_N \leq \liminf_{q_i\to N}\|\varphi_{q_i}\|_{q_i}^{q_i/N}$ by Theorem 3.17. Hence $\|\varphi_0\|_N = 1$ and $J(\varphi_0) = \mu$.
Moreover by Radon's theorem, 3.47, $\varphi_{q_i} \to \varphi_0$ strongly in H_1 because $\|\varphi_{q_i}\|_{\mathrm{H}_1} \to \|\varphi_0\|_{\mathrm{H}_1}$ since $\mu_{q_i} \to \mu$. Therefore by the Sobolev imbedding theorem $\varphi_{q_i} \to \varphi_0$ strongly in L_N.

ε) In fact, when $\mu < n(n-1)\,\omega_n^{2/n}$, it is possible to prove directly that the functions φ_q $q \in]2, N[$ are uniformly bounded and we can proceed as in the negative case, without using Trüdinger's theorem. ■

6.8 More generally, let us consider the equation

$$(10)\qquad 4\frac{n-1}{n-2}\Delta\varphi + h(x)\varphi = \lambda f(x)\varphi^{N-1},$$

with $h \in C^\infty$, $f \in C^\infty$ given ($f > 0$), and λ ($=0, 1$, or -1) to be determined as in 5.3. Let

$$I(\varphi) = \left[4\frac{n-1}{n-2}\int \nabla^i\varphi\nabla_i\varphi\, dV + \int h\varphi^2\, dV\right] \Big/ \left[\int f\varphi^N\, dV\right]^{-2/N}$$

and define $\nu = \inf I(\varphi)$ for all $\varphi \in \mathsf{H}_1$, $\varphi \not\equiv 0$. Using the same method one can prove:

Theorem. $\nu \le n(n-1)\omega_n^{2/n}[\sup f]^{-2/N}$. *If* $\nu < n(n-1)\omega_n^{2/n}[\sup f]^{-2/N}$, *Equation* (10) *has a* C^∞ *strictly positive solution.*

6.9 Now we have to investigate when the inequalities of Theorems 6.7 and 6.8 are strictly satisfied. For this, consider the sequence of functions ψ_k ($k \in \mathbb{N}$):

$$\psi_k(Q) = \left(\frac{1}{k} + r^2\right)^{1-n/2} - \left(\frac{1}{k} + \delta^2\right)^{1-n/2}, \quad \text{for } r < \delta,$$

and $\psi_k(Q) = 0$ for $r \ge \delta$, with δ the injectivity radius, $r = d(P, Q)$, P fixed. A computation shows that $\lim_{k\to\infty} I(\psi_k) = n(n-1)\omega_n^{2/n}[f(P)]^{-2/N}$. Pick a point P where $f(P)$ has its maximum. In order to see if equality in Theorem 6.8 does not hold, we compute an asymptotic expansion.
If $n \ge 4$, the coefficient of the second term has the sign of

$$h(P) - R(P) + \frac{n-4}{2}\frac{\Delta f(P)}{f(P)}.$$

More precisely, the asymptotic expansion for $n > 4$ is

$$I(\psi_k) = n(n-1)\omega_n^{2/n}[f(P)]^{-2/N} \times \left\{1 + [n(n-4)k]^{-1}\left[h(P) - R(P) + \frac{n-4}{2}\frac{\Delta f(P)}{f(P)}\right]\right\} + o\left(\frac{1}{k}\right),$$

and for $n = 4$

$$I(\psi_k) = 12[\omega_4/f(P)]^{1/2}\left\{1 + [h(P) - R(P)]\frac{\operatorname{Log} k}{8k}\right\} + o\left(\frac{\operatorname{Log} k}{k}\right).$$

Proposition (Aubin [14] p. 286). *If, at a point* P *where* f *is maximum,* $h(P) - R(P) + ((n-4)/2)(\Delta f(P)/f(P)) < 0$, *Equation* (10) *has a* C^∞ *strictly positive solution when* $n \ge 4$.

2.1. Geometrical Applications

6.10 Let us return to Yamabe's equation (1) with $R' = \text{Const}$. That is equation (10) with $f(P) = 1$ and $h(P) = R(P)$. We cannot apply Proposition 6.9. Hence Yamabe's equation is a limiting case in two ways: first with the exponent $(n+2)/(n-2)$ and second with the function R. An asymptotic expansion of $J(\psi_k)$ yields nothing. It is necessary to consider other functions, those of the sphere of radius $1/\alpha$.

$$\tilde{\psi}_k(Q) = \left(\frac{1}{k} + \frac{1 - \cos \alpha r}{\alpha^2}\right)^{1-n/2} - \left(\frac{1}{k} + \frac{1 - \cos \alpha\delta}{\alpha^2}\right)^{1-n/2} \quad \text{for } r < \delta$$

and $\tilde{\psi}_k(Q) = 0$ for $r \geq \delta$ with $R(P) = n(n-1)\alpha^2$ and $r = d(P, Q)$, P fixed.

After a change of conformal metric, the calculation of an asymptotic expansion yields a second term whose sign is $-\mathsf{W}_{ijkl}\mathsf{W}^{ijkl}$, when $n \geq 6$, where W_{ijkl} is the Weyl tensor at P. Because P is arbitrary we obtain:

6.11 Theorem (Aubin [14] p. 292). *If M_n $(n \geq 6)$ is a compact nonlocally conformally flat Riemannian manifold, then $\mu < n(n-1)\omega_n^{2/n}$. Hence the minimum μ is attained and equation* (1) *has a C^∞ strictly positive solution with $R' = \mu$, so Yamabe's problem is solved in this case.*

Proof. A manifold M_n is locally conformally flat if each point has a neighborhood where there exists a conformal metric whose curvature vanishes. When $n > 3$, a necessary and sufficient condition for a manifold to be locally conformally flat is that the Weyl tensor is zero.

By definition the Weyl tensor is

$$\mathsf{W}_{ijkl} = R_{ijkl} - \frac{1}{n-2}(R_{ik}g_{jl} - R_{il}g_{jk} + R_{jl}g_{ik} - R_{jk}g_{il})$$
$$+ \frac{R}{(n-1)(n-2)}(g_{jl}g_{ik} - g_{jk}g_{il}),$$

and $\mathsf{W}_{i\ kl}^{\ j}$ is a conformal invariant.

If M_n is not locally conformally flat, there exists a point P where the Weyl tensor is not zero. The computation of the asymptotic expansion in question in 6.10 shows that $\mu < n(n-1)\omega_n^{2/n}$. Then, according to Theorem 6.7, Yamabe's problem is solved. ∎

6.12 Theorem (Aubin [13] p. 588 and Aubin [14] p. 293). *For the sphere $\mathbb{S}_n$, $(n \geq 3)$, $\mu = n(n-1)\omega_n^{2/n}$ and Equation* (1) *with $R' = R$ has infinitely many solutions. In fact, the functions $\varphi(r) = (\beta - \cos \alpha r)^{1-n/2}$, with $1 < \beta$ a real number and $\alpha^2 = R/n(n-1)$, are solutions of* (1) *with $R' = R(\beta^2 - 1)$. Moreover on the sphere $\mathbb{S}_n$ with $\int dV = 1$, all $\varphi \in \mathsf{H}_1$ satisfy:*

$$\|\varphi\|_N^2 \leq \mathsf{K}^2(n, 2)\|\nabla\varphi\|_2^2 + \|\varphi\|_2^2. \tag{11}$$

Proof. i) Recall r is the distance to a fixed point P. According to Theorem 6.13 below, Equation (1) has no solution on the sphere when $R' = 1 + \varepsilon \cos \alpha r$, $\varepsilon \neq 0$. Indeed, $F = \cos \alpha r$ are the spherical harmonics of degree 1: $\Delta F = \lambda_1 F$, with $\lambda_1 = R/(n-1)$ the first nonzero eigenvalue.
If $\mu < n(n-1)\omega_n^{2/n}$, we can choose ε small enough to apply Theorem 6.8 with $h = R$ and $f = 1 + \varepsilon \cos \alpha r$ since $\nu \leq \mu(1-\varepsilon)^{-2/N}$. This contradicts the nonsolvability of Theorem 6.13

ii) We easily verify that $\varphi(r)$ satisfies Equation (1):

$$-4\frac{n-1}{n-2}(\sin \alpha r)^{1-n}\frac{\partial}{\partial r}\left[(\sin \alpha r)^{n-1}\frac{\partial}{\partial r}\varphi\right] + n(n-1)\alpha^2\varphi$$
$$= n(n-1)\alpha^2(\beta^2 - 1)\varphi^{(n+2)/(n-2)}.$$

(See 4.9 for the expression of the Laplacian in polar geodesic coordinates.) For the sphere $r^{n-1}\sqrt{|g|} = (\sin \alpha r)^{n-1}$.
These functions $\varphi(r)$ correspond to the changes of metrics $g' = (\beta - \cos \alpha r)^{-2} g$. We can verify that the manifold (M_n', g') is the sphere, because it has constant curvature. Indeed, $R'_{ij} = (\beta^2 - 1)(R/n)g'_{ij}$.

iii) Writing $\mu = n(n-1)\omega_n^{2/n}$ yields (11) when $\int dV = 1$. (11) is an improvement of the inequality of Theorem 2.28. Both constants are optimal, the second, $A_q(0) \geq 1$, since the inequality must be satisfied by the function $\varphi \equiv 1$.

6.13 Theorem (Kazdan and Warner [160] p. 130). *If u is a positive solution of equation* (1) *on the standard n-sphere* ($n \geq 3$), *then*

$$\int_{S_n} u^N \nabla^i R' \nabla_i F \, dV = 0$$

for all spherical harmonics F of degree 1. *Therefore on $\mathbb{S}_n$ there is no positive solution of* (1) *if $R' = R + \varepsilon F$ for any constant $\varepsilon \neq 0$.*

The proof uses the fact that all spherical harmonics F of degree 1 satisfy

$$\nabla_{ij} F = -\frac{\lambda_1 F}{n} g_{ij} = -\alpha^2 F g_{ij}.$$

6.14 Theorem (Aubin [14] p. 291). *A compact locally conformally flat manifold M_n ($n \geq 3$) with finite fundamental group has the Yamabe property: there exists a conformal metric g' with $R' =$* Const. *If M_n is not simply connected, $\mu < n(n-1)\omega_n^{2/n}$.*

For the proof we consider $\tilde{M}_n$ the universal covering of M_n. $\tilde{M}_n$ is compact, locally conformally flat and simply connected.

Kuiper's theorem [172] then implies that $\tilde{M}_n$ is conformally equivalent to the sphere $\mathbb{S}_n$. Hence Equation (1) has a solution with $R' = \mu$.

6.15 Theorem. *If* $\int R\, dV \leq n(n-1)\omega_n^{2/n}$, *Equation* (1) *has a* C^∞ *strictly positive solution with* $R' = \mu$.

Proof. $\mu \leq J(1) = \int R\, dV$. If $\mu = J(1)$, $\varphi_0 \equiv 1$ is a solution of Equation (1) with $R' = \mu$. And if $\mu < J(1)$ we can apply Theorem 6.7.

6.16 Proposition. *When the minimum* μ *is attained, let* $J(\varphi_0) = \mu$. *In the corresponding metric* g_0 *whose scalar curvature* R_0 *is constant, the first non-zero eigenvalue of the Laplacian* $\lambda_1 \geq R_0/(n-1)$.

For the proof one computes the second variation of $J(\varphi)$ (see Aubin [14] p. 292).

2.2. Open Questions

6.17 For the compact manifolds M_n ($n \geq 6$), there remains to be discussed the case of locally conformally flat manifolds with infinite fundamental group for which $\int R\, dV > n(n-1)\omega_n^{2/n}$.
$M_n = \mathrm{C} \times \mathbb{S}_{n-1}$, the product of a circle with the sphere, is of this kind (as one can easily verify). But this manifold has constant scalar curvature. To prove that μ is attained is stronger than proving the existence of a conformal metric with constant scalar curvature. This is the reason for which we propose the following two conjectures:

Conjecture 1. *All compact Riemannian manifolds* M_n ($n \geq 3$) *satisfy* $\mu < n(n-1)\omega_n^{2/n}$, *except manifolds which are simultaneously simply connected and locally conformally flat.*

For these latter $\mu = n(n-1)\omega_n^{2/n}$ (Theorem 6.12). Recall that always $\mu \leq n(n-1)\omega_n^{2/n}$. When $n \geq 6$ this conjecture is proved except for the locally conformally flat manifold with infinite fundamental group. Note that some of these manifolds do already satisfy the assumptions of Theorem 6.15. Conjecture 1 implies:

Conjecture 2. *On a compact Riemannian manifold* M_n ($n \geq 3$), *there exists a* C^∞ *strictly positive function* φ *such that* $J(\varphi) = \mu$.

This conjecture implies Yamabe's conjecture.

Remark. Now there exist direct proofs for solving Equation (1) without considering equation (3). See that of Inouë [150] and that of Vaugon [265].

§3. Other Problems

6.18 About uniqueness. α) In the negative and null cases ($\mu \leq 0$) two solutions of (1) with $R' = \text{Const}$ are proportional.
Let φ_0 be a solution of (1) with $R' = \mu$. In the corresponding metric g_0 the Yamabe equation is always of the type of Equation (1), since Yamabe's problem is conformally invariant.
So let φ_1 be a solution of Equation (1) with $R = R' = \mu$.
If $\mu = 0$, $\Delta\varphi_1 = 0$, thus $\varphi_1 \equiv \text{Const}$.
If $\mu < 0$, at a point P where φ_1 is maximum, $\Delta\varphi_1 \geq 0$, thus $[\varphi_1(P)]^{N-2} \leq 1$ and at a point Q where φ_1 is minimum, $\Delta\varphi_1 \leq 0$ thus $[\varphi_1(Q)]^{N-2} \geq 1$.
Consequently $\varphi_0 \equiv 1$ is the unique solution of (1) when $R = R' = \mu < 0$.

β) In the positive case, we do not have uniqueness generally.

Examples.

The sphere $\mathbb{S}_n$ (Theorem 6.12).

$M_n = \mathsf{T}_2 \times \mathbb{S}_{n-2}$ with T_2 the torus and $n \geq 6$ when $R(\int dV)^{2/n} > n(n-1)\omega_n^{2/n}$. Indeed in this case there exists on M_n at least two solutions of Equation (1) with $R' = R = \text{Const}$. First, $\varphi_1 \equiv 1$ and second, φ_0 for which $J(\varphi_0) = \mu < n(n-1)\omega_n^{2/n} < J(1)$ (Theorem 6.11).

γ) On the other hand, according to Obata [225], we have uniqueness for Einstein manifolds other than the sphere. Thus $J(1) = \mu$ and we find again (Proposition 6.16) the well-known result of Lichnerowicz (Theorem 1.78):

$$\lambda_1 \geq R/(n-1).$$

But we have more for Einstein manifolds:

$$J(1) = R\left(\int dV\right)^{2/n} = \mu \leq n(n-1)\omega_n^{2/n},$$

with equality only for the sphere.

3.1. Topological Meaning of the Scalar Curvature

6.19 We have seen that μ is a conformal invariant (Proposition 6.4), but does its sign have a topological meaning? Considering a change of metric (a non-conformal one, obviously) it is possible to prove:

Theorem (Aubin [11] p. 388). *A compact Riemannian manifold M_n ($n \geq 3$) carries a metric whose scalar curvature is a negative constant.*

Proof. According to Theorem 6.5 if we are not in the negative case, there exists a metric g with $R \geq 0$. Then we consider a change of metric of the kind:

$\tilde{g}_{ij} = \psi g_{ij} + \partial_i \psi \, \partial_j \psi$ with $\psi > 0$ a C^∞ function. It is possible to determine ψ such that the corresponding functional $\tilde{J}(u)$ is negative for some u. Hence the result follows by Theorem 6.5. ∎

Since on every compact manifold M_n ($n \geq 3$) there is some metric with $\mu < 0$, there is no topological significance to having negative scalar curvature. In contrast to this, Lichnerowicz (186) has proved that there are topological obstructions to admitting a metric with $\mu > 0$, that is, to positive scalar curvature. He showed that if there is a metric with nonnegative scalar curvature (not identically zero), then the Hirzebruch Â-genus of M must be zero. This work was extended by Hitchin (145), who proved that certain exotic spheres do not admit metrics with positive scalar curvature—and hence certainly have no metrics with positive sectional curvature.

In a related work, Kazdan and Warner [161] proved that there are also topological obstructions to admitting metrics with identically zero scalar curvature, that is, to $\mu = 0$. Thus there are obstructions to $\mu > 0$ and $\mu = 0$, but not to $\mu < 0$.

More recently, Gromov and Lawson [136] and [137] proved that every compact simply-connected manifold M_n ($n \geq 5$), which is not spin, carries a metric of positive scalar curvature. For the spin manifolds they generalize Hirzebruch's Â-genus in order to obtain almost necessary and sufficient conditions for a compact manifold to carry a metric of positive scalar curvature. In particular, the tori T_n, $n \geq 3$, do not admit metrics of positive scalar curvature. For the details see the article in references [136] and [137] or Bourbaki [34].

3.2. Kazdan and Warner's Problem

6.20 These authors ([159], and [160]), discuss the problem of describing the set of scalar curvature functions associated with Riemannian metrics on a given manifold of dimension $n \geq 3$. They proved, among other things, the following result:

Theorem. *Let M_n be a compact manifold of dimension $n \geq 3$, and let $R \in C^\infty(M_n)$. Then R is the scalar curvature of some Riemannian metric on M_n in the following cases:*

i) *if R is negative somewhere on M_n*

or

ii) *in the case $R \geq 0$ on M_n if there exists a metric with constant positive scalar curvature on M_n.*

Thus, every $R \in C^\infty(M_n)$ is a scalar curvature if and only if M_n admits a metric of positive constant scalar curvature.

Using this result, the problem of describing the set of scalar curvature functions on M_n is completely solved if $n \geq 6$. To see this, note that the topological obstructions mentioned above show that there are three cases.

The first case: M does not admit any metrics with $\mu = 0$ or $\mu > 0$. Then $\mu < 0$ for every metric so the scalar curvature functions are precisely those which are negative somewhere.

The second case: M does not admit a metric with $\mu > 0$, but does admit metrics with $\mu = 0$ and $\mu < 0$. This is identical with the first case except that the zero function is also a scalar curvature.

The third case: M has some metric g with $\mu > 0$.

If (M_n, g) is not locally conformally flat at some point, then, since $n \geq 6$, we can use Theorem 6.11 to obtain a conformal metric with positive constant scalar curvature. The above theorem then proves that every function is a scalar curvature. If (M_n, g) is everywhere locally conformally flat, one merely perturbs g slightly to obtain a new metric g', which is not locally conformally flat at some point and for which μ is still positive. The previous reasoning then applies to (M_n, g').

To prove the above theorem, it is natural to fix a metric g_0 on M with scalar curvature R_0 and seek a solution $u > 0$ of

$$4\frac{n-1}{n-2}\Delta u + R_0 u = f u^{N-1},$$

where $N = 2n/(n-2)$ as before. If there is a solution of this equation with $f = R$, then $g = u^{4/(n-2)} g_0$ is the desired metric with scalar curvature R. Unfortunately, as these authors showed, one may not always be able to solve this equation. Instead, they introduce a diffeomorphism Φ of M_n and seek a solution of the same equation, except that now f is replaced by $R \circ \Phi$. If $u > 0$ is a solution of this modified equation, then the metric $g_1 = (\Phi^{-1})^* g$ will have scalar curvature R. Their strategy is thus to solve the above equation for a small class of functions f and then use the group of diffeomorphisms to complete the proof of the theorem. We refer to their paper for more details.

6.21 Remark. J. Kazdan pointed out to me that we can solve Equation (1) with $R' = \text{Const}$ in the negative case by using the method of sub and supersolution. We only sketch the proof. It will be a good exercise for the reader to fill in the details.

Let ν be the first eigenvalue of the operator $L\varphi = 4((n-1)/(n-2))\Delta\varphi + R\varphi$. The corresponding eigenspace is of dimension 1 and the eigenfunctions do not change sign. Let $\psi > 0$ be one of them. These facts hold in general (see Krasnoselski [171]) but here we can give a short proof using the method of Section 4.4).

Since we are in the negative case, v is negative. If $c > 0$ is a constant we can write

$$L(c\psi) + (c\psi)^{(n+2)/(n-2)} = c\psi[v + (c\psi)^{4/(n-2)}].$$

So obviously for $c_+ > 0$, a large constant $u_+ = c_+\psi$ is a supersolution, while for $c_- > 0$, a small constant $u_- = c_-\psi$ is a subsolution.
Therefore there is a solution u of (1) with $R' = 1$ satisfying $c_-\psi \le u \le c_+\psi$. For a construction of a solution using the method of sub and supersolutions, see Section 7.25.

Chapter 7

Complex Monge–Ampère Equation on Compact Kähler Manifolds

7.1 Introduction. In this chapter we shall use the continuity method and the method of upper and lower solutions to solve *complex Monge–Ampère equations.*

But they can also be solved by the variational method. The difficulty is to obtain the *a priori estimates*; either method can be used indiscriminately.

These equations arise in some geometric problems which will be explained. The results and proofs appeared in Aubin [11], [18] and [20], and Yau [277]. An exposition can also be found in Bourguignon [59] and [60].

We introduce some notation. Let g, ω, Ψ (respectively, g', ω', Ψ') denote the metric, the first fundamental form 7.2, and the Ricci form 7.4. For a compact manifold, $V = \int dV$. In complex coordinates, d' and d'' are defined by $d'\varphi = \partial_\lambda \varphi \, dz^\lambda$ and $d''\varphi = \partial_{\bar\mu} \varphi \, dz^{\bar\mu}$. Also, let $d^c\varphi = (d' - d'')\varphi$. Then $dd^c\varphi = -2\, \partial_{\lambda\bar\mu}\varphi \, dz^\lambda \wedge dz^{\bar\mu}$.

7.2 First definitions. Let M_{2m} be a manifold of real even dimension $2m$. We consider only local charts (Ω, φ), where Ω is considered to be homeomorphic by a map φ to an open set of $\mathbb{C}^m$: $\varphi(\Omega)$.

The complex coordinates are $\{z^\lambda\}$, $(\lambda = 1, 2, \ldots, m)$. We write $z^{\bar\lambda} = \overline{z^\lambda}$. A *complex manifold* is a manifold which admits an atlas whose changes of coordinate charts are holomorphic. A complex manifold is analytic. A *Hermitian metric* g is a Riemannian metric whose components in a local chart satisfy for all ν, μ:

$$g_{\nu\mu} = g_{\bar\nu\bar\mu} = 0, \qquad g_{\nu\bar\mu} = g_{\bar\mu\nu} = \bar g_{\mu\bar\nu}.$$

The *first fundamental form* of the Hermitian manifold is $\omega = (i/2\pi) g_{\lambda\bar\mu}\, dz^\lambda \wedge dz^{\bar\mu}$, where g is a Hermitian metric,

§1. Kähler Manifolds

7.3 A Hermitian metric g is said to be *Kähler*: if the first fundamental form is closed: $d\omega = 0$. A necessary and sufficient condition for g to be Kähler is that its components in a local chart satisfy, for all λ, μ, ν,

(1) $$\partial_\lambda g_{\nu\bar\mu} = \partial_\nu g_{\lambda\bar\mu}.$$

On a Kähler manifold we consider the Riemannian connection (Lichnerowicz [184] and Kobayashi–Nomizu [167]).

It is easy to verify that Christoffel's symbols of mixed type vanish. Only $\Gamma^{\nu}_{\lambda\mu} = \overline{\Gamma^{\bar\nu}_{\bar\lambda\bar\mu}}$ may be nonzero. Thus, if $f \in C^2$, then $\nabla_{\lambda\bar\mu} f = \partial_{\lambda\bar\mu} f$. On a Kähler manifold we will write $\Delta f = -g^{\lambda\bar\mu}\,\partial_{\lambda\bar\mu} f$, which is half of the real Laplacian (warning!).

Only the components of mixed type $R_{\alpha\bar\beta\lambda\bar\mu}$ of the curvature tensor may be nonzero. It is easy to verify that the components of the Ricci tensor satisfy $R_{\lambda\mu} = R_{\bar\lambda\bar\mu} = 0$ and

$$R_{\lambda\bar\mu} = -\partial_{\lambda\bar\mu} \log|g|, \tag{2}$$

where $|g|$ is the determinant of the metric,

$$g = \begin{vmatrix} g_{1\bar 1} & \cdots & g_{1\bar m} \\ \vdots & & \vdots \\ g_{m\bar 1} & \cdots & g_{m\bar m} \end{vmatrix}.$$

In the real case we used the square of this determinant.

$\eta = (i/2)^m |g|\, dz^1 \wedge dz^{\bar 1} \wedge \cdots \wedge dz^\lambda \wedge dz^{\bar\lambda} \wedge \cdots \wedge dz^m \wedge dz^{\bar m}$ defines a global $2m$-form. A complex manifold is orientable.

1.1. First Chern Class

7.4 $\Psi = (i/2\pi) R_{\lambda\bar\mu}\, dz^\lambda \wedge dz^{\bar\mu}$ is called the *Ricci form*. According to (2), Ψ is closed: $d\Psi = 0$. Hence Ψ defines a cohomology class called the *first Chern class*: $\mathbf{C}_1(M)$. Recall that the cohomology class of Ψ is the set of the forms homologous to Ψ. Chern [91] defined the classes $\mathbf{C}_r(M)$ in an intrinsic way. For our purpose we only need to verify that $\mathbf{C}_1(M)$, so defined, does not depend on the metric. Indeed, let g' be another metric and Ψ' the corresponding Ricci form, let us prove that $\Psi' - \Psi$ is homologous to zero.

Since η and η' are positive $2m$-forms, there exists f, a strictly positive function, such that $\eta' = f\eta$. Hence, according to (2),

$$\Psi' - \Psi = -\frac{i}{2\pi}\,\partial_{\lambda\bar\mu} \log f\, dz^\lambda \wedge dz^{\bar\mu},$$

and the result follows from the following:

7.5 Lemma. *A 1-1 form $\gamma = a_{\lambda\bar\mu}\, dz^\lambda \wedge dz^{\bar\mu}$ is homologous to zero if and only if there exists a function h such that $a_{\lambda\bar\mu} = \partial_{\lambda\bar\mu} h$. For the necessity we suppose the manifold is compact.*

Proof. The sufficiency is established at once:

$$\gamma = \partial_{\lambda\bar{\mu}} h \, dz^\lambda \wedge dz^{\bar{\mu}} = dd''h, \quad \text{where } d''h = \partial_{\bar{\mu}} h \, dz^{\bar{\mu}}.$$

Now let us consider γ, a $1-1$ form homologous to zero.
Pick a function h such that $\Delta h = -g^{\lambda\bar{\mu}} a_{\lambda\bar{\mu}} + \text{Const}$ (in fact the constant is zero), and define $\tilde{\gamma} = \tilde{a}_{\lambda\bar{\mu}} \, dz^\lambda \wedge dz^{\bar{\mu}}$ with $\tilde{a}_{\lambda\bar{\mu}} = \partial_{\lambda\bar{\mu}} h$.

$$g^{\lambda\bar{\mu}}(\tilde{a}_{\lambda\bar{\mu}} - a_{\lambda\bar{\mu}}) = \text{Const, so } \nabla_\nu[g^{\lambda\bar{\mu}}(\tilde{a}_{\lambda\bar{\mu}} - a_{\lambda\bar{\mu}})] = g^{\lambda\bar{\mu}}\nabla_\nu(\tilde{a}_{\lambda\bar{\mu}} - a_{\lambda\bar{\mu}}) = 0$$

and $\delta''(\tilde{\gamma} - \gamma) = g^{\lambda\bar{\mu}}\nabla_\lambda(\tilde{a}_{\nu\bar{\mu}} - a_{\nu\bar{\mu}}) \, dz^\nu = 0$, since $d\tilde{\gamma} = d\gamma = 0$ implies $\nabla_\lambda a_{\nu\bar{\mu}} = \nabla_\nu a_{\lambda\bar{\mu}}$ and $\nabla_\lambda \tilde{a}_{\nu\bar{\mu}} = \nabla_\nu \tilde{a}_{\lambda\bar{\mu}}$. Likewise, $\delta'(\tilde{\gamma} - \gamma) = -g^{\lambda\bar{\mu}}\nabla_{\bar{\mu}}(\tilde{a}_{\lambda\bar{\nu}} - a_{\lambda\bar{\nu}}) \, dz^{\bar{\nu}} = 0$.

$\tilde{\gamma} - \gamma$ is homologous to zero and coclosed, so it vanishes (de Rham's theorem 1.72). On p-forms, the operators δ' and δ'' are defined by $\delta' = (-1)^{p-1} *^{-1} d' *$ and $\delta'' = (-1)^{p-1} *^{-1} d'' *$, (see 1.69); they are, respectively, of type $(-1, 0)$ and $(0, -1)$. ∎

1.2. Change of Kähler Metrics. Admissible functions

7.6 Let us consider the change of Kähler metric:

$$g'_{\lambda\bar{\mu}} = g_{\lambda\bar{\mu}} + \partial_{\lambda\bar{\mu}}\varphi, \tag{3}$$

where $\varphi \in C^\infty$ is said to be admissible (so that g' is positive definite). Obviously g' is a Kähler metric, since (1) is satisfied.
Let $\mathsf{M}(\varphi) = |g'||g|^{-1}$. Then $dV' = \mathsf{M}(\varphi) \, dV$. Since

$$\mathsf{M}(\varphi) = |g' \circ g^{-1}| = \begin{vmatrix} 1 + \nabla_1^1\varphi & \nabla_2^1\varphi & \cdots & \nabla_m^1\varphi \\ \nabla_1^2\varphi & 1 + \nabla_2^2\varphi & & \\ \vdots & & \ddots & \\ \nabla_1^m\varphi & & & 1 + \nabla_m^m\varphi \end{vmatrix}, \tag{4}$$

by expanding the determinant we find

$$\mathsf{M}(\varphi) = 1 + \nabla_\nu^\nu\varphi + \frac{1}{2}\begin{vmatrix} \nabla_\nu^\nu\varphi & \nabla_\mu^\nu\varphi \\ \nabla_\nu^\mu\varphi & \nabla_\mu^\mu\varphi \end{vmatrix} + \cdots + \frac{1}{m!}\begin{vmatrix} \nabla_\nu^\nu\varphi & \nabla_\mu^\nu\varphi & \cdots & \nabla_\lambda^\nu\varphi \\ \nabla_\nu^\mu\varphi & \nabla_\mu^\mu\varphi & & \\ \vdots & & \ddots & \\ \nabla_\nu^\lambda\varphi & & & \nabla_\lambda^\lambda\varphi \end{vmatrix}, \tag{5}$$

where the last determinant has m rows and m columns.

7.7 Remark. The first fundamental forms ω' corresponding to the metrics g' defined by (3) belong to the same cohomology class (Lemma 7.5). Conversely, if two first fundamental forms belong to the same cohomology class, there exists a function φ such that the corresponding metrics satisfy (3).

A cohomology class γ is said to be positive definite if there exists in γ a Hermitian form $(i/2\pi)C_{\lambda\bar{\mu}}\, dz^\lambda \wedge dz^{\bar{\mu}} \in \gamma$ such that everywhere $C_{\lambda\bar{\mu}}\xi^\lambda\xi^{\bar{\mu}} > 0$ for all vectors $\xi \neq 0$. A Kähler manifold M has at least one positive definite cohomology class defined by ω. Thus the second Betti number, $b_2(M)$, is nonzero.
If a Kähler manifold has only one positive definite cohomology class up to a proportionality constant, in particular if $b_2(M) = 1$, then all Kähler metrics are proportional to one of the form (3).

7.8 Lemma. *The Kähler manifolds* (M, g') *with* M *compact and* g' *defined by* (3) *have the same volume.*

Proof. The determinants in (5) are divergences

$$\nabla^\lambda \begin{vmatrix} \nabla_\lambda \varphi & \nabla^\mu_\lambda \varphi & \cdots & \nabla^\nu_\lambda \varphi \\ \nabla_\mu \varphi & \nabla^\mu_\mu \varphi & & \\ \vdots & & \ddots & \\ \nabla_\nu \varphi & \nabla^\mu_\nu \varphi & & \nabla^\nu_\nu \varphi \end{vmatrix} = \begin{vmatrix} \nabla^\lambda_\lambda \varphi & \nabla^\mu_\lambda \varphi & \cdots & \nabla^\nu_\lambda \varphi \\ \nabla^\lambda_\mu \varphi & \nabla^\mu_\mu \varphi & & \\ \vdots & & \ddots & \\ \nabla^\lambda_\nu \varphi & & & \nabla^\nu_\nu \varphi \end{vmatrix}.$$

Indeed the differentiation of the other columns gives zero, because on a Kähler manifold $\nabla^\lambda \nabla^\mu_\lambda \varphi = \nabla^\mu \nabla^\lambda_\lambda \varphi$.

So integrating (5) yields: $V' = \int_M dV' = \int_M \mathsf{M}(\varphi)\, dV = \int_M dV = V$. ∎

We can prove Lemma 7.8 by another method. Denote by ω^m (respectively, ω'^m) the m-fold tensor product of ω (respectively, ω'). $\omega' = \omega - (i/4\pi)\, dd^c\varphi$ and

$$\omega^m = \left(\frac{i}{2\pi}\right)^m m!(-2i)^m |g|\, dx^1 \wedge dy^1 \wedge dx^2 \wedge dy^2 \wedge \cdots \wedge dx^m \wedge dy^m.$$

Since $d\omega = 0$, then by Stokes' formula, $\int \omega'^m = \int \omega^m$. Hence

$$V' = \frac{\pi^m}{m!} \int \omega'^m = \frac{\pi^m}{m!} \int \omega^m = V.$$

§2. Calabi's Conjecture

7.9 The *Calabi conjecture* [(73] and [74]). which is proved in 7.19, asserts that every form representing the first Chern class $\mathsf{C}_1(M)$ is the Ricci form Ψ' of some Kähler metric on a compact Kähler manifold (M, g).

Let $(i/2\pi)C_{\lambda\bar{\mu}}\, dz^{\lambda} \wedge dz^{\bar{\mu}}$ belong to $\mathsf{C}_1(M)$. According to Lemma 7.5, there exists an $f \in C^{\infty}$ such that $C_{\lambda\bar{\mu}} = R_{\lambda\bar{\mu}} - \partial_{\lambda\bar{\mu}} f$.
Consider a change of metric of type (3), the components of the corresponding Ricci tensor in a local chart are:

$$R'_{\lambda\bar{\mu}} = \partial_{\lambda\bar{\mu}} \log|g'| = -\partial_{\lambda\bar{\mu}} \log \mathsf{M}(\varphi) + R_{\lambda\bar{\mu}}.$$

So we shall have $R'_{\lambda\bar{\mu}} = C_{\lambda\bar{\mu}}$, if there is an admissible function $\varphi \in C^{\infty}$ that satisfies

(6) $$\log \mathsf{M}(\varphi) = f + k, \text{ with } k \text{ a constant.}$$

By Lemma 7.8, we can compute k, $k = \log V - \log \int e^f \, dV$.

§3. Einstein–Kähler Metrics

7.10 Suppose g' is an Einstein–Kähler metric. Then by definition this means there exists a real number λ such that $\Psi' = -\lambda\omega'$. Since ω' is positive definite, the first Chern class is positive, negative, or zero. This is a necessary condition for a manifold to have an Einstein–Kähler metric.

The case $\lambda = 0$ ($\mathsf{C}_1(M)$ is zero) is a special case of Calabi's conjecture. We want to find a Kähler metric whose Ricci tensor vanishes, the zero-form belongs to $\mathsf{C}_1(M)$.

If $\mathsf{C}_1(M)$ is positive or negative ($\lambda \neq 0$), we can choose the metric g such that $-\lambda\omega \in \mathsf{C}_1(M)$. Then the Einstein–Kähler metric g', if it exists, is of the type (3). Hence

$$R'_{\lambda\bar{\mu}} = -\partial_{\lambda\bar{\mu}} \log \mathsf{M}(\varphi) + R_{\lambda\bar{\mu}}.$$

By Lemma 7.5, since $-\lambda\omega \in \mathsf{C}_1(M)$, there exists $f \in C^{\infty}$ such that $R_{\lambda\bar{\mu}} = -\lambda g_{\lambda\bar{\mu}} + \partial_{\lambda\bar{\mu}} f$. So

$$R'_{\lambda\bar{\mu}} = -\partial_{\lambda\bar{\mu}} \log \mathsf{M}(\varphi) - \lambda(g'_{\lambda\bar{\mu}} - \partial_{\lambda\bar{\mu}}\varphi) + \partial_{\lambda\bar{\mu}} f.$$

g' will be Einstein if there is an admissible function $\varphi \in C^{\infty}$ that satisfies

(7) $$\log \mathsf{M}(\varphi) = \lambda\varphi + f + k,$$

with k a constant. Here we can drop the constant k since $\tilde{\varphi} = \varphi + k\lambda^{-1}$ satisfies the equation

(8) $$\log \mathsf{M}(\varphi) = \lambda\varphi + f.$$

According to Lemma 7.5, when $\lambda \neq 0$, there exists a *Kähler–Einstein metric* if and only if Equation (8) has a C^∞ admissible solution.

§4. Complex Monge–Ampère Equation

7.11 More generally, we can consider an equation of the type

$$\mathsf{M}(\varphi) = \exp[F(\varphi, x)], \tag{9}$$

where $I \times M \ni (t, x) \to F(t, x)$ is a C^∞ function on $I \times M$ (or only C^3), with I an interval of $\mathbb{R}$.

(9) is called a Monge–Ampère Equation of complex type.

4.1. About Regularity

7.12 Proposition. *If F is in C^∞, then a solution of (9) is C^∞ admissible. If F is only $C^{r+\alpha}$ $r \geq 1, 0 < \alpha < 1$, the solution is $C^{2+r+\alpha}$.*

Proof. At Q a point of M, where φ, a C^2 solution of (9), has a minimum, $\partial_{\lambda\bar{\lambda}}\varphi(Q) \geq 0$ for all directions λ. So at Q, g' is positive definite. By continuity, no eigenvalue of g' can be zero since $\mathsf{M}(\varphi) > 0$. Hence φ is admissible.

Consider the following mapping of the C^2 admissible functions to C^0:

$$\Gamma\colon \varphi \to F(\varphi, x) - \log \mathsf{M}(\varphi). \tag{10}$$

Γ is continuously differentiable. Let $d\Gamma_\varphi$ denote its differential at φ:

$$d\Gamma_\varphi(\psi) = F'_t(\varphi, x)\psi + \Delta'_\varphi\psi. \tag{11}$$

Δ'_φ is the Laplacian in the metric g'; $\Delta'_\varphi = -g'^{\nu\bar{\mu}}\,\partial_{\nu\bar{\mu}}$, where $g'^{\nu\bar{\mu}}$ denotes the components of the inverse matrix of $g'_{\nu\bar{\mu}}$. F'_t means $\partial F/\partial t$.
Since φ is admissible, Equation (9) is elliptic at φ. Hence, by Theorem 3.56, $\varphi \in C^2$ implies $\varphi \in C^\infty$. If F is only $C^{r+\alpha}(r \geq 1)$, φ belongs to $C^{2+r+\alpha}$. ■

4.2. About Uniqueness

7.13 Proposition. *Equation (9) has at most one C^2 solution, possibly up to a constant, if $F'_t(t, x) \geq 0$ for all $(t, x) \in I \times M$.*
In particular, Equation (8) has at most one C^2 solution when $\lambda > 0$, while the solution is unique up to a constant if $\lambda = 0$.

Proof. This follows from the maximum principle (Theorem 3.74). Let φ_1 and φ_2 be two solutions of (9). According to the mean value Theorem 3.6, there exists a function θ $(0 < \theta < 1)$ such that $\psi = \varphi_2 - \varphi_1$ satisfies

$$(12) \qquad \Delta'_\gamma \psi + F'_t(\gamma, x)\psi = 0 \quad \text{with} \quad \gamma = \varphi_1 + \theta(\varphi_2 - \varphi_1).$$

Since $F'_t \geq 0$, Equation (12) has at most the constant solution.
If $F'_t > 0$, (12) has no solution except zero. ∎

§5. Theorem of Existence (the Negative Case)

7.14 *On a compact Kähler manifold, Equation* (8) *has a unique admissible* $C^{5+\alpha}$ *solution if* $\lambda > 0$ *and* $f \in C^{3+\alpha}$. *The solution is* C^∞ *if* $f \in C^\infty$.

Proof. We shall use the continuity method. For $t \geq 0$ a parameter, let us consider the equation:

$$(13) \qquad \log \mathsf{M}(\varphi) = \lambda\varphi + tf,$$

with $f \in C^{3+\alpha}$. If for some t, Equation (13) has a C^2 solution φ_t, then φ_t is unique, admissible, and belongs to $C^{5+\alpha}$, by Propositions 7.12 and 7.13.

a) The set of functions f, for which Equation (8) has a $C^{5+\alpha}$ solution is open in $C^{3+\alpha}$.
To prove this, let us consider Γ, the mapping of the set Θ of the $C^{5+\alpha}$ admissible functions in $C^{3+\alpha}$ defined by:

$$C^{5+\alpha} \supset \Theta \ni \varphi \xrightarrow{\Gamma} \lambda\varphi - \log \mathsf{M}(\varphi) \in C^{3+\alpha}.$$

$\log \mathsf{M}(\varphi) \in C^{3+\alpha}$ since φ is admissible and $\mathsf{M}(\varphi)$ involves only the second derivatives of φ.
Γ is continuously differentiable; its differential at φ is

$$d\Gamma_\varphi(\psi) = \lambda\psi + \Delta'_\varphi \psi.$$

Indeed for φ given, $\|d\Gamma_\varphi(\psi)\|_{3+\alpha} \leq \text{Const} \times \|\psi\|_{5+\alpha}$ and $C^{5+\alpha} \supset \Theta \ni \varphi \to d\Gamma_\varphi \in \mathscr{L}(C^{5+\alpha}, C^{3+\alpha})$ is continuous since $C^{5+\alpha} \supset \Theta \ni \varphi \to g'^{\lambda\bar{\mu}} \in C^{3+\alpha}$ is continuous. By Theorem 4.18 the operator $d\Gamma_\varphi$ is invertible since $\lambda > 0$. Indeed, we can write the equation $d\Gamma_\varphi(\Psi) = \tilde{f}$ in the form:

$$-\nabla_\nu[\mathsf{M}(\varphi)g'^{\nu\bar{\mu}}\nabla_{\bar{\mu}}\psi] + \lambda\mathsf{M}(\varphi)\psi = \tilde{f}\mathsf{M}(\varphi).$$

Since $C^{5+\alpha}$ and $C^{3+\alpha}$ are Banach spaces, we can use the inverse function Theorem 3.10. Thus if $\tilde{\varphi} \in C^{5+\alpha}$ satisfies

$$\log \mathsf{M}(\tilde{\varphi}) = \lambda\tilde{\varphi} + \tilde{f},$$

there exists $\mathscr{V}$, a $C^{3+\alpha}$ neighborhood of $\tilde{f}$, such that Equation (8) has a $C^{5+\alpha}$ solution when $f \in \mathscr{V}$.

Return to Equation (13), where f is given. Because $\varphi_0 = 0$ is the solution of (13) for $t = 0$, (13) has a solution for some interval $t \in [0, \tau[$, where $\tau > 0$. Let τ be the largest real number such that Equation (13) has a solution for all $t \in [0, \tau[$. If $\tau > 1$, then φ_1 is the solution of (8) and Theorem 7.14 is proved. So suppose $\tau \leq 1$, and come to a contradiction.

b) We claim that the set $\mathscr{B}$ of functions φ_t, $t \in [0, \tau[$, is bounded in $C^{2+\alpha}$, $(0 < \alpha < 1)$.
If φ_t has a maximum at P, then $\mathsf{M}(\varphi_t) \leq 1$. Indeed in a local chart for which $g_{\lambda\bar{\mu}}(P) = \delta^{\mu}_{\lambda}$ (δ^{μ}_{λ} the Kronecker tensor), and $\partial_{\lambda\bar{\mu}}\varphi_t = 0$ for $\lambda \neq \mu$, at P, we have $\mathsf{M}(\varphi_t) = \prod_{\lambda=1}^{m} (1 + \partial_{\lambda\bar{\lambda}}\varphi) \leq 1$, since all the terms are less than or equal to 1. Thus $\lambda\varphi_t(P) + tf(P) \leq 0$.
Similarly, we prove that if φ_t has a minimum at Q, then $\mathsf{M}(\varphi_t) \geq 1$ and

$$\lambda\varphi_t(Q) + tf(Q) \geq 0.$$

Hence $\sup|\varphi_t| \leq (\tau/\lambda)\sup|f|$. The set $\mathscr{B}$ is bounded in C^0. According to Proposition 7.23 below, $\mathscr{B}$ is bounded in $C^{2+\alpha}$.

c) We now show that (13) has a solution for $t = \tau$ and hence for some $t > \tau$ by a). This will give the desired contradiction.
According to Ascoli's theorem 3.15, the imbedding $C^{2+\alpha} \subset C^2$ is compact. Thus there exists $\varphi_\tau \in C^2$ and $t_i \to \tau$ an increasing sequence such that φ_{t_i} converges to φ_τ in C^2.
Letting $i \to \infty$ in $\log \mathsf{M}(\varphi_{t_i}) = \lambda\varphi_{t_i} + t_i f$ we prove that φ_τ is the solution of (13) for $t = \tau$. According to the regularity theorem $\varphi_\tau \in C^{5+\alpha}$, and the contradiction follows from a), since (13) has a solution for t in a neighborhood of τ. ∎

§6. Existence of Kähler–Einstein Metric

7.15 Theorem (Aubin [18]). *A compact Kähler manifold with negative first Chern class has an Einstein–Kähler metric (all the Einstein–Kähler metrics are proportional).*

Proof. According to 7.10, finding an Einstein–Kähler metric when $\mathsf{C}_1(M) < 0$ is equivalent to solving Equation (8) with $\lambda > 0$. By Theorem 7.14, Equation (8) has a unique solution. Thus there exists a unique Einstein Kähler metric whose Ricci curvature is equal to $-\lambda$ (we must choose g such that $-\lambda\omega \in \mathsf{C}_1(M)$).

7.16 An application of the preceding theorem is the proof of the following, which is equivalent to the Poincaré conjecture in the case of a compact Kähler manifold of dimension 4:

Theorem. *A compact Kähler manifold homeomorphic to* $\mathsf{P}_2(\mathbb{C})$, *the complex projectif space of dimension 2, is biholomorphic to* $\mathsf{P}_2(\mathbb{C})$.

In their proof, Hirzebruch–Kodaira [143] supposed that the first Chern class is nonnegative. This extra hypothesis can be removed, as Yau [276] recently pointed out.
If $\mathsf{C}_1(M) < 0$, by Theorem 7.15 there exists an Einstein–Kähler metric. Some computations done with this metric (see Yau [276]) lead to a contradiction: the manifold would be covered by the ball and could not be simply connected.

§7. Theorem of Existence (the Null Case)

7.17 *On a compact Kähler manifold, Equation* (6) *has, up to a constant, a unique admissible* $C^{r+2+\alpha}$ *solution (respectively,* C^∞*) if* $f \in C^{r+\alpha}$, $r \geq 3$ *(respectively,* $f \in C^\infty$*).*

Proof. We shall use the continuity method. For $t \geq 0$ a parameter, let us consider the equation:

$$\mathsf{M}(\varphi) - 1 = t(e^f - 1) \tag{14}$$

with $f \in C^{3+\alpha}$ satisfying $\int e^f\, dV = \int dV$.

If for some t $(0 \leq t \leq 1)$, Equation (14) has a C^2 solution φ_t, then it is unique up to a constant, admissible, and belongs to $C^{5+\alpha}$. Indeed $\mathsf{M}(\varphi_t) = (1 - t) + te^f$, so for t in a neighborhood of $[0, 1]$, $\mathsf{M}(\varphi_t)$ is strictly positive and we can apply Propositions 7.12 and 7.13.
Set $\tilde{C}^{r+\alpha} = \{f \in C^{r+\alpha} / \int f\, dV = 0\}$.

a) The set of the functions $h \in \tilde{C}^{3+\alpha}$ for which the equation

$$\mathsf{M}(\varphi) - 1 = h$$

has a $C^{5+\alpha}$ admissible solution is open in $\tilde{C}^{3+\alpha}$.
Let us consider the mapping Γ of the set Θ of the admissible functions belonging to $\tilde{C}^{5+\alpha}$ in $\tilde{C}^{3+\alpha}$ defined by

$$\tilde{C}^{5+\alpha} \supset \Theta \ni \varphi \xrightarrow{\Gamma} \mathsf{M}(\varphi) - 1 \in \tilde{C}^{3+\alpha}.$$

Γ is continuously differentiable; its differential at φ

$$d\Gamma_\varphi(\psi) = \mathsf{M}(\varphi)\Delta'_\varphi \psi,$$

is invertible. Indeed, $\int \mathsf{M}(\varphi)\Delta'_\varphi \psi\, dV = \int \Delta'_\varphi \psi\, dV' = 0$.
Since $\tilde{C}^{5+\alpha}$ and $\tilde{C}^{3+\alpha}$ are Banach spaces, we can use the inverse function Theorem 3.10. Thus if $\tilde{\varphi} \in \Theta$ satisfies $\mathsf{M}(\tilde{\varphi}) - 1 = \tilde{h}$ there exists $\mathscr{V}$, a $C^{3+\alpha}$

neighborhood of $\tilde{h}$ in $\tilde{C}^{3+\alpha}$, such that equation $\mathsf{M}(\varphi) - 1 = h$ has a solution in Θ when $h \in \mathscr{V}$.

Return to Equation (14). Because $\varphi_0 = 0$ is the solution in Θ of (14) for $t = 0$, (14) has an admissible solution for $t \in [0, \tau[$, $\tau > 0$. Let τ be the largest real number such that Equation (14) has an admissible solution $\varphi_t \in \Theta$ for all $t \in [0, \tau[$. If $\tau > 1$, then φ_1 is the desired solution of (6) in Θ.
So we suppose $\tau \leq 1$ and come to a contradiction.

b) We claim that the set $\mathscr{B} \subset \tilde{C}^{2+\alpha}$ of the functions φ_t, $t \in [0, \tau[$, is bounded in $C^{2+\alpha}$.
Let us prove that $\mathscr{B}$ is bounded in C^0. Then by Proposition 7.23 below, $\mathscr{B}$ is bounded in $C^{2+\alpha}$. Repeating the proof in 7.14c then establishes Theorem 7.17. The idea is to find a bound, uniform in t and $p \geq 2$, of $\|\varphi_t\|_p$ for $0 < t < \tau \leq 1$. Then $\|\varphi_t\|_p \leq \gamma$ and letting $p \to \infty$ will imply $\sup|\varphi_t| \leq \gamma$. For simplicity we drop the subscript t.

Setting $h(\varphi) = \varphi|\varphi|^{p-2}$ in Proposition 7.18 below yields:

$$(15) \qquad 4\frac{p-1}{mp^2}\int \nabla^\nu |\varphi|^{p/2} \nabla_\nu |\varphi|^{p/2}\, dV \leq \int [1 - \mathsf{M}(\varphi)]\varphi|\varphi|^{p-2}\, dV.$$

According to the Sobolev imbedding theorem, there is a constant, independent of p, such that

$$\| |\varphi|^p \|_{m/(m-1)} = \| |\varphi|^{p/2} \|_{2m/(m-1)}^2 \leq \text{Const} \times (\|\nabla|\varphi|^{p/2}\|_2^2 + \|\varphi\|_p^p)$$

This inequality together with (15) leads to

$$(16) \qquad \| |\varphi|^p \|_{m/(m-1)} \leq \tilde{C}\left(p \int |\varphi|^{p-1}\, dV + \int |\varphi|^p\, dV\right), \qquad (p > 1)$$

where $\tilde{C}$ is a constant, since $\mathsf{M}(\varphi)$ is uniformly bounded; we pick $\tilde{C} \geq 1$. The desired result, $\|\varphi\|_p \leq \gamma$ for all p and $\varphi \in \mathscr{B}$, will follow from:

Lemma. *There exists a constant γ such that for all real numbers $p \geq 1$ and all $\varphi \in \mathscr{B}$:*

$$(17) \qquad \|\varphi\|_p \leq \gamma(\alpha^{m-1} C p)^{-m/p}$$

with $\alpha = m/(m-1)$ and $C = \tilde{C}(1 + V^{1/p})$, $\tilde{C}$ the constant of (16).

Proof. Because φ is admissible, then $\Delta\varphi < m$. Thus $\|\Delta\varphi\|_1 < 2m \int dV$, since $\int \Delta\varphi\, dV = 0$. According to Theorem 4.13, as $\int \varphi\, dV = 0$, there exists a constant C_0 such that $\|\varphi\|_1 \leq C_0\|\Delta\varphi\|_1 \leq 2mC_0 V$. Picking $p = 2$ in (15) gives $\|\nabla\varphi\|_2 \leq \text{Const}$, since $\mathsf{M}(\varphi)$ is uniformly bounded. Hence $\|\varphi\|_2 \leq \text{Const}$, by Corollary 4.3 because $\int \varphi\, dV = 0$.

Choosing $p = 2$ in (16) yields $\|\varphi\|_{2m/(m-1)} \leq \text{Const}$. By the interpolation inequality 3.69 there exists a constant k such that $\|\varphi\|_q \leq k$ for $1 \leq q \leq 2m/(m-1)$.
Set $\gamma = k\alpha^{m(m-1)}C^m e^{m/e}$. Then we can verify that inequality (17) is satisfied for $1 \leq p \leq 2m/(m-1)$. Either $\|\varphi\|_p$ is always smaller than 1, and there is nothing to prove (we pick $k \geq 1$ and (17) is satisfied); or else, for some p, $\|\varphi\|_p > 1$ and then by Hölder's inequality $\int |\varphi|^{p-1}\, dV \leq \|\varphi\|_p^{p-1}(\int dV)^{1/p} \leq V^{1/p} \int |\varphi|^p\, dV$.
Inequality (16) becomes

$$\left(\int |\varphi|^{p\alpha}\, dV\right)^{1/\alpha} \leq Cp \int |\varphi|^p\, dV$$

and inequality (17) follows by induction:

$$\int |\varphi|^{p\alpha}\, dV \leq (Cp)^\alpha \gamma^{p\alpha}(\alpha^{m-1}Cp)^{-m\alpha} = \gamma^{p\alpha}(\alpha^{m-1}Cp\alpha)^{-m}$$

since $\alpha(m-1) = m$, $(\alpha - 1)(m-1) = 1$. ■

7.18 Proposition. *Let $h(t)$ be a C^1 increasing function on $\mathbb{R}$. Then all C^2 admissible functions φ satisfy:*

$$\text{(18)} \qquad \int [1 - \mathsf{M}(\varphi)]h(\varphi)\, dV \geq \frac{1}{m} \int h'(\varphi)\nabla^\nu\varphi\nabla_\nu\varphi\, dV.$$

Proof. In the notation of 7.8,

$$\int [1 - \mathsf{M}(\varphi)]h(\varphi)\, dV = \frac{\pi^m}{m!} \int h(\varphi)(\omega^m - \omega'^m).$$

But $\omega^m - \omega'^m = (i/4\pi)\, dd^c\varphi \wedge (\omega^{m-1} + \omega^{m-2} \wedge \omega' + \cdots + \omega'^{m-1})$. Applying Stokes' formula leads to

$$\int h(\varphi)(\omega^m - \omega'^m)$$

$$= \frac{-i}{4\pi} \int h'(\varphi)\, d\varphi \wedge d^c\varphi \wedge (\omega^{m-1} + \omega^{m-2} \wedge \omega' + \cdots + \omega'^{m-1})$$

$$\geq \frac{(m-1)!}{\pi}\left(\frac{i}{2\pi}\right)^{m-1}(-2i)^{m-1} \int h'(\varphi)\nabla^\nu\varphi\nabla_\nu\varphi\, dV,$$

which gives (18). ■

§8. Proof of Calabi's Conjecture

7.19 Theorem. *On a compact Kähler manifold, every form representing* $\mathsf{C}_1(M)$ *is the Ricci form of some Kähler metric.*
To each positive cohomology class there corresponds one and only one metric. In particular, if $b_2(M) = 1$, *the solution is unique up to a homothetic change of metric.*

Proof. Let g be the Kähler metric, ω its first fundamental form, and $\gamma \in \mathsf{C}_1(M)$. According to 7.7, to find g', with $\omega' - \omega$ homologous to zero and $\Psi' = \gamma$, is equivalent to solving equation (6).
By Theorem 7.17, Equation (6) has a unique solution up to a constant. Thus in each positive cohomology class we find a unique ω' whose Ψ' equals γ. ∎

§9. The Positive Case

7.20 *According to* (7.10), *in the case* $\mathsf{C}_1(M) > 0$ *there exists a Kähler–Einstein metric, if and only if Equation* (8) *with* $\lambda < 0$ *has a* C^∞ *admissible solution.*

This problem is not yet solved. It is more difficult than the two preceding cases. First, since the linear map $d\Gamma_\varphi$ of 7.14 is not necessarily invertible, it is not obvious how to use the continuity method. Then we must find a C^0 estimate in order to use Proposition 7.23. On the other hand, Equation (8) with $\lambda < 0$ may have many solutions (see Aubin [20] pp. 85 and 86). For instance, on the complex projective space, $\log \mathsf{M}(\varphi) = -\lambda_1 \varphi$ has many solutions; these solutions come from the infinitesimal holomorphic transformations which are not isometries. Worse, we know that some Equations (8) have no solution, since if we blow up one or two points of projective space, the manifold obtained cannot carry an Einstein–Kähler metric according to a theorem of Lichnerowicz [185] p. 156 (see Yau [275]).

§10. A Priori Estimate for $\Delta\varphi$

7.21 Notations. On a compact Kähler manifold, let $\mathscr{B}$ be a set of $C^{5+\alpha}$ admissible functions and λ real numbers satisfying $|\lambda| \leq \lambda_0$. We suppose that $\mathscr{C}$, the set of the corresponding functions $f = \log \mathsf{M}(\varphi) - \lambda\varphi$, is bounded in $C^{3+\alpha}$.
F_0 and F_1 are real numbers such that everywhere for all $\varphi \in \mathscr{B}$,

$$\Delta f \leq F_1 \quad \text{and} \quad f \leq F_0.$$

Proposition (Aubin [18]). *There exist two constants k and K depending only on λ_0, F_0, F_1, and the curvature, such that all $\varphi \in \mathscr{B}$ satisfy:*

$$0 < m - \Delta\varphi \leq K e^{k\varphi - (k - \lambda/m)\inf\varphi}.$$

Moreover, if $\mathscr{B}$ is bounded in C^0, all corresponding metrics g' are equivalent.

Proof. The first inequality is obvious. Since φ is admissible, for all directions μ, $g_{\mu\bar{\mu}} + \partial_{\mu\bar{\mu}}\varphi > 0$; thus by summing over μ we obtain $m - \Delta\varphi > 0$.
To prove the second inequality set

$$A = \log(m - \Delta\varphi) - k\varphi,$$

where k is a real number that we will choose later. Let us compute $\Delta'_\varphi A = -g'^{\lambda\bar{\mu}}_\varphi \partial_{\lambda\bar{\mu}} A$ (we will omit the index φ below).

(19) $$\Delta' A = -(m - \Delta\varphi)^{-1}\Delta'\Delta\varphi - k\Delta'\varphi + (m - \Delta\varphi)^{-2} g'^{\lambda\bar{\mu}} \nabla_\lambda \Delta\varphi \nabla_{\bar{\mu}} \Delta\varphi.$$

Recall $g'^{\lambda\bar{\mu}}$ are the components of the inverse matrix of $((g_{\lambda\bar{\mu}} + \partial_{\lambda\bar{\mu}}\varphi))$. Differentiating (8) yields:

$$\lambda\nabla_\nu\varphi + \nabla_\nu f = \nabla_\nu \log M(\varphi) = g'^{\alpha\bar{\beta}}\nabla_\nu\nabla_{\alpha\bar{\beta}}\varphi$$

(20) $$-\lambda\Delta\varphi - \Delta f = g'^{\alpha\bar{\beta}}\nabla^\nu\nabla_\nu\nabla_{\alpha\bar{\beta}}\varphi - g'^{\alpha\bar{\mu}} g'^{\gamma\bar{\beta}} \nabla^\nu\nabla_{\gamma\bar{\mu}}\varphi\nabla_\nu\nabla_{\alpha\bar{\beta}}\varphi$$

But from 1.13,

(21) $$\Delta'\Delta\varphi - g'^{\alpha\bar{\beta}}\nabla^\nu\nabla_\nu\nabla_{\alpha\bar{\beta}}\varphi = R_{\alpha\bar{\beta}\lambda\bar{\mu}}\nabla^{\alpha\bar{\beta}}\varphi g'^{\lambda\bar{\mu}} - R_{\lambda\bar{\mu}}\nabla^\lambda_\nu\varphi g'^{\nu\bar{\mu}} = E$$

and there exists a constant C such that E satisfies

(22) $$|E| \leq C(m - \Delta\varphi) g'^{\lambda\bar{\mu}} g_{\lambda\bar{\mu}}.$$

Write $\Delta'\varphi = -g'^{\lambda\bar{\mu}}(g'_{\lambda\bar{\mu}} - g_{\lambda\bar{\mu}}) = g'^{\lambda\bar{\mu}} g_{\lambda\bar{\mu}} - m$ and observe that

(23) $$g'^{\alpha\bar{\beta}} g'^{\lambda\bar{\mu}} \nabla^\nu\nabla_{\alpha\bar{\mu}}\varphi\nabla_\nu\nabla_{\lambda\bar{\beta}}\varphi \geq (m - \Delta\varphi)^{-1} g'^{\lambda\bar{\mu}}\nabla_\lambda\Delta\varphi\nabla_{\bar{\mu}}\Delta\varphi.$$

To verify this inequality, we have only to expand

$$[(m - \Delta\varphi)\nabla_\nu\nabla_{\lambda\bar{\beta}}\varphi + \nabla_\lambda\Delta\varphi g'_{\nu\bar{\beta}}] \times [(m - \Delta\varphi)\nabla_{\bar{\gamma}}\nabla_{\alpha\bar{\mu}}\varphi + \nabla_{\bar{\mu}}\Delta\varphi g'_{\alpha\bar{\gamma}}] g'^{\alpha\bar{\beta}} g'^{\lambda\bar{\mu}} g^{\nu\bar{\gamma}} \geq 0.$$

(19)–(23) lead to

(24) $$\Delta' A \leq k(m - g'^{\lambda\bar{\mu}} g_{\lambda\bar{\mu}}) - (m - \Delta\varphi)^{-1}(E - \lambda\Delta\varphi - \Delta f).$$

At a point P where A has a maximum, $\Delta' A \geq 0$. We find, using (22),

$$(25) \qquad (k - C)g'^{\nu\bar{\mu}}g_{\nu\bar{\mu}} \leq (m - \Delta\varphi)^{-1}(\lambda\Delta\varphi + \Delta f) + mk.$$

Since the arithmetic mean is greater than or equal to the geometric mean,

$$1 - \Delta\varphi/m \geq [\mathsf{M}(\varphi)]^{1/m} = e^{(\lambda\varphi + f)/m}.$$

Thus at P inequality (25) yields

$$(26) \qquad (k - C)g'^{\nu\bar{\mu}}g_{\nu\bar{\mu}} \leq mk - \lambda + (\lambda + \Delta f/m)e^{-(\lambda\varphi + f)/m}.$$

However, $g'^{\nu\bar{\mu}}g_{\nu\bar{\mu}} \geq m[\mathsf{M}(\varphi)]^{-1/m}$, so that

$$(27) \qquad m(k - C) - \lambda - \Delta f/m \leq (mk - \lambda)e^{(\lambda\varphi + f)/m}.$$

Pick k such that $m(k - C) \geq 1 + \sup(\lambda, 0) + \sup(\Delta f)/m$, expressions (26) and (27) lead to: There exists a constant K_0 such that at P

$$(28) \qquad (g'^{\nu\bar{\mu}}g_{\nu\bar{\mu}})_P \leq K_0.$$

K_0 depends on λ_0, F_1, and the curvature through C.

In an orthonormal chart at P for which $\partial_{\nu\bar{\mu}}\varphi = 0$ if $\nu \neq \mu$

$$1 + \partial_{\nu\bar{\nu}}\varphi \leq \mathsf{M}(\varphi)\prod_{\mu\neq\nu} g'^{\mu\bar{\mu}} \leq \mathsf{M}(\varphi)\left[\frac{1}{m-1}\sum_{\mu\neq\nu} g'^{\mu\bar{\mu}}\right]^{m-1}.$$

Taking the sum and using (28) yields at P

$$(m - \Delta\varphi)_P \leq m[K_0/(m-1)]^{m-1}e^{\lambda\varphi(P) + f(P)}.$$

Hence everywhere

$$(29) \qquad (m - \Delta\varphi)e^{-k\varphi} \leq (m - \Delta\varphi)_P e^{-k\varphi(P)} \leq Ke^{-(k-\lambda/m)\varphi(P)},$$

where K is a constant depending on K_0 and F_0.

The inequality of Proposition 7.21 now follows since $k \geq \lambda/m$.

If $\mathscr{B}$ is bounded in C^0 that is $|\varphi| \leq k_0$, using (29) for all $\varphi \in \mathscr{B}$ we have $\Delta\varphi$ uniformly bounded: $|\Delta\varphi| \leq k_1$. Therefore, in an orthonormal chart adapted to φ ($\partial_{\nu\bar{\mu}}\varphi = 0 \quad$ if $\nu \neq \mu$),

$$\text{as } \partial_{\mu\bar{\mu}}\varphi > -1, \ \partial_{\nu\bar{\nu}}\varphi < m - 1 + k_1,$$

and $(1 + \partial_{\mu\bar{\mu}}\varphi)^{-1} \leq (m + k_1)^{m-1}[\mathsf{M}(\varphi)]^{-1} < (m + k_1)^{m-1}e^{k_0|\lambda|+f}$. Thus the metrics g'_φ, $\varphi \in \mathscr{B}$, are equivalent to g; for all directions μ

$$e^{-k_0|\lambda| - \sup f}(m + k_1)^{1-m}g_{\mu\bar{\mu}} \leq g'_{\mu\bar{\mu}} \leq (m + k_1)g_{\mu\bar{\mu}}.$$ ∎

§11. A Priori Estimate for the Third Derivatives of Mixed Type

7.22 Once we have uniform bounds for $|\varphi|$ and $|\Delta\varphi|$, to obtain estimates for the third derivatives of mixed type, consider

$$(30) \qquad |\psi|^2 = g'^{\alpha\bar{\beta}}g'^{\lambda\bar{\mu}}g'^{\nu\bar{\gamma}}\nabla_\alpha\nabla_{\nu\bar{\mu}}\varphi\nabla_{\bar{\beta}}\nabla_{\lambda\bar{\gamma}}\varphi.$$

The choice of this norm instead of a simpler equivalent norm (in the metric g, for instance) imposes itself on those who make the computation. We now give the result; the reader can find the details of the calculation in Aubin (11) pp. 410 and 411.

Lemma.

$$\begin{aligned}
-\Delta'|\varphi|^2 = {} & g'^{\lambda\bar{\mu}}g'^{\alpha\bar{\beta}}g'^{a\bar{b}}g'^{c\bar{d}}[(\nabla_{\bar{\mu}\alpha\bar{b}c}\varphi - \nabla_{\bar{\mu}\gamma\bar{b}}\varphi\nabla_{\alpha\bar{\delta}c}\varphi g'^{\gamma\bar{\delta}})\text{ (conjugate expression)} \\
& + (\nabla_{\lambda\alpha\bar{b}c}\varphi - \nabla_{\lambda\bar{b}\rho}\varphi\nabla_{\alpha\bar{\nu}c}\varphi g'^{\rho\bar{\nu}} - \nabla_{\lambda\bar{\nu}c}\varphi\nabla_{\rho\bar{b}\alpha}\varphi g'^{\rho\bar{\nu}}) \\
& \times \text{(conjuguate expression)}] - g'^{c\bar{d}}(2g'^{\alpha\bar{\delta}}g'^{\gamma\bar{\beta}}g'^{a\bar{b}} \\
& + g'^{\alpha\bar{\beta}}g'^{a\bar{\delta}}g'^{\gamma\bar{b}})\nabla_{\alpha\bar{b}c}\varphi\nabla_{\bar{\beta}a\bar{d}}\varphi[\nabla_{\gamma\bar{\delta}}(\lambda\varphi + f) - R_{\gamma\bar{\delta}}] \\
& + g'^{\alpha\bar{\beta}}g'^{a\bar{b}}g'^{c\bar{d}}[\nabla_{\bar{\beta}a\bar{d}}\varphi\nabla_{\alpha\bar{b}c}(\lambda\varphi + f) + \nabla_{\alpha\bar{b}c}\varphi\nabla_{\bar{\beta}a\bar{d}}(\lambda\varphi + f)] \\
& + g'^{\lambda\bar{\mu}}g'^{\alpha\bar{\beta}}g'^{a\bar{b}}g'^{cd}[\nabla_{\bar{\beta}a\bar{d}}\varphi(R^{\nu}_{c\lambda\bar{b}}\nabla_{\alpha\bar{\mu}\nu}\varphi + R^{\bar{\rho}}_{\bar{b}\bar{\mu}\alpha}\nabla_{\lambda\bar{\rho}c}\varphi \\
& + R^{\nu}_{c\bar{\mu}\alpha}\nabla_{\lambda\bar{b}\nu}\varphi) + \text{conjuguate expression} \\
& + g'^{\alpha\bar{\beta}}g'^{c\bar{d}}[\nabla_{\bar{\beta}a\bar{d}}\varphi(g'^{\lambda\bar{\mu}}\nabla_\lambda R^{a}_{c\bar{\mu}\alpha} - g'^{a\bar{b}}\nabla_\alpha R_{c\bar{b}}) \\
& + \text{conjuguate expression}].
\end{aligned}$$

Hence there exists a constant k_2 which depends on λ_0, $\|\mathscr{B}\|_{C^0}$, $\|\mathscr{C}\|_{C^3}$, and the curvature such that

$$(31) \qquad \Delta'|\psi|^2 \leq k_2(|\psi|^2 + |\psi|).$$

Proposition. *There exists a constant k_3, depending only on λ_0, $\|\mathscr{B}\|_{C^0}$, $\|\mathscr{C}\|_{C^3}$, and the curvature, such that $\nabla_{\lambda\bar{\mu}\nu}\varphi\nabla^{\lambda\bar{\mu}\nu}\varphi \leq k_3$, for all $\varphi \in \mathscr{B}$.*

Proof. Equations (20) and (21) give

$$(32) \qquad \Delta'\Delta\varphi = g'^{\alpha\bar{\mu}}g'^{\gamma\bar{\beta}}\nabla^\nu\nabla_{\gamma\bar{\mu}}\varphi\nabla_\nu\nabla_{\alpha\bar{\beta}}\varphi - \lambda\Delta\varphi - \Delta f + E.$$

As all metrics g' are equivalent (Proposition 7.21), there exists a constant $B > 0$ such that

$$g'^{\alpha\bar{\mu}} g'^{\gamma\bar{\beta}} \nabla^\nu \nabla_{\gamma\bar{\mu}} \varphi \nabla_\nu \nabla_{\alpha\bar{\beta}} \varphi \geq B|\psi|^2.$$

Let $h > 0$ be a real number. According to (31),

$$\Delta'(|\psi|^2 - h\Delta\varphi) \leq k_2(|\psi|^2 + |\psi|) - hB|\psi|^2 + h(\lambda\Delta\varphi + \Delta f - E).$$

Picking $h = 2k_2 B^{-1}$, we get

$$(33) \quad \Delta'(|\psi|^2 - h\Delta\varphi) \leq -(k_2/2)|\psi|^2 + k_2/2 + 2k_2 B^{-1}(\lambda\Delta\varphi + \Delta f - E).$$

At a point P where $|\psi|^2 - h\Delta\varphi$ has a maximum, the first member of (33) is nonnegative. Thus

$$|\psi(P)|^2 \leq 1 + 4B^{-1}(\lambda\Delta\varphi(P) + \Delta f(P) - E(P)).$$

So by Proposition 7.21, $|\psi(P)|^2 \leq$ Const. Hence everywhere, $|\psi|^2 \leq$ Const.

7.23 Proposition. *On a compact Kähler manifold, let $\mathcal{B}$ be a set of C^5 admissible functions, $\mathcal{B}$ bounded in C^0. Let $(\varphi, \lambda) \in \mathcal{B} \times [-\lambda_0, \lambda_0]$ with λ_0 a constant, and let*

$$f = \log \mathsf{M}(\varphi) - \lambda\varphi.$$

If the set $\mathscr{C}$ of corresponding functions f is bounded in C^3, then $\mathcal{B}$ is bounded in $C^{2+\alpha}$ for all $\alpha \in]0, 1[$.

Proof. According to Proposition 7.22, the third derivatives of mixed type of the functions $\varphi \in \mathcal{B}$ are uniformly bounded. Hence there exists a constant k such that for all $\varphi \in \mathcal{B}$ $|\nabla\Delta\varphi| \leq k$ since the gradient of $\Delta\varphi$ involves only third derivatives of mixed type. By the properties of Green's function (Theorem 4.13), for any $\alpha \in]0, 1[$, $\mathcal{B}$ is bounded in $C^{2+\alpha}$. ∎

§12. The Method of Lower and Upper Solutions

7.24 Suppose we have to solve an elliptic differential equation $\mathscr{E}$. If there exist a lower solution u and an upper solution v satisfying $u \leq v$ we can hope to use the method of lower and upper solutions. But for this we also need to be able to solve an equation "close" to equation $\mathscr{E}$. Thus the method of lower and upper solutions requires an additional basic step. It is simpler to give an example.

7.25 Return to Equation (9),

$$\log \mathsf{M}(\varphi) = F(\varphi, x), \tag{34}$$

and set $S(t) = \sup_{x \in M} F(t, x)$ and $P(t) = \inf_{x \in M} F(t, x)$. Recall that $F(t, x)$ is a C^∞ function on $I \times M$ where $I =]\alpha, \beta[$ is an interval of $\mathbb{R}$. We will prove:

Theorem 7.25. *Equation* (9) *has a* C^∞ *solution if there exist two real numbers* a *and* b *belonging to* I, $(a \leq b)$ *such that* $S(a) = P(b) = 0$.

In this problem a is a lower solution of Equation (9). Indeed, $\log \mathsf{M}(a) \geq F(a, x)$, since $F(a, x) \leq S(a) = 0$; and b is an upper solution, $\log \mathsf{M}(b) \leq F(b, x)$, since $F(b, x) \geq P(b) = 0$. Moreover, $a \leq b$. Before giving the proof of this theorem, let us establish the following:

Corollary 7.25. *The equation* $\log \mathsf{M}(\varphi) - F(\varphi, x) = \psi(x)$ *has a* C^∞ *solution for any function* $\psi \in C^\infty$ *if* $P(t) \to +\infty$ *as* $t \to \beta$, *and* $S(t) \to -\infty$ *as* $t \to \alpha$.

Proof. Corollary 7.25 follows from Theorem 7.25. Indeed set $P_0(t) = \inf_{x \in M}[F(t, x) + \psi(x)]$ and $S_0(t) = \sup_{x \in M}[F(t, x) + \psi(x)]$. These functions are continuous and obviously $P(t) + \inf \psi \leq P_0(t) \leq S_0(t) \leq S(t) + \sup \psi$. Thus, when $t \to \alpha$, $S_0(t)$ and $P_0(t)$ go to $-\infty$ and when $t \to \beta$, $S_0(t)$ and $P_0(t)$ go to $+\infty$. So there exists $a \in I$ such that $S_0(a) = 0$. But as $P_0(a) \leq 0$ there exists also $b \in [a, \beta[$ such that $P_0(b) = 0$. The hypotheses of Theorem 7.25 are satisfied. ∎

If one solves the equation under the assumptions of Corollary 7.25, note that only the behavior of $F(t, x)$ as t goes to α and β is important.

Now we give the proof of Theorem 7.25.

α) By Theorem 7.14, the equation $\log \mathsf{M}(\varphi) - \lambda\varphi = f$ has a unique solution when $\lambda > 0$. With this result we will consider an increasing sequence of functions converging to a solution of Equation (9). Pick $\lambda > \sup[0, F'_t(t, x)]$ for all $(t, x) \in [a, b] \times M$. According to Theorem 7.14 we can define the sequence of functions φ_j by $\varphi_0 = a$ and

$$\log \mathsf{M}(\varphi_j) - \lambda\varphi_j = F(\varphi_{j-1}, x) - \lambda\varphi_{j-1} \quad \text{for } j \geq 1. \tag{35}$$

This sequence of C^∞ functions is increasing and satisfies $a \leq \varphi_j \leq b$. The proof proceeds by induction by using the maximum principle. We suppose that for all $i \leq j$, $a \leq \varphi_{i-1} \leq \varphi_i \leq b$ and we write:

$$\begin{aligned} &\log \mathsf{M}(\varphi_{j+1}) - \log \mathsf{M}(\varphi_j) - \lambda(\varphi_{j+1} - \varphi_j) \\ &\quad = [F(\varphi_j, x) - \lambda\varphi_j] - [F(\varphi_{j-1}, x) - \lambda\varphi_{j-1}] \leq 0. \end{aligned}$$

The last inequality is obtained by applying the mean value theorem. The maximum principle implies $\varphi_{j+1} - \varphi_j \geq 0$. Moreover we can start the induction because $\log \mathsf{M}(\varphi_1) - \lambda(\varphi_1 - a) = F(a, x) \leq S(a) = 0$. Similarly

$$\log \mathsf{M}(\varphi_{j+1}) - \lambda(\varphi_{j+1} - b) = [F(\varphi_j, x) - \lambda\varphi_j] + \lambda b \geq F(b, x) \geq P(b) = 0$$

shows that $\varphi_{j+1} - b \leq 0$. Therefore the sequence $\{\varphi_j\}$, which is increasing and bounded, converges pointwise to a function ψ which satisfies $a \leq \psi \leq b$. It remains to establish the regularity of ψ. For this we need estimates on the functions φ_j.

β) We will prove that the set of the functions φ_j is bounded in $C^{2+\alpha}$ $(0 < \alpha < 1)$, because then, by Ascoli's theorem, there exists a subsequence $\{\varphi_i\}$ of the sequence $\{\varphi_j\}$ which converges in C^2 to a function which cannot be different from ψ. Then $\psi \in C^{2+\alpha}$ and by the regularity theorem 3.56 $\psi \in C^\infty$. We already have the C^0-estimate $a \leq \varphi_j \leq b$. To prove that $|\Delta\varphi_j|$ is smaller than a constant independent of j, we compute $\Delta'_j B$ where

$$B = \log(m - \Delta\varphi_j) - k\varphi_j + \sigma\varphi_{j-1} + \nu\varphi_{j-1}^2,$$

k, σ, and ν being real numbers which we will choose later. Recall that $\Delta'_j = -g_j'^{\lambda\bar{\mu}}\nabla_{\lambda\bar{\mu}}$ and $((g_j'^{\lambda\bar{\mu}}))$ is the inverse matrix of $((g_{\lambda\bar{\mu}} + \partial_{\lambda\bar{\mu}}\varphi_j))$. At a point P where B has a maximum, we note that $\Delta'_j B \geq 0$. First of all we choose ν so that the terms in $\Delta'_j B$ which involve the first derivatives of φ_{j-1} are less than some constant. Then we pick σ large enough to control the terms with the Laplacian of φ_{j-1}. Finally k is chosen sufficiently large (in particular $k > \sigma$) so that the inequality $\Delta'_j B \geq 0$ at P becomes $g_j'^{\alpha\bar{\beta}} g_{\alpha\bar{\beta}} \leq \text{Const}$. (For the details of this computation see Aubin [20] p. 89). In 7.21 we saw that $g_j'^{\alpha\bar{\beta}} g_{\alpha\bar{\beta}} \leq \text{Const}$ at a point where B is maximum implies that the functions $|\Delta\varphi_j|$ are uniformly bounded, and consequently that the metrics g'_j are uniformly equivalent to g.

γ) Finally we prove that the set $\{\Delta\varphi_j\}$ is bounded in H_1^q for all $q \geq 1$ (see Aubin [20] p. 90) by using the following inequality instead of inequality (31):

$$|\Delta'_j|\psi_j|^2 + \Gamma_j^2| \leq k_2(|\psi_j|^2 + |\psi_j||\psi_{j-1}| + |\psi_j|),$$

where $|\psi_j|$ is defined by (30) with $\varphi = \varphi_j$, and Γ_j^2 is a positive term which involves the fourth derivatives of φ_j.
Hence the set of the functions $\{\varphi_j\}$ is bounded in $C^{2+\alpha}$ for all α $(0 < \alpha < 1)$.

Chapter 8

Monge–Ampère Equations

§1. Monge–Ampère Equations on Bounded Domains of $\mathbb{R}^n$

8.1 In this chapter we study the *Dirichlet Problem* for real Monge–Ampère equations.

Let B be the ball of radius 1 in $\mathbb{R}^n$ and let I be a closed interval of $\mathbb{R}$. $f(x, t)$ will denote a C^∞ function on $\bar{B} \times I$ and g a C^∞ Riemannian metric on $\bar{B}$. Consider $\text{u}(x)$ a C^∞ function on $S = \partial B$ with values in I, defined as the restriction to S of a C^∞ function γ on $\bar{B}$.

The problem is to prove the existence of a function $\varphi \in C^\infty(\bar{B})$ satisfying:

$$\text{(1)} \qquad \log \det((\nabla_{ij}\varphi + a_{ij})) = f(x, \varphi), \qquad \varphi/S = \text{u},$$

where $a_{ij}(x) = a_{ji}(x)(1 \le i, j \le n)$ are $n(n + 1)/2$ C^∞ functions on $\bar{B}$. This problem is not yet solved, except for dimension two under some additional hypotheses. The reason for the difficulty is the following: for the present it is possible to obtain *a priori* estimates up to the second derivatives but not for the third derivatives in the general case. We need such estimates to exhibit a subsequence which converges in $C^2(\bar{B})$ to a C^2 function which will be a solution of (1). Then according to Nirenberg [217] the solution is C^∞. In the special case when $n = 2$, Nirenberg [216] found an estimate for the third derivatives in terms of a bound on the second derivatives. When $n \ge 3$ this estimate depends in addition on the modulus of continuity of the second derivatives.

1.1. The Fundamental Hypothesis

8.2 The hypothesis that B is convex in the metric g is fundamental: there exists $h \in C^\infty(\bar{B})$, $h/S = 0$ satisfying $\nabla_{ij}h(x)\xi^i\xi^j > 0$ for all vectors $\xi \neq 0$ and all points x in $\bar{B}$.

Proposition. *Under the hypothesis of convexity, there exists a lower solution of* (1): $\gamma_1 \in C^\infty(\bar{B})$ *if the right-hand side satisfies* $\lim_{t \to -\infty}[|t|^{-n} \exp f(x, t)] = 0$.

Proof. Consider the functions $\varphi_\alpha = \gamma + \alpha h$ for $\alpha > 0$.
They are equal to u on S and when $\alpha \to \infty$, $\det((\nabla_{ij}\varphi_\alpha + a_{ij}))$ converges to $\alpha^n \det((\nabla_{ij}h))$. Thus, for α large enough $\alpha^{-n}\det((\nabla_{ij}\varphi_\alpha + a_{ij})) \geq \text{Const} > 0$ and $\exp f(x, \varphi_\alpha) \leq \det((\nabla_{ij}\varphi_\alpha + a_{ij}))$.
Hence there exists $\gamma_1 \in C^\infty(\bar{B})$ satisfying

$$(2) \qquad \log \det((\nabla_{ij}\gamma_1 + a_{ij})) \geq f(x, \gamma_1), \qquad \gamma_1/S = \mathrm{u}. \qquad \blacksquare$$

Remark. An open question: can one remove the hypothesis of convexity for some problems?

8.3 The problem. For simplicity we are going to consider the more usual Dirichlet problem for Monge–Ampère equations.

Let Ω be a bounded strictly convex domain in $\mathbb{R}^n$ $(n \geq 2)$ defined by a C^∞ strictly convex function h on $\bar{\Omega}$ satisfying $h/\partial\Omega = 0$. Given $\mathrm{u}(x)$ *a* C^∞ function on $\partial\Omega$ which is the restriction to $\partial\Omega$ of a C^∞ function γ on $\bar{\Omega}$, we consider the equation:

$$(3) \qquad \log \det((\partial_{ij}\varphi)) = f(x, \varphi), \qquad \varphi/\partial\Omega = \mathrm{u},$$

where $f(x, t) \in C^\infty(\bar{\Omega} \times \mathbb{R})$.

This equation was studied by Alexandrov [5], Pogorelov [235], and Cheng and Yau [89]. These authors all use the same method, that of Alexandrov, while the ideas for the estimates are due to Pogorelov. Under some hypotheses they prove the existence of a "generalized" solution of (3) (see 8.13 below) and then they try to establish its regularity. The result obtained is the following:

If $f'_t(x, t) \geq 0$ for $x \in \Omega$ and $t \leq \sup_{\partial\Omega} \mathrm{u}$, then there exists $\varphi \in C^\infty(\Omega)$, a strictly convex solution of (3), which is Lipschitz continuous on $\bar{\Omega}$ (Ω is strictly convex).

Here we will use the continuity method advocated by Nirenberg [222]. The continuity method is simpler and allows us to prove the existence of a solution of (3) which belongs to $C^\infty(\bar{\Omega})$ if there exists a strictly convex upper solution of (3) (Theorem 8.5). Unfortunately the proof is complete only in dimension two. When $n \geq 3$, estimates for the third derivatives is still an open question.
We are going to show, among other things, how to obtain the estimates by using the continuity method. Pogorelov's estimates are different.

Notation. Henceforth we set $M(\varphi) = \det((\partial_{ij}\varphi))$.

1.2. Extra Hypothesis

8.4 For the continuity method we suppose that $f'_t(x, t) \geq 0$ on $\bar{\Omega} \times \mathbb{R}$. We will remove this hypothesis later by using the method of lower and upper solution. But we must suppose (the obviously necessary condition) that there exists a strictly convex upper solution of (3), $\gamma_0 \in C^2(\bar{\Omega})$, which satisfies:

(4) $$M(\gamma_0) = \det((\partial_{ij}\gamma_0) \leq \exp f(x, \gamma_0), \qquad \gamma_0/\partial\Omega = \text{u}.$$

This hypothesis will be used to estimate the second derivatives on the boundary.
If there exists a convex function $\psi \in C^2(\bar{\Omega})$ satisfying:

$$M(\psi) = 0 \qquad \psi/\partial\Omega = \text{u}$$

(ψ exists, in particular, if u is constant), then ψ satisfies (4) strictly for any function f and we can choose for γ_0 a function of the form $\gamma_0 = \psi + \beta h$ with $\beta > 0$.

1.3. Theorem of Existence

8.5 *The Dirichlet problem* (3) *has a unique strictly convex solution belonging to* $C^\infty(\bar{\Omega})$, *when* $n = 2$, *if there exists a strictly convex upper solution* $\gamma_0 \in C^2(\bar{\Omega})$ *satisfying* (4) *and if* $f_t'(x, t) \geq 0$ *for all* $x \in \Omega$ *and* $t \leq \sup_{\partial\Omega} \text{u}$ (*we assume* Ω *is strictly convex*).

Proof. If $n > 2$ only inequality (23) is missing; otherwise the whole proof works for any dimension. This is why we give the proof for arbitrary n.

Let us consider the equations:

(5) $$\log M(\varphi) = f(x, \varphi) + (1 - \sigma)[\log M(\gamma_1) - f(x, \gamma_1)], \qquad \varphi/\partial\Omega = \text{u},$$

where $\sigma \geq 0$ is a real number and γ_1 is a lower solution, the existence of which was proved in Proposition 8.2. Thus γ_1 satisfies:

(6) $$\log M(\gamma_1) \geq f(x, \gamma_1), \qquad \gamma_1/\partial\Omega = \text{u}.$$

Let $\mathscr{A}$ be the set of strictly convex functions belonging to $C^{2+\alpha}(\bar{\Omega})$ with $\alpha \in]0, 1[$ which are equal to u on $\partial\Omega$. The operator

$$\Gamma : \mathscr{A} \ni \varphi \to f(x, \varphi) - \log M(\varphi) \in C^\alpha$$

is continuously differentiable,

$$(d\Gamma_\varphi)(\psi) = f_t'(x, \varphi)\psi - g_\varphi^{ij}\,\partial_{ij}\psi,$$

$\mathscr{A} \ni \varphi \to d\Gamma_\varphi \in \mathscr{L}(C_0^{2+\alpha}(\bar{\Omega}), C^\alpha(\bar{\Omega}))$ is continuous, and $d\Gamma_\varphi$ is invertible because $f_t' \geq 0$ (Theorem 6.14, p. 101 of Gilbarg and Trüdinger [125]). $C_0^r(\bar{\Omega})$

denotes the functions of $C^r(\overline{\Omega})$ which vanish on the boundary $\partial\Omega$, and g_φ^{ij} are the components of the inverse matrix of $((\partial_{ij}\varphi))$.

Thus we can apply the inverse function theorem. If $\tilde{\varphi} \in \mathscr{A}$ satisfies $\log M(\tilde{\varphi}) = f(x, \tilde{\varphi}) + f_0(x)$, there exists $\mathscr{V}$, a neighborhood of f_0 in $C^\alpha(\overline{\Omega})$, such that the equation $\Gamma(\varphi) = -f_1(x)$ has a solution $\psi \in \mathscr{A}$ when $f_1 \in \mathscr{V}$. If $f_1 \in C^\infty(\overline{\Omega})$, $\psi \in C^\infty(\overline{\Omega})$, according to the regularity theorem 3.55. Moreover, ψ is strictly convex, because it is so at a minimum of ψ and remains so by continuity since $M(\psi) > 0$. Lastly, the solution is unique since $f'_t \geq 0$ (Theorem 3.74).

At $\sigma = 0$ Equation (5) has a solution $\varphi_0 = \gamma_1$; therefore there exists $\sigma_0 > 0$ such that (5) has a strictly convex solution $\varphi_\sigma \in C^\infty(\overline{\Omega})$ when $\sigma \in [0, \sigma_0[$. Let σ_0 be the largest real number having this property.

If $\sigma_0 > 1$, Equation (3) has a solution and Theorem 8.5 is proved. If $\sigma_0 \leq 1$, let us suppose for a moment the following, which we will prove shortly: the set of the functions φ_σ for $\sigma \in [0, \sigma_0[$ is bounded in $C^3(\overline{\Omega})$. Then there exist $\varphi_{\sigma_0} \in C^{2+\alpha}(\overline{\Omega})$ for some $\alpha \in]0, 1[$ and a sequence $\sigma_i \to \sigma_0$ such that $\varphi_{\sigma_i} \to \varphi_{\sigma_0}$ in $C^{2+\alpha}(\overline{\Omega})$.
Since φ_{σ_i} satisfies (5), letting $i \to \infty$, we see that φ_{σ_0} satisfies (5) with $\sigma = \sigma_0$. But now we can apply the inverse function theorem at φ_{σ_0} and find a neighborhood $\mathfrak{J}$ of σ_0 such that Equation (5) has a solution when $\sigma \in \mathfrak{J}$. This contradicts the definition of σ_0.

Now we have to establish the estimates, the hardest part of the proof.

§2. The Estimates

2.1. The First Estimates

8.6 $\mathbf{C^0}$ and $\mathbf{C^1}$ estimates. Henceforth, when no confusion is possible, we drop the subscript σ. Then $\varphi = \varphi_\sigma \in C^\infty(\overline{\Omega})$ is the solution of (5) with $\sigma \in [0, 1]$. We have

$$\log M(\varphi) - f(x, \varphi) \leq \log M(\gamma_1) - f(x, \gamma_1),$$

since the second term is positive. Thus by the maximum principle (Theorem 3.74), $\varphi - \gamma_1 \geq 0$ on Ω because $f'_t \geq 0$ and $\varphi - \gamma_1 = 0$ on $\partial\Omega$. This implies that on $\partial\Omega$: $\partial_\nu \varphi \leq \partial_\nu \gamma_1$, where ∂_ν denotes the exterior normal derivative.

We have thus proved the C^0 estimate:

$$\inf_\Omega \gamma_1 \leq \varphi \leq \sup_{\partial\Omega} u. \tag{7}$$

Since φ is convex, the gradient of φ attaines its maximum on the boundary. Let P be a point of $\partial\Omega$. The tangential derivatives of φ at P are bounded since $u \in C^2(\partial\Omega)$. On the other hand we previously saw that $\partial_\nu \varphi \leq \partial_\nu \gamma_1$. It remains to establish an inequality in the other direction. The normal at P intersects $\partial\Omega$ at one other point, which we call Q.

On the straight line $\mathfrak{D}$, through P and Q, let $\mathfrak{w}$ be the linear function equal to $\mathfrak{u}$ at P and Q. Let $\mu(P)$ be the gradient of $\mathfrak{w}$. Since φ is convex on $\mathfrak{D}$, $\varphi \leq \mathfrak{w}$ on $\Omega \cap \mathfrak{D}$ and $(\partial_\nu \varphi)_P \geq \mu(P)$. $\mu(P)$ is a continuous function on $\partial\Omega$; let μ be its minimum on the compact set $\partial\Omega$. Hence $\partial_\nu \varphi \geq \mu$.

2.2. C^2-Estimate

8.7 C^2-estimate on the boundary. Let $P \in \partial\Omega$ and Y be a vector field on $\mathbb{R}^n$ which is tangent to $\partial\Omega$. Suppose $\|Y(P)\| = 1$ and choose on $\mathbb{R}^n$ orthonormal coordinates such that $\nu = (1, 0, 0, \ldots, 0)$ is the unit exterior normal at P and $Y(P) = (0, 1, 0, \ldots, 0)$. We will estimate $\partial^2_{x_2x_2}\varphi$, $\partial^2_{x_1x_2}\varphi$ and $\partial^2_{x_1x_1}\varphi$ at P.

a) Let R_2 be the radius of curvature of $\partial\Omega$ at P in the direction Y. Since $\partial\Omega$ is strictly convex, $R_2 \geq R_0 > 0$ (R_0 a real number independent of P and Y). At P:

$$\partial^2_{x_2x_2}\varphi = \frac{1}{R_2^2}\,\partial^2_{YY}\varphi + \frac{1}{R_2}\,\partial_\nu\varphi = \frac{1}{R_2^2}\,\partial^2_{YY}\mathfrak{u} + \frac{1}{R_2}\,\partial_\nu\varphi. \tag{8}$$

Therefore $\partial^2_{x_2x_2}\varphi$ is estimated

b) Let us consider a family $\mathfrak{g}$ of vector fields on $\mathbb{R}^n$ tangent to $\partial\Omega$ and bounded in $C^2(\overline{\Omega})$; thus the components $X^i(x)$ of the vector field $X \in \mathfrak{g}$ are uniformly bounded in C^2 on $\overline{\Omega}$.
Set $\psi = \varphi - \gamma$ and $L = X^i(x)\,\partial_{x_i}$ for $X \in \mathfrak{g}$.
Differentiating the equation

$$\log M(\varphi) = F(x, \varphi) \tag{9}$$

yields

$$LF = g^{ij}X^k\,\partial_{ijk}\varphi, \tag{10}$$

where $F(x, \varphi)$ is the right-hand side of (5) (recall that $((g^{ij}))$ is the inverse matrix of $((g_{ij}))$ with $g_{ij} = \partial_{ij}\varphi$. We will compute $B = g^{ij}\,\partial_{ij}(L\psi + \alpha h + \beta\psi)$, where α and β are two real numbers which we will choose later.

$$B = g^{ij}X^k\,\partial_{ijk}\psi + 2g^{ij}\,\partial_i X^k\,\partial_{jk}\psi + g^{ij}\,\partial_{ij}X^k\,\partial_k\psi$$
$$+ \alpha g^{ij}\,\partial_{ij}h + \beta(g^{ij}\,\partial_{ij}\varphi - g^{ij}\,\partial_{ij}\gamma).$$

Since $g^{ij}\,\partial_{ik}\varphi = \delta^j_k$, using (10) we obtain:

$$B = LF + \beta n + 2\,\partial_i X^i + g^{ij}(\mathfrak{m}_{ij} + \alpha\,\partial_{ij}h),$$

with

$$\mathfrak{m}_{ij} = -\beta\,\partial_{ij}\gamma - X^k\,\partial_{ijk}\gamma - 2\,\partial_i X^k\,\partial_{jk}\gamma + \partial_{ij}X^k\,\partial_k\psi.$$

At first we pick $\beta = \beta_0 \geq -(1/n)\inf(LF + 2\,\partial_i X^i)$, where the inf is taken for all $x \in \Omega$ and all functions φ_σ. Note that this inf is finite since the functions φ_σ are already estimated in C^1.

Then we choose $\alpha = \alpha_0$ large enough so that

$$(\mathfrak{m}_{ij} + \alpha_0\,\partial_{ij}h)g^{ij} \geq 0.$$

The real numbers α_0 and β_0 can be chosen independent of $X \in \mathfrak{g}$. This is possible by our hypothesis. Thus

$$g^{ij}\,\partial_{ij}(L\psi + \alpha_0 h + \beta_0\psi) \geq 0.$$

Likewise, let $\beta = \beta_1 \leq -(1/n)\sup(LF + 2\,\partial_i X^i)$, where the sup is taken for all $x \in \Omega$ and all φ_σ, and let α_1 be such that $g^{ij}(\mathfrak{m}_{ij} + \alpha_1\,\partial_{ij}h) \leq 0$. β_1 and α_1 are chosen independent of $X \in \mathfrak{g}$. Thus

$$g^{ij}\,\partial_{ij}(L\psi + \alpha_1 h + \beta_1\psi) \leq 0.$$

Since $L\psi$, h, and ψ vanish on $\partial\Omega$, by the maximum principle:

$$-(\alpha_1 h + \beta_1\psi) \leq L\psi \leq -(\alpha_0 h + \beta_0\psi)$$

and

$$-\partial_\nu(\alpha_0 h + \beta_0\psi) \leq \partial_\nu L\psi \leq -\partial_\nu(\alpha_1 h + \beta_1\psi).$$

These inequalities yield the estimate of $\partial_{\nu X}\psi$.

In order for the family $\mathfrak{g}$ to be large enough so that, for all pairs (P, Y) with $P \in \partial\Omega$, and $Y \in T_P(\partial\Omega)$ a unit vector, there exists $X \in \mathfrak{g}$ such that $X(P) = Y$, we define $\mathfrak{g}$ as follows.
Let B be the unit ball of $\mathbb{R}^n$ and Φ be a C^3-diffeomorphism from $\bar{B}$ to $\bar{\Omega}$. Then $\mathfrak{v}_{ij} = x_i\,\partial_j - x_j\,\partial_i$ are vector fields tangent to ∂B. Consider the family $\mathfrak{F} = \{\mathfrak{V} = \sum a_{ij}\mathfrak{v}_{ij}$, where a_{ij} are real numbers with $|a_{ij}| \leq \|(\Phi^{-1})_*\|\}$. Then $\mathfrak{g} = \Phi_*\mathfrak{F}$ has the desired property.

c) To estimate $\partial^2_{\nu\nu}\varphi$, we need to know a strictly convex upper solution γ_0 satisfying (4).
Since φ_σ satisfies (5), $\log M(\varphi_\sigma) \geq f(x, \varphi_\sigma)$. According to the maximum principle, since $f'_t \geq 0$, then as in 8.6, $\varphi_\sigma \leq \gamma_0$ and $\partial_\nu\varphi_\sigma \geq \partial_\nu\gamma_0$ on $\partial\Omega$.
Since γ_0 is strictly convex, there exists an $\varepsilon > 0$ such that for all $x \in \bar{\Omega}$ and all $i = 1, 2, \ldots, n$, $\partial^2_{x_i x_i}\gamma_0 \geq \varepsilon$. From (8) it follows for $i \geq 2$:

$$\partial^2_{x_i x_i}\varphi = \partial^2_{x_i x_i}\gamma_0 + \frac{1}{R_i}(\partial_\nu\varphi - \partial_\nu\gamma_0) \geq \varepsilon.$$

Suppose we choose the orthonormal frame in $T_P(\partial\Omega)$ such that for $j \geq i \geq 2$, $\partial^2_{x_i x_j}\varphi = 0$ if $i \neq j$. Then (9) implies

$$\partial^2_{x_1x_1}\varphi \prod_{p=2}^{n} \partial^2_{x_px_p}\varphi = \exp F(x, \varphi) + \sum_{p=2}^{n}(-1)^p\, \partial^2_{x_1x_p}\varphi\mu_{1p}, \tag{11}$$

where μ_{1p}, the minor of $\partial^2_{x_1x_p}\varphi$ in the determinant $M(\varphi)$, does not contain $\partial^2_{x_1x_1}\varphi$. Therefore μ_{1p} can be estimated by the inequalities in the preceding paragraphs.

Thus by (11) $\partial^2_{x_1x_1}\varphi$ is estimated on $\partial\Omega$ since $\partial^2_{x_px_p}\varphi \geq \varepsilon$.

8.8 C^2-estimate on $\bar{\Omega}$. Since φ is convex, $\partial_{yy}\varphi \geq 0$ for all directions y. Thus an upperbound for $\sum_{k=1}^{n}\partial_{kk}\varphi$ is enough to yield the C^2-estimate. Computing the Laplacian of (9) leads to:

$$\sum_{k=1}^{n} g^{ij}\,\partial_{ijkk}\varphi = \sum_{k=1}^{n} g^{im}g^{jl}\,\partial_{ijk}\varphi\,\partial_{mlk}\varphi + \sum_{k=1}^{n}\partial_{kk}F(x, \varphi). \tag{12}$$

Let $\beta_2 \leq 1/n \inf \sum_{k=1}^{n}\partial_{kk}F(x, \varphi_\sigma)$, where the inf is taken over Ω and for all functions φ_σ. It is finite since all of the terms have been estimated except the term which involves $\Delta\varphi$, and that term is positive: $F'_t(x, \varphi)\sum_{k=1}^{n}\partial_{kk}\varphi \geq 0$. β_2 can be chosen negative and independent of σ. Hence

$$g^{ij}\,\partial_{ij}\left(\sum_{k=1}^{n}\partial_{kk}\varphi - \beta_2\varphi\right) \geq -n\beta_2 + \sum_{k=1}^{n}\partial_{kk}F(x, \varphi) \geq 0.$$

By the maximum principle $\sum_{k=1}^{n}\partial_{kk}\varphi - \beta_2\varphi$ attains its maximum on $\partial\Omega$. But by (8.8) $\Delta\varphi$ is bounded on $\partial\Omega$. Hence the C^2-estimate follows:

$$0 \leq \partial_{yy}\varphi \leq 2|\beta_2|\sup_{\Omega}|\varphi| + \sup_{\partial\Omega}\sum_{k=1}^{n}\partial_{kk}\varphi.$$

Consequently the metrics $(g_\sigma)_{ij} = \partial_{ij}\varphi_\sigma$ are equivalent for $\sigma \in [0, 1]$. Indeed, according to the preceding inequality $(g_\sigma)_{yy} \leq C$, where C is a constant, and (5) implies

$$C^{n-1}(g_\sigma)_{xx} \geq B > 0,$$

where B is a constant. Thus for all vectors ξ and $\sigma \in [0, 1]$

$$C^{1-n}B\|\xi\|^2 \leq (g_\sigma)_{ij}\xi^i\xi^j \leq C\|\xi\|^2. \tag{13}$$

2.3. C^3-Estimate

8.9 Proposition. *$R = \frac{1}{4}g^{\alpha\beta}g^{ik}g^{j\ell}\,\partial_{\alpha ij}\varphi_{\beta k\ell}\varphi$ satisfies*

$$g^{ij}\nabla_{ij}R \geq \frac{2}{n-1}R^2 + CR^{1/2}, \tag{14}$$

where ∇ is the covariant derivative with respect to g and C is a constant which depends on the function $F(x, t)$ and on an upper bound of $\|\varphi\|_{C^2}$ (as do the constants introduced in the proof).

Proof. Calabi [75] p. 113 establishes the following inequality in the special case $F(x, \varphi) = 0$:

$$g^{ij}\nabla_{ij}R \geq \frac{2(n+1)}{n(n-1)}R^2.$$

He introduced $A_{ijk} = \Gamma_{ijk} = \frac{1}{2}\,\partial_{ijk}\varphi$. A_{ijk} is symmetric with respect to its subscripts and we can verify the following equalities:

$$g^{ij}A_{ijk} = \tfrac{1}{2}\,\partial_k F \quad \text{and} \quad \nabla_\ell A_{ijk} = \nabla_i A_{\ell jk} \tag{15}$$

where F is written in place of $F(x, \varphi)$ for simplicity.

A computation similar to that of Calabi (see Pogorelov [235] p. 39) leads to

$$g^{ij}\nabla_{ij}R \geq A^{ijk}\nabla_{jk}\nabla_i F + 2\frac{n+1}{n(n-1)}R^2 + C_1R^{3/2} + C_2R + 2\nabla^\ell A^{ijk}\nabla_\ell A_{ijk}.$$

C_1 and C_2 are two constants and the indices are raised using g^{ij}, for instance $\Gamma^j_{ik} = g^{j\ell}\Gamma_{i\ell k}$ are Christoffel's symbols of the Riemannian connection. Moreover:

$$\begin{aligned} A^{ijk}\nabla_{ijk}F = {} & A^{ijk}\,\partial_i(\partial_{jk}F - \Gamma^\ell_{jk}\,\partial_\ell F) \\ & - A^{ijk}\Gamma^\ell_{ij}(\partial_{\ell k}F - \Gamma^m_{\ell k}\,\partial_m F) \\ & - A^{ijk}\Gamma^\ell_{ik}[\partial_{j\ell}F - \Gamma^m_{j\ell}\,\partial_m F]. \end{aligned}$$

Thus

$$A^{ijk}(\nabla_{ijk}F + \nabla^\ell F\nabla_i A_{jk\ell}) \leq \text{Const} \times (1 + R)\sqrt{R}.$$

According to (15) for some constant C_3 we get:

$$g^{ij}\nabla_{ij}R \geq 2\frac{n+1}{n(n-1)}R^2 + C_3(1+R)\sqrt{R}$$
$$+ 2(\nabla^\ell A^{ijk} - \tfrac{1}{4}A^{ijk}\nabla^\ell F)(\nabla_\ell A_{ijk} - \tfrac{1}{4}A_{ijk}\nabla_\ell F),$$

and inequality (14) follows. ■

8.10 Interior C^3-estimate. In this paragraph we assume that the derivatives of φ up to order two are estimated. A term which involves only derivatives of φ of order at most two is called a bounded term. First of all note that $\partial_i[g^{ij}M(\varphi)] = 0$. Indeed,

$$\partial_i[g^{ij}M(\varphi)] = M(\varphi)[g^{ij}g^{k\ell}\,\partial_{k\ell i}\varphi - g^{ik}g^{j\ell}\,\partial_{k\ell i}\varphi].$$

Interchanging ℓ and i in the last term, we obtain the result. Multiply (12) by $h^2M(\varphi)$ and integrate over Ω. Since $\partial_i[g^{ij}M(\varphi)] = 0$, integrating by parts twice leads to:

$$\int_\Omega h^2 \sum_{k=1}^n g^{im}g^{j\ell}\,\partial_{ijk}\varphi\,\partial_{m\ell k}\varphi M(\varphi)\,dx \leq \text{Const.}$$

Set $R = \frac{1}{4}g^{\alpha\beta}g^{ik}g^{j\ell}\,\partial_{\alpha ij}\varphi\,\partial_{\beta k\ell}\varphi$. Since the metrics g_σ are equivalent, the preceding inequality implies

$$\int_\Omega h^2R\,dx \leq \text{Const.} \tag{16}$$

It is possible to show that $\int_\Omega R\,dx \leq \text{Const}$, but that yields nothing more here. Let us prove by induction that for all integer p:

$$\int_\Omega h^{2p}R^p\,dx \leq \text{Const.} \tag{17}$$

Assume (17) holds for a given p. Multiplying (14) by $h^{2p+2}R^{p-1}\sqrt{M(\varphi)}$ and integrating by parts over Ω lead to:

$$(1-p)\int_\Omega h^{2p+2}g^{ij}R^{p-2}\,\partial_iR\,\partial_jR\sqrt{M(\varphi)}\,dx$$
$$-\int_\Omega g^{ij}\,\partial_ih^{2p+2}R^{p-1}\,\partial_jR\sqrt{M(\varphi)}\,dx$$
$$\geq \frac{2}{n-1}\int_\Omega h^{2p+2}R^{p+1}\sqrt{M(\varphi)}\,dx + C\int_\Omega h^{2p+2}R^{p-1/2}\sqrt{M(\varphi)}\,dx.$$

Integrating the second integral by parts again gives, by (15),

$$\frac{2}{n-1}\int_\Omega h^{2p+2}R^{p+1}\sqrt{M(\varphi)}\,dx \leq \frac{1}{p}\int_\Omega R^p g^{ij}\nabla_{ij}h^{2p+2}\sqrt{M(\varphi)}\,dx + \text{Const}$$

$$\leq \text{Const} \times \left[1 + \frac{1}{p}\int_\Omega h^{2p}R^p\,dx\right].$$

Thus, since (17) holds for $p = 1$ (inequality 16), (17) holds for all p. Accordingly, for any compact set $K \subset \Omega$ and any integer p, $\|\varphi_\sigma\|_{H_3^p(K)} \leq \text{Const}$ for all $\sigma \in [0, 1]$.

By the regularity theorem (3.56), for all $r > 0$:

$$\|\varphi_\sigma\|_{C^r(K)} \leq \text{Const}.$$

In particular the third derivatives of φ_σ are uniformly bounded on K.

8.11 C^3**-estimate on the boundary.**

a) Recall (8.7), where we defined $L = X^k\,\partial_k$ with $X \in \mathfrak{g}$. Differentiating (9) twice with respect to L gives: $L^2F(x, \varphi) = -g^{i\ell}g^{jk}L(\partial_{\ell k}\varphi)L(\partial_{ij}\varphi) + g^{ij}L^2(\partial_{ij}\varphi)$. Next we compute $g^{ij}\,\partial_{ij}L^2\varphi$. Since

$$(18) \qquad L^2\varphi = L(X^k\,\partial_k\varphi) = X^l X^k\,\partial_{lk}\varphi + X^l\,\partial_l X^k\,\partial_k\varphi,$$

then

$$g^{ij}(\partial_{ij}L^2\varphi) = g^{ij}L^2(\partial_{ij}\varphi) + 4g^{ij}(\partial_i X^\ell)X^k\,\partial_{j\ell k}\varphi + g^{ij}\,\partial_{ij}(X^\ell X^k)\,\partial_{\ell k}\varphi$$
$$+ g^{ij}\,\partial_{ij}(X^\ell\,\partial_\ell X^k)\,\partial_k\varphi + 2\,\partial_i(X^k\,\partial_k X^i).$$

Thus

$$(19) \qquad g^{ij}\,\partial_{ij}L^2\varphi = (X^k\,\partial_{j\ell k}\varphi g^{ij} + 2\,\partial_\ell X^i)(X^\lambda\,\partial_{\lambda\beta i}\varphi g^{\beta\ell} + 2\,\partial_i X^\ell) + \text{bounded terms}.$$

Consequently, there exists a constant α such that

$$g^{ij}\,\partial_{ij}(L^2\psi + \alpha h) \geq 0.$$

Since $L^2\psi$ and h vanish on the boundary $\partial\Omega$, by the maximum principle $L^2\psi + \alpha h \leq 0$ on Ω and on the boundary

$$(20) \qquad \partial_\nu L^2\psi \geq -\alpha\,\partial_\nu h.$$

b) If we get an inequality in the opposite direction, the third derivatives will be estimated on the boundary. Indeed, then we have on the boundary:

$$|\partial_\nu L^2\varphi| < \text{Const.} \tag{21}$$

Consider $P \in \partial\Omega$ and use the coordinates in (8.7). Differentiating (18) with respect to L yields $L^3\varphi = X^iX^jX^k\,\partial_{ijk}\varphi +$ bounded terms.

On $\partial\Omega$, $L^3\varphi = L^3(\psi + \gamma) = L^3\gamma$, so $(\partial^3_{222}\varphi)_P$ is estimated. Likewise the third derivatives with respect to coordinates x_i with $i \geq 2$ are estimated. By (21), $(\partial_{1ij}\varphi)_P$ is estimated for i and $j \geq 2$.

To get an estimate for $(\partial_{11j}\varphi)_P, j > 1$, we differentiate (9) with respect to x_j. This yields:

$$g^{11}\,\partial_{11j}\varphi = \text{bounded terms.}$$

Because the metrics g_{ij} are equivalent, there exists a constant $\eta > 0$ such that $g^{11} \geq \eta > 0$ and the estimate of $(\partial_{11j}\varphi)_P$ follows.

Finally, differentiating (9) with respect to x_1 yields $g^{11}\,\partial^3_{111}\varphi =$ bounded terms. Hence all the third derivatives are estimated on the boundary.

c) It remains to find an upper bound for $\partial_\nu L^2\psi$ on the boundary. For the present such a bound is established only in the case $n = 2$.

From now on, $n = 2$; consequently the dimension of $\partial\Omega$ is equal to one. Consider a vector field X tangent to $\partial\Omega$ and of norm one on $\partial\Omega$. Since the second derivatives of φ are estimated, there exists M such that $|L^2\psi| \leq M$ on $\partial\Omega$. Recall that $\psi = \varphi - \gamma$.

Let p be an integer that we will choose later and set

$$\mathfrak{E} = (1 + M + L^2\psi)^{-p}.$$

Let $K \subset \Omega$ be a compact set such that $\|X\| > 1/2$ on $\Omega - K$. Compute $g^{\alpha\beta}\,\partial_{\alpha\beta}\mathfrak{E}$ on $\Omega - K$ to obtain

$$\begin{aligned} g^{\alpha\beta}\,\partial_{\alpha\beta}\mathfrak{E} = &-p(1 + M + L^2\psi)^{-p-1}g^{\alpha\beta}\,\partial_{\alpha\beta}L^2\psi \\ &+ p(p + 1)(1 + M + L^2\psi)^{-p-2}g^{\alpha\beta}\,\partial_\alpha L^2\psi\,\partial_\beta L^2\psi. \end{aligned}$$

Using (19) gives

$$\begin{aligned} g^{\alpha\beta}\,\partial_{\alpha\beta}\mathfrak{E} = p(1 + M + L^2\psi)^{-p-2}&[(p + 1)g^{\alpha\beta}\,\partial_\alpha L^2\psi\,\partial_\beta L^2\psi \\ &- (1 + M + L^2\psi)(X^k\,\partial_{jlk}\varphi g^{ij} + 2\,\partial_l X^i) \\ &\times (X^\lambda\,\partial_{\lambda\beta i}\varphi g^{\beta l} + 2\,\partial_i X^l)] + \text{bounded terms.} \end{aligned} \tag{21}$$

At $Q \in \Omega - K$ suppose that X is in the direction of x_2. According to (10), there exists a constant k_0 such that

$$|\partial_{211}\varphi| \leq k_0(1 + |\partial_{221}\varphi| + |\partial_{222}\varphi|).$$

Thus there is a constant k_1 such that

$$(X^k \partial_{jlk}\varphi g^{ij} + 2\,\partial_l X^i)(X^\lambda \partial_{\lambda\beta i}\varphi g^{\beta l} + 2\,\partial_i X^l) \\ \leq k_1(1 + |\partial_{221}\varphi|^2 + |\partial_{222}\varphi|^2).$$

Moreover,

$$g^{\alpha\beta}\,\partial_\alpha L^2\psi\,\partial_\beta L^2\psi \geq k_2(|\partial_{221}\varphi|^2 + |\partial_{222}\varphi|^2 - 1),$$

where k_2 is a constant which depends on K and on the preceding estimates, much as did k_0 and k_1.

Pick p such that $(p + 1)k_2 \geq (1 + 2M)k_1$. Since the third derivatives are bounded on K (by 8.9), (21) shows that $\mathfrak{E}$ satisfies the following inequality on Ω:

$$g^{\alpha\beta}\,\partial_{\alpha\beta}\mathfrak{E} \geq \text{Const.}$$

Hence there exists a constant $\mathfrak{k}$ such that

$$g^{\alpha\beta}\,\partial_{\alpha\beta}(\mathfrak{E} + \mathfrak{k}h) \geq 0 \tag{22}$$

since the metrics g_σ are equivalent. Because $\mathfrak{E}$ and $\mathfrak{h}$ are constant on $\partial\Omega$, according to the maximum principle.

$$\mathfrak{E} + \mathfrak{k}h \leq (1 + M)^{-p},$$

while on $\partial\Omega$: $\partial_\nu\mathfrak{E} + \mathfrak{k}\,\partial_\nu h \geq 0$. This gives

$$-p(1 + M + L^2\psi)^{-p-1}\,\partial_\nu L^2\psi \geq -\mathfrak{k}\,\partial_\nu h.$$

Hence we obtain the desired inequality

$$\partial_\nu L^2\psi \leq \frac{\mathfrak{k}}{p}(1 + 2M)^{p+1}\sup_{\partial\Omega}\partial_\nu h. \tag{23}$$

C^3-estimate on $\overline{\Omega}$.

By inequality (14) there exist two positive constants a and α such that:

$$g^{ij}\nabla_{ij}R \geq \frac{2}{n-1}[(R - a)^2 - \alpha^2].$$

Two cases can occur: R attains its maximum on $\partial\Omega$ or in Ω. In the first case we saw that R is bounded, while if R attains its maximum at $P \in \Omega$, according to the preceding inequality:

$$R \leq \sup_\Omega R \leq a + \alpha$$

since $g^{ij}\nabla_{ij}R \leq 0$ at P.

8.12 Using the method of lower and upper solutions (as in 7.25), and using Aubin [23] p. 374 for the estimates, we can prove the following.

Theorem. *Let $\Omega \subset \mathbb{R}^n$ be strictly convex. If there exist a strictly convex upper solution $\gamma_0 \in C^2(\overline{\Omega})$ satisfying (4) and a strictly convex lower solution $\gamma_1 \in C^\infty(\overline{\Omega})$ satisfying (6) with $\gamma_1 \leq \gamma_0$, the Dirichlet problem (3) has a strictly convex solution φ belonging to $C^\infty(\overline{\Omega})$ when $n = 2$, and φ satisfies $\gamma_1 \leq \varphi \leq \gamma_0$.*

Corollary. *If there exists γ_0 as above and if $\lim_{t \to -\infty}[|t|^{-n} \exp f(x, t)] = 0$, then the Dirichlet problem (3) has a strictly convex solution $\varphi \in C^\infty(\overline{\Omega})$ when $n = 2$.*

Proof. By Proposition (8.2), for α large enough $\gamma_1 = \gamma + \alpha h$ is a strictly convex lower solution satisfying (6). Thus we can choose α so that $\gamma_1 \leq \gamma_0$. Moreover $\gamma_1 \in C^\infty(\overline{\Omega})$.

The preceding theorem now implies the stated result.

§3. The Radon Measure $\mathcal{M}(\varphi)$

8.13 Definition. For a C^2 convex function on Ω, we set

$$\mathcal{M}(\varphi) = M(\varphi)\, dx^1 \wedge dx^2 \wedge \cdots \wedge dx^n$$

and we define the Radon measure:

$$C_0(\Omega) \ni \psi \to \int_\Omega \psi \mathcal{M}(\varphi),$$

where $C_0(\Omega)$ denote the set of continuous functions with compact support.

This definition extends to convex functions in general, according to Alexandrov. Here we will follow the analytic approach of Rauch and Taylor [242].

First let us remark that for a C^2 convex function

$$\mathcal{M}(\varphi) = d(\partial_1 \varphi) \wedge d(\partial_2 \varphi) \wedge \cdots \wedge d(\partial_n \varphi).$$

We are going to show by induction that $d(\partial_{i_1}\varphi) \wedge d(\partial_{i_2}\varphi) \wedge \cdots \wedge d(\partial_{i_m}\varphi)$ defines a current when φ is a convex function and $1 \leq i_1 < i_2 < \cdots < i_m \leq n$. Let us begin with $m = 1$. Let ω be a continuous $(n-1)$-form with compact support $K \subset \Omega$, and $\omega = \sum_{l=1}^n f_l(x)\, dx^1 \wedge \cdots \wedge \widehat{dx^l} \wedge \cdots \wedge dx^n$. For such ω we set $\|\omega\|_0 = \sup_{1 \leq l \leq n} \sup_{x \in K} |f_l(x)|$. When φ is a convex function we will define

$$\int_\Omega d(\partial_i \varphi) \wedge \omega.$$

Choose a C^∞ positive function γ with compact support $\tilde{K} \subset \Omega$ which is equal to one on K.

Now consider a sequence of convex C^∞ functions φ_j which converges to φ uniformly on $\tilde{K}$. We have

$$\left|\int_\Omega f_1\, d(\partial_i\varphi_j) \wedge dx^2 \wedge \cdots \wedge dx^n\right| \leq \sup|f_1| \int_K |\partial_{1i}\varphi_j|\, dx^1 \wedge dx^2 \wedge \cdots \wedge dx^n.$$

But since φ_j is convex,

(24) $$2|\partial_{1i}\varphi_j| \leq \partial_{11}\varphi_j + \partial_{ii}\varphi_j$$

and

$$2\int_K |\partial_{1i}\varphi_j|\, dx \leq \int_\Omega \gamma(\partial_{11}\varphi_j + \partial_{ii}\varphi_j)\, dx = \int_\Omega \varphi_j(\partial_{11}\gamma + \partial_{ii}\gamma)\, dx.$$

Thus there exists a constant C_1 such that

(25) $$\left|\int_\Omega d(\partial_i\varphi_j) \wedge \omega\right| \leq C_1\|\omega\|_0 \sup|\varphi_j|.$$

The constants C_α ($\alpha \in \mathbb{N}$) introduced depend on K and γ only. In particular, they are independent of φ_j and ω.
Let ω_k be a sequence of C^∞ $(n-1)$-forms which converges uniformly to ω. We choose ω_k such that supp $\omega_k \subset K$.
We define $\int_\Omega d(\partial_i\varphi) \wedge \omega_k$ as a distribution and let $j \to \infty$,

$$\int_\Omega d(\partial_i\varphi_j) \wedge \omega_k \to \int_\Omega d(\partial_i\varphi) \wedge \omega_k.$$

Then by (25)

$$\left|\int_\Omega d(\partial_i\varphi) \wedge \omega_k\right| \leq C_1\|\omega_k\|_0 \sup|\varphi|,$$

so that $\int_\Omega d(\partial_i\varphi) \wedge \omega_k$ is a Cauchy sequence which converges to a limit independent of the sequence ω_k. We call this limit $\int_\Omega d(\partial_i\varphi) \wedge \omega$. It satisfies:

$$\left|\int_\Omega d(\partial_i\varphi) \wedge \omega\right| \leq C_1\|\omega\|_0 \sup|\varphi|.$$

Now suppose we have defined $\int_\Omega d(\partial_{i_1}\varphi) \wedge d(\partial_{i_2}\varphi) \wedge \cdots \wedge d(\partial_{i_m}\varphi) \wedge \breve{\omega}$ for all families of integers $1 \leq i_1 < i_2 < \cdots < i_m \leq n$, $(m < n)$ and all continuous

$(n - m)$-forms $\check{\omega}$ with support contained in K. Suppose we have proved that there exists a constant C_2 such that:

$$(26)\quad \left|\int_\Omega d(\partial_{i_1}\varphi) \wedge d(\partial_{i_2}\varphi) \wedge \cdots \wedge d(\partial_{i_m}\varphi) \wedge \check{\omega}\right| \le C_2 \|\check{\omega}\|_0[\sup|\varphi|]^m.$$

As before, $\|\check{\omega}\|_0$ is equal to the sup of the components of $\check{\omega}$.

Let $\tilde{\omega}$ be a continuous $(n - m - 1)$-form with support contained in K, and let $\tilde{\omega}_k$ be a sequence of C^∞ $(n - m - 1)$-forms which converge uniformly to $\tilde{\omega}$: $\|\tilde{\omega}_k - \tilde{\omega}\|_0 \to 0$ when $k \to \infty$. We pick $\tilde{\omega}_k$ such that supp $\tilde{\omega}_k \subset K$. Consider $(m + 1)$ integers $1 \le i_1 < i_2 < \cdots < i_{m+1} \le n$. We want to define $\Gamma(\tilde{\omega}) = \int_\Omega d(\partial_{i_1}\varphi) \wedge \cdots \wedge d(\partial_{i_{m+1}}\varphi) \wedge \tilde{\omega}$. First we define $\int_\Omega d(\partial_{i_1}\varphi) \wedge \cdots \wedge d(\partial_{i_{m+1}}\varphi) \wedge \tilde{\omega}_k = \Gamma(\tilde{\omega}_k)$ by integrating by parts. For instance, if $f \in \mathscr{D}(\Omega)$, then by definition:

$$(m + 1)\int_\Omega f(x)\, d(\partial_{i_1}\varphi) \wedge \cdots \wedge d(\partial_{i_{m+1}}\varphi) \wedge dx^{m+2} \wedge dx^{m+3} \wedge \cdots \wedge dx^n$$

$$= \sum_{q=1}^{m+1} \int_\Omega \varphi\, d(\partial_{i_1}\varphi) \wedge \cdots \wedge d(\partial_{i_{q-1}}\varphi) \wedge d(\partial_{i_q} f) \wedge d(\partial_{i_{q+1}}\varphi) \wedge \cdots \wedge$$

$$\times\, d(\partial_{i_{m+1}}\varphi) \wedge dx^{m+2} \wedge \cdots \wedge dx^n.$$

This equality holds if $\varphi \in C^\infty$. When φ is only convex the integrals of the second term make sense by our assumption, with the continuous $(n - m)$-forms

$$\check{\omega}_q = (-1)^{m+1-q}\varphi\, d(\partial_{i_q} f) \wedge dx^{m+2} \wedge \cdots \wedge dx^n.$$

Then letting $j \to \infty$,

$$(27)\quad \Gamma_j(\tilde{\omega}_k) = \int_\Omega d(\partial_{i_1}\varphi_j) \wedge d(\partial_{i_2}\varphi_j) \wedge \cdots \wedge d(\partial_{i_{m+1}}\varphi_j) \wedge \tilde{\omega}_k \to \Gamma(\tilde{\omega}_k)$$

On the other hand,

$$\left|\int_\Omega f\, d(\partial_{i_1}\varphi_j) \wedge \cdots \wedge d(\partial_{i_{m+1}}\varphi_j) \wedge dx^{m+2} \wedge \cdots \wedge dx^n\right|$$

$$\le \sup|f| \int_K |D(x)|\, dx,$$

where $D(x) = \det((\partial_{li_k}\varphi_j))$ is the $(m + 1)$-determinant with $1 \le l, k \le m + 1$. But because φ_j is convex, $2|D(x)| \le D_1(x) + D_2(x)$ where $D_1(x) = \det((\partial_{lk}\varphi_j))$ and $D_2 = \det((\partial_{i_k i_l}\varphi_j))$.

Indeed this inequality is the same as (24) on $\Lambda^{m+1}(\mathbb{R}^n)$. The extension to Λ^{m+1} of the quadratic form defined by $((\partial_{ih}\varphi_j))$ is a positive quadratic form whose components are the $(m+1)$-determinants extracted from $((\partial_{ih}\varphi_j))$ with $1 \le i, h \le n$.
Thus there exists a constant C_3 such that

$$2\int_K |D(x)|\,dx \le \int_\Omega \gamma[D_1(x) + D_2(x)]\,dx \le C_3[\sup|\varphi_j|]^{m+1}.$$

For instance, according to (26):

$$\begin{aligned}\int_\Omega \gamma D_1(x)\,dx &= \int_\Omega \gamma\, d(\partial_1\varphi_j) \wedge d(\partial_2\varphi_j) \wedge \cdots \wedge d(\partial_{m+1}\varphi_j) \wedge dx^{m+2} \wedge \cdots \wedge dx^n \\ &= \int_\Omega \varphi_j\, d(\partial_1\varphi_j) \wedge d(\partial_2\varphi_j) \wedge \cdots \wedge d(\partial_m\varphi_j) \\ &\quad \wedge d(\partial_{m+1}\gamma) \wedge dx^{m+2} \wedge \cdots \wedge dx^n \le C_3[\sup|\varphi_j|]^{m+1}.\end{aligned}$$

Consequently, there exists a constant C_4 such that

$$|\Gamma_j(\tilde{\omega}_k)| \le C_4\|\tilde{\omega}_k\|_0[\sup|\varphi_j|]^{m+1}.$$

Letting $j \to \infty$, by (27):

$$|\Gamma(\tilde{\omega}_k)| \le C_4\|\tilde{\omega}_k\|_0[\sup|\varphi|]^{m+1}.$$

Therefore $\Gamma(\tilde{\omega}_k)$ is a Cauchy sequence which converges to $\Gamma(\tilde{\omega})$ and this limit is independent of the sequence $\tilde{\omega}_k$.

By induction we have thus defined the Radon measure $\mathcal{M}(\varphi)$ when φ is a convex function. Moreover, for any compact set $K \subset \Omega$ there exists a constant C_5 such that

$$\left|\int_\Omega \psi\mathcal{M}(\varphi)\right| \le C_5 \sup|\psi|[\sup|\varphi|]^n \tag{28}$$

for all $\psi \in C(\Omega)$ with supp $\psi \subset K$.

8.14 Proposition. *Let $\{\varphi_p\}$ be a sequence of convex functions on Ω which converges uniformly to φ on Ω. Then $\mathcal{M}(\varphi_p) \to \mathcal{M}(\varphi)$ vaguely (i.e. for all $\psi \in C_0(\Omega)$, $\int_\Omega \psi\mathcal{M}(\varphi_p) \to \int_\Omega \psi\mathcal{M}(\varphi)$).*

Proof. As in the preceding paragraph, the proof proceeds by induction. We use the same notation.

Let ω be a continuous $(n-1)$-form and let $\{\omega_k\}$ be a sequence of C^∞ $(n-1)$-forms with support contained in K, and which converges uniformly to ω. Obviously

$$\int_\Omega d(\partial_i \varphi_p) \wedge \omega_k \to \int_\Omega d(\partial_i \varphi) \wedge \omega_k.$$

By definition

$$\int_\Omega d(\partial_i \varphi_p) \wedge \omega = \lim_{k\to\infty} \int_\Omega d(\partial_i \varphi_p) \wedge \omega_k.$$

But according to (25) this convergence is uniform in p:

$$\left| \int_\Omega d(\partial_i \varphi_p) \wedge (\omega - \omega_k) \right| \leq C_1 \|\omega - \omega_k\|_0 \sup|\varphi_p| \leq \tilde{C}_1 \|\omega_k - \omega\|_0$$

Thus we can interchange the limits in p and in k:

$$\left| \int_\Omega d(\partial_i \varphi) \wedge \omega - \int_\Omega d(\partial_i \varphi_p) \wedge \omega \right| \leq \left| \int_\Omega d(\partial_i \varphi) \wedge (\omega - \omega_k) \right|$$
$$+ \left| \int_\Omega d(\partial_i \varphi) \wedge \omega_k - \int_\Omega d(\partial_i \varphi_p) \wedge \omega_k \right| + \left| \int_\Omega d(\partial_i \varphi_p) \wedge (\omega - \omega_k) \right|.$$

These three terms are smaller than $\varepsilon > 0$ if we choose k such that $\|\omega_k - \omega\|_0 < \varepsilon/\tilde{C}_1$, and then p large enough.

This proves the result for $m = 1$. We suppose now that it is true for some $m < n$, so we assume:

for all continuous $(n-m)$-form $\check{\omega}$ with support included in K,

$$\int_\Omega d(\partial_{i_1} \varphi_p) \wedge d(\partial_{i_2} \varphi_p) \wedge \cdots \wedge d(\partial_{i_m} \varphi_p) \wedge \check{\omega}$$
$$\to \int_\Omega d(\partial_{i_1} \varphi) \wedge d(\partial_{i_2} \varphi) \wedge \cdots \wedge d(\partial_{i_m} \varphi) \wedge \check{\omega}$$

We then prove the result for $m + 1$ in a similar way to the $m = 1$ case. ■

§4. The Functional $\mathscr{I}(\varphi)$

8.15 Definition. For a C^2 convex function which is zero on the boundary $\partial\Omega$ we set:

$$\mathscr{I}(\varphi) = -n \int_\Omega \varphi M(\varphi)\,dx. \tag{29}$$

This definition makes sense if the convex function is not C^2. Indeed, for $a < 0$ set $\Omega_a = \{x \in \Omega / \varphi(x) \leq a\}$. According to the preceding paragraph

$$-\int_{\Omega_a} \varphi \mathscr{M}(\varphi) = \tau(a)$$

makes sense. Since $\mathscr{M}(\varphi)$ is a non-negative Radon measure and $\varphi \leq 0$, $\tau(a)$ is an increasing function. We define

$$\mathscr{I}(\varphi) = -n \lim_{a\to 0} \int_{\Omega_a} \varphi \mathscr{M}(\varphi).$$

Of course $\mathscr{I}(\varphi)$ may be infinite. The set of the convex functions for which $\mathscr{I}(\varphi)$ is finite will play surely an important role.

4.1. Properties of $\mathscr{I}(\varphi)$

8.16 Let us suppose that $\varphi \in C^2(\overline{\Omega})$ is a strictly convex function. As before, set $g_{ij} = \partial_{ij}\varphi$ and let g^{ij} be the components of the inverse matrix of $((g_{ij}))$. We will prove that

$$\mathscr{I}(\varphi) = \int_\Omega g^{ij}\,\partial_i\varphi\,\partial_j\varphi M(\varphi)\,dx. \tag{30}$$

$$M(\varphi) = \begin{vmatrix} \partial_{11}\varphi & \partial_{12}\varphi & \cdots & \partial_{1n}\varphi \\ \vdots & & & \vdots \\ \partial_{1n}\varphi & & & \partial_{nn}\varphi \end{vmatrix} = \frac{1}{n!}\sum \begin{vmatrix} \partial_{ii}\varphi & \partial_{ij}\varphi & \cdots & \partial_{ik}\varphi \\ \partial_{ij}\varphi & \partial_{jj}\varphi & \cdots & \\ \vdots & & & \\ \partial_{ik}\varphi & \cdots & & \partial_{kk}\varphi \end{vmatrix},$$

where the sum $\sum$ is extended to all $n \times n$ determinants obtained when the n subscripts $i, j, \ldots, k$ run from 1 to n. If two subscripts are equal the determinant is zero, otherwise it is equal to $M(\varphi)$. Thus we see that $M(\varphi)$ is a divergence:

$$M(\varphi) = \frac{1}{n!}\sum \partial_i \begin{vmatrix} \partial_i\varphi & \partial_{ij}\varphi & \cdots & \partial_{ik}\varphi \\ \partial_j\varphi & \partial_{jj}\varphi & & \vdots \\ \vdots & & & \\ \partial_k\varphi & \partial_{jk}\varphi & & \partial_{kk}\varphi \end{vmatrix},$$

$$nM(\varphi) = \partial_i(g^{ij}M(\varphi)\,\partial_j\varphi).$$

When we differentiate a column other than the first, we get a determinant which is zero. Thus integrating (29) by parts leads to (30).
For the preceding proof we supposed that $\varphi \in C^3$, but the result is true if φ is only C^2. In this case we approximate φ in C^2 by strictly convex C^∞ functions $\tilde{\varphi}$. We have

$$-n \int_\Omega \varphi M(\tilde{\varphi})\, dx = \int_\Omega \tilde{g}^{ij}\, \partial_i \varphi\, \partial_j \tilde{\varphi} M(\tilde{\varphi})\, dx.$$

By passing to the limit we get the result.

We will now show that the functional $\mathscr{I}(\varphi)$ is *convex* on the set of the strictly convex functions $\varphi \in C^2(\overline{\Omega})$. Let $\psi \in C^2(\overline{\Omega})$ be a function which vanishes on the boundary. $\varphi + t\psi$ is strictly convex for $|t|$ small enough and the second derivative of $\mathscr{I}(\varphi + t\psi)$ with respect to t at $t = 0$ is equal to

$$n(n+1) \int_\Omega g^{ij}\, \partial_i \psi\, \partial_j \psi M(\varphi)\, dx \geq 0.$$

Indeed, we have n terms of the form

$$\frac{1}{(n-1)!} \sum \int_\Omega \psi \begin{vmatrix} \partial_{ii}\psi & \partial_{ij}\varphi & \cdots & \partial_{ik}\varphi \\ \partial_{ij}\psi & \partial_{jj}\varphi & & \vdots \\ \vdots & & & \\ \partial_{ik}\psi & \cdots & & \partial_{kk}\varphi \end{vmatrix} dx = \int_\Omega \psi\, \partial_i[g^{ij} M(\varphi)\, \partial_j \psi]\, dx$$

and $n(n-1)/2$ terms of the form

$$\frac{1}{(n-1)!} \sum \int_\Omega \varphi \begin{vmatrix} \partial_{ii}\psi & \partial_{ij}\psi & \partial_{il}\varphi & \cdots & \partial_{ik}\varphi \\ \partial_{ij}\psi & \partial_{jj}\psi & \partial_{jl}\varphi & \cdots & \\ \vdots & & & & \\ \partial_{ik}\psi & \partial_{jk}\psi & \partial_{kl}\varphi & \cdots & \partial_{kk}\varphi \end{vmatrix} dx,$$

which are equal to

$$\frac{1}{(n-1)!} \sum \int_\Omega \partial_i \varphi \begin{vmatrix} \partial_i\psi & \partial_{ij}\psi & \partial_{il}\varphi & \cdots & \partial_{ik}\varphi \\ \partial_j\psi & \partial_{jj}\psi & \partial_{jl}\varphi & & \partial_{jk}\varphi \\ \vdots & \vdots & & & \vdots \\ \partial_k\psi & \partial_{jk}\psi & \partial_{kl}\varphi & & \partial_{kk}\varphi \end{vmatrix} dx.$$

This may be rewritten as

$$\frac{1}{(n-1)!} \sum \int_\Omega \partial_i \psi \begin{vmatrix} \partial_i\varphi & \partial_{ij}\psi & \partial_{il}\varphi & \cdots & \partial_{ik}\varphi \\ \partial_j\varphi & \partial_{jj}\psi & \partial_{jl}\varphi & & \vdots \\ \vdots & \vdots & \vdots & & \\ \partial_k\varphi & \partial_{jk}\psi & \partial_{kl}\varphi & \cdots & \partial_{kk}\varphi \end{vmatrix} dx,$$

which is equal to $\int_\Omega \psi\, \partial_i[g^{ij}M(\varphi)\, \partial_j\psi]\, dx$. Altogether we have $n(n+1)/2$ terms. Integrating by parts gives the result.

8.17 Remark. If Ω is a ball of radius τ and φ a C^2 radially symmetric function vanishing on the boundary: $\varphi(x) = g(\|x\|)$ with $g(\tau) = 0$.
The integrand in (30) is equal to $|g'|^{n+1}\rho^{1-n}$ with $\rho = \|x\|$. Thus in this case

$$\mathscr{I}(\varphi) = \omega_{n-1} \int_0^\tau |g'|^{n+1}\, d\rho.$$

Let f be a function belonging to $H_1^{n+1}([0, \tau])$. f is Hölder continuous. Indeed:

$$(31)\; |f(b) - f(a)| = \left| \int_a^b f'(s)\, ds \right| \leq \left| \int_a^b |f'(s)|^{n+1}\, ds \right|^{1/(n+1)} |b - a|^{n/(n+1)}$$

Thus, $f(\tau) = 0$ makes sense, and if $\tilde{\varphi}(x) = f(\|x\|)$ we can define $\mathscr{I}(\tilde{\varphi})$ to be equal to $\omega_{n-1} \int_0^\tau |f'|^{n+1}\, d\rho$.

8.18 Proposition. *Let $\{\varphi_p\}_{p\in\mathbb{N}}$ be a sequence of convex functions on $\overline{\Omega}$, vanishing on $\partial\Omega$, which converges uniformly to φ on $\overline{\Omega}$. Then*

$$\mathscr{I}(\varphi) \leq \liminf_{p\to\infty} \mathscr{I}(\varphi_p).$$

Thus, $\mathscr{I}(\varphi)$ is lower semi-continuous.

Proof. We use the notation of 8.14. For $a < 0$, according to (28):

$$\left| \int_{\Omega(a)} (\varphi - \varphi_p)\mathscr{M}(\varphi_p) \right| \leq C(a) \sup|\varphi - \varphi_p|[\sup|\varphi_p|]^n$$

$$\leq \text{Const} \times \|\varphi - \varphi_p\|_0;$$

thus $\int_{\Omega(a)} (\varphi - \varphi_p)\mathscr{M}(\varphi_p) \to 0$ when $p \to \infty$. But we can write

$$-n \int_{\Omega(a)} (\varphi_p - \varphi)\mathscr{M}(\varphi_p) - n \int_{\Omega(a)} \varphi\mathscr{M}(\varphi_p) = -n \int_{\Omega(a)} \varphi_p\mathscr{M}(\varphi_p) \leq \mathscr{I}(\varphi_p).$$

Taking the lim inf when $p \to \infty$ leads to

$$-n \int_{\Omega(a)} \varphi\mathscr{M}(\varphi) \leq \liminf_{p\to\infty} \mathscr{I}(\varphi_p),$$

by Proposition 8.14. Then letting $a \to 0$ we get the desired result. ∎

8.19 In this paragraph we will recall some properties of convex bodies which will be useful in 8.20.

As previously, Ω is a strictly convex bounded set of $\mathbb{R}^n$. If $Q \in \partial\Omega$, $\omega(Q)$ will denote the direction of the vector $\overrightarrow{PH}$ where P is the origin of the coordinates of $\mathbb{R}^n$ (which we assume is inside Ω) and $H(Q)$ the projection of P on the tangent plane of $\partial\Omega$ at Q.
We identify ω with a point on the sphere $\mathbb{S}_{n-1}(1)$. We let $\ell(Q)$ be the length of $\overrightarrow{PH}$.

For the proof of the next theorem, we are going to define a symmetrization procedure which is not the usual one of 2.17. It will be a consequence of the general inequality of Minkowski (see Buseman [71] p. 48) which applies to the convex bodies Ω and B and asserts that

$$\left(\frac{1}{n}\int_{\partial\Omega} \ell(Q)\, d\omega\right)^n \geq \mu(\Omega)[\mu(B)]^{n-1}, \tag{32}$$

where μ denote the Euclidean measure and $d\omega$ the element of measure on the unit sphere. Since equality holds for the ball, a consequence of (32) is:

Proposition 8.19. *Among the convex bodies with $\int_{\partial\Omega} \ell\, d\omega$ given, the ball has the greatest volume.*

8.20 Let d and D be the inradius and circumradius of Ω.

Theorem. *Let $\varphi \in C^2(\overline{\Omega})$ be a convex function which vanishes on $\partial\Omega$. There exists a radially symmetric function $\tilde{\varphi} \in C^1(\bar{B}_\tau)$ ($d \leq \tau \leq D$), vanishing for $\|x\| = \tau$, with the following properties:*

(33)
- α) *$\tilde{\varphi}$ has the same extrema as φ;*
- β) $\mathscr{I}(\tilde{\varphi}) \leq \mathscr{I}(\varphi)$;
- γ) $\mu(\Omega_a) \leq \mu(\tilde{\Omega}_a)$ for all $a < 0$,

where $\tilde{\Omega}_a = \{x \in B_\tau \mid \tilde{\varphi}(x) \leq a\}$.

Proof. Let m be the minimum of φ. On $[m, 0]$ we define the function ρ by $\rho(0) = (1/\omega_{n-1}) \int_{\partial\Omega} \ell\, d\omega$ and $\rho(a) = (1/\omega_{n-1}) \int_{\partial\Omega_a} \ell\, d\omega$ for $m \leq a < 0$. ρ is strictly increasing and C^1 on $]m, 0[$. Indeed,

$$\rho'(a) = \frac{1}{\omega_{n-1}} \int_{\partial\Omega_a} \frac{d\omega}{|\nabla\varphi|} \quad \text{for } a \in]m, 0[. \tag{34}$$

$\rho'(a)$ is a continuous strictly positive function which goes to infinity as $a \to m$ because $|\nabla\varphi(x)| \to 0$ as $d(x, \Omega_m) \to 0$. Moreover, when $a \to 0$,

$$\rho'(a) \to \frac{1}{\omega_{n-1}} \int_{\partial\Omega} \frac{d\omega}{|\nabla\varphi|}.$$

Thus $a \to \rho(a)$ is invertible. Let g be its inverse function: $a = g(\rho)$.

On $\partial\Omega$ ℓ satisfies $d \le \ell \le D$; hence $d \le \tau = \rho(0) \le D$. If $\mu(\Omega_m) \ne 0$, $\rho(m) > 0$. In this case we set $g(\rho) = m$ for $0 \le \rho \le \rho(m)$. Thus $g \in C^1([0, \tau])$, $g'(0) = 0$, and $g' \ge 0$.

Consider now the radially symmetric function $\tilde{\varphi}(x) = g(\|x\|)$. Obviously α) is true and γ) holds by Proposition 8.19 since

$$\int_{\partial\Omega_a} \ell \, d\omega = \rho(a)\omega_{n-1} = \int_{\partial\tilde{\Omega}_a} \ell \, d\omega. \tag{35}$$

It remains to prove β). By (30)

$$\mathscr{I}(\varphi) = \int_m^0 da \int_{\partial\Omega_a} |\nabla\varphi| \prod_{i=1}^{n-1} \left(\frac{|\nabla\varphi|}{R_i}\right) d\sigma,$$

where $R_i(Q)$ are the principal radii of curvature of $\partial\Omega_a$ at $Q \in \partial\Omega_a$. Thus

$$\mathscr{I}(\varphi) = \int_m^0 da \int_{\partial\Omega_a} |\nabla\varphi|^n \, d\omega$$

because $d\sigma = \prod_{i=1}^{n-1} R_i \, d\omega$ and

$$\begin{aligned} \mathscr{I}(\tilde{\varphi}) &= \int_m^0 da \int_{\partial\tilde{\Omega}_a} |\nabla\tilde{\varphi}|^n \, d\omega = \omega_{n-1} \int_m^0 |g'[\rho(a)]|^n \, da \\ &= \omega_{n-1} \int_m^0 [\rho'(a)]^{-n} \, da. \end{aligned} \tag{36}$$

Applying Hölder's inequality yields

$$\omega_{n-1} = \int_{\partial\Omega_a} d\omega \le \left(\int_{\partial\Omega_a} |\nabla\varphi|^n \, d\omega\right)^{1/(n+1)} \left(\int_{\partial\Omega_a} \frac{d\omega}{|\nabla\varphi|}\right)^{n/(n+1)}.$$

Consequently by (34):

$$\int_{\partial\Omega_a} |\nabla\varphi|^n \, d\omega \ge \omega_{n-1}[\rho'(a)]^{-n}.$$

Integrating with respect to a over $[m, 0]$ gives β).

8.21 Theorem. *All convex functions φ, which are zero on $\partial\Omega$, satisfy:*

$$\left|\inf_\Omega \varphi\right|^{n+1} \le D^n \omega_{n-1}^{-1} \mathscr{I}(\varphi), \tag{37}$$

where D is the circumradius of Ω.

Proof. If $\varphi \in C^2(\bar{\Omega})$, Theorem 8.20 shows that it is enough to prove the result for radially symmetric function $\tilde{\varphi}$.
Set $\tilde{\varphi}(x) = f(\|x\|)$. We have $f(0) = \inf \tilde{\varphi}$ and $f(\tau) = 0$ with $\tau \leq D$. Using (31) gives

$$|f(0)|^{n+1} \leq \tau^n \int_0^\tau |f'(s)|^{n+1}\, ds = \tau^n \omega_{n-1}^{-1} \mathscr{I}(\tilde{\varphi}).$$

And we get the result from Theorem 8.20.

If φ is a convex function but not necessarily C^2, we consider a sequence of convex C^∞ functions ψ_i which converges uniformly to φ on $\bar{\Omega}_a$, where $a < 0$ is close to zero. We can suppose that on $\partial\Omega_a$, $\psi_i > 2a$. By Proposition 8.14, $\int_{\Omega_a} \psi_i \mathscr{M}(\psi_i) \to \int_{\Omega_a} \varphi \mathscr{M}(\varphi)$.
Applying (37) to the function $\inf(0, \psi_i - 2a)$ leads to

$$|\inf(\psi_i - 2a)|^{n+1} \leq D^n \omega_{n-1}^{-1} \left[-n \int_{\Omega_a} \psi_i \mathscr{M}(\psi_i) \right];$$

letting $i \to \infty$ and then $a \to 0$ yields (37). ∎

§5. Variational Problem

8.22 Let $\tilde{f}(t) \in C^k(]-\infty, 0])$ $(k \geq 0)$ be a strictly positive function when $t \neq 0$ and greater than some $\varepsilon > 0$ for $t < t_0$, t_0 some real number. Set $F(t) = \int_{-|t|}^{0} \tilde{f}(u)\, du$ and consider the functional Γ defined on the set of continuous functions on the unit ball $\bar{B}$ by:

$$\Gamma(\psi) = \int_B F(\psi(x))\, dx.$$

We are interested in the following problem:
Minimize $\mathscr{I}(\psi)$ over the set $\mathscr{A}$ of convex functions which are zero on ∂B and which satisfy $\Gamma(\psi) = \mathscr{k}$, for $\mathscr{k} > 0$ some given real number.

Theorem 8.22. *The* inf *of $\mathscr{I}(\psi)$ for all $\psi \in \mathscr{A}$, which we call m, is attained by a radially symmetric convex function $\psi_0 \in C^{k+2}(\bar{B})$ which vanishes on ∂B and which satisfies $\mathscr{I}(\psi_0) = m$, $\Gamma(\psi_0) = \mathscr{k}$, and for some $\nu > 0$*

(38) $$M(\psi_0) = \nu \tilde{f}(\psi_0).$$

Proof. α) First of all $\mathscr{A}$ is not empty. More precisely if $\psi \leq 0$ $(\psi \not\equiv 0)$ is a continuous function on B there exists a unique real number $\mu_0 > 0$ for which $\Gamma(\mu_0 \psi) = \mathscr{k}$. Indeed for $\mu > 0$

(39) $$\partial_\mu \Gamma(\mu\psi) = -\int_B \psi \tilde{f}(\mu\psi)\, dx > 0$$

because the integrand is strictly negative somewhere. It is easy to verify that the hypotheses imply $\Gamma(\mu\psi) \to \infty$ as $\mu \to \infty$ and $\Gamma(\mu\psi) \to 0$ as $\mu \to 0$. Hence μ_0 exists and is unique.

β) One can show that the inf of $\mathscr{I}(\psi)$ for $\psi \in \mathscr{A}$ and that for $\psi \in \mathscr{A} \cap C^2(\bar{B})$ are equal. The reader will find the details in Aubin [23] p. 370.

γ) Now let $\psi \in \mathscr{A} \cap C^2(\bar{B})$ and $\tilde{\psi}$ be the corresponding radially symmetric functions introduced in 8.20. Then $\Gamma(\tilde{\psi}) \geq \Gamma(\psi)$. Indeed consider $\check{\psi}$, the radially symmetric function such that $\mu(\Omega_a) = \mu(\check{\Omega}_a)$ for all $a < 0$, where $\check{\Omega}_a = \{x \in B \,|\, \check{\psi}(x) \leq a\}$.
By Theorem 8.19, $\mu(\check{\Omega}_a) \leq \mu(\tilde{\Omega}_a)$. Thus $\tilde{\psi} \leq \check{\psi}$ on B and therefore $\Gamma(\tilde{\psi}) \geq \Gamma(\check{\psi})$ since Γ is decreasing in ψ. See (39). Moreover obviously, $\Gamma(\check{\psi}) = \Gamma(\psi)$.

Hence there exists $\mu_0 \leq 1$ such that $\Gamma(\mu_0 \tilde{\psi}) = \Gamma(\psi) = k$. See α). But according to Theorem 8.20, $\mathscr{I}(\tilde{\psi}) \leq \mathscr{I}(\psi)$. Thus $\mathscr{I}(\mu_0 \tilde{\psi}) = \mu_0^{n+1} \mathscr{I}(\tilde{\psi}) \leq \mathscr{I}(\psi)$. Therefore m is equal to the inf of $\mathscr{I}(\psi)$ for all radially symmetric functions $\psi \in \mathscr{A} \cap C^2(B)$.

δ) It remains for us to solve a variational problem in one dimension. That is the aim of the following.

8.23 Theorem. *The* inf *of* $\mathscr{I}(g) = \omega_{n-1} \int_0^1 |g'(r)|^{n+1}\, dr$ *for all nonpositive functions* $g \in H_1^{n+1}([0, 1])$ *which vanish at* $r = 1$ *and which satisfy* $\Gamma(g) = \omega_{n-1} \int_0^1 F(g) r^{n-1}\, dr = k$ *is attained by a convex function* $g_0 \in C^{k+2}([0, 1])$ *which is a solution of Equation* (38) *with* $v > 0$, g_0 *satisfying* $g_0'(0) = 0$ *and* $g_0(1) = 0$.

Proof. Since this is similar to several proofs done previously, we only sketch it. We already saw that this problem makes sense: g is Hölder continuous on $[0, 1]$ (see 8.17), and there exist functions g satisfying $\Gamma(g) = k$ (see 8.22, α)).

Let $\{g_i\}$ be a minimizing sequence. These functions are equicontinuous by (31). Applying Ascoli's theorem 3.15 there exists a subsequence of the $\{g_i\}$ which converges uniformly to a continuous function g_0. Thus $\Gamma(g_0) = k$, $g_0 \leq 0$, and $g_0(1) = 0$. Moreover a subsequence converges weakly to g_0 in $H_1^{n+1}([0, 1])$. Thus by 3.17 the inf of $I(g)$ is attained by g_0. Writing the Euler equation yields

$$\int_0^1 \psi' |g_0'|^{n-1} g_0'\, dr = -v \int_0^1 \psi f(g_0) r^{n-1}\, dr \tag{40}$$

for some real v and all function $\psi \in H_1^{n+1}([0, 1])$ vanishing at $r = 1$. Picking $\psi = g_0$ we see that $v > 0$ ($v = 0$ is impossible because this would imply $\mathscr{I}(g_0) = 0$ and consequently $g_0 \equiv 0$).

We now prove that $g_0(r)$ is equal to $\tilde{g}(r) = \int_1^r [v \int_0^u \tilde{f}(g_0(t))t^{n-1}\,dt]^{1/n}\,du$. $\tilde{g}(1) = 0$, $\tilde{g} \in C^1([0, 1])$, and $(\tilde{g}'^n(r))' = v\tilde{f}(g_0(r))r^{n-1}$. Thus for all integrable functions γ on $[0, 1]$:

$$\int_0^1 \gamma(|g_0'|^{n-1}g_0' - \tilde{g}'^n)\,dr = 0.$$

This implies $|g_0'|^{n-1}g_0' = \tilde{g}'^n$, so $g_0' = \tilde{g}'$ since $\tilde{g}' \geq 0$. Hence $g_0 = \tilde{g}$. Considering the expression

$$g_0'(r) = \left[v \int_0^r \tilde{f}(g_0(t))t^{n-1}\,dt\right]^{1/n},$$

we see that $g_0'(1) > 0$ since $g_0 \not\equiv 0$ and therefore $g_0(r) < 0$ for $r < 1$. Thus $g_0'(r) \neq 0$ for $r > 0$ and g_0 is C^2 on $]0, 1]$ where $g_0'' > 0$. Moreover as $r \to 0$, $g_0'(r) \sim [(v/n)\tilde{f}(g_0(0))]^{1/n}r$. Thus $g_0 \in C^2([0, 1])$ and it is convex. If $\tilde{f} \in C^k$, $g_0 \in C^{2+k}$. ■

8.24 Corollary. *Let $f(x, t)$ be a C^∞ function on $\bar{B} \times]-\infty, 0]$. There exists a real number $v_0 > 0$ such that the equation*

(41) $$M(\varphi) = v \exp f(x, \varphi), \qquad \varphi/\partial B = 0$$

has a strictly convex solution $\tilde{\varphi} \in C^\infty(\bar{B})$ when $n = 2$ and $0 < v \leq v_0$.

Proof. For some $\varepsilon > 0$ define $\tilde{f}(t) = \varepsilon + \sup_{x \in \bar{B}} \exp f(x, t)$. Consider $F(t) = \int_t^0 \tilde{f}(u)\,du$ and the functional $\Gamma(\psi) = \int_B F(\psi(x))\,dx$ as in 8.22.

By Theorem 8.22, there exist $v_0 > 0$ and a convex function $\psi_0 \in C^\infty(\bar{B})$ satisfying

$$M(\psi_0) = v_0 \tilde{f}(\psi_0), \qquad \psi_0/\partial B = 0.$$

Obviously ψ_0 is a strictly convex lower solution of (41) for $v \leq v_0$:

$$M(\psi_0) \geq v_0 \exp f(x, \psi_0(x))$$

and ψ_0 is strictly convex since $M(\psi_0) > 0$.

Then we choose $\beta > 0$ small enough so that $\gamma_0 = \beta(\|x\|^2 - 1)$ is an upper solution of (41) greater than ψ_0, where $v \leq v_0$ is given. Using Theorem 8.12 we obtain the stated result.

§6. The Complex Monge–Ampère Equation

8.25 The problem. We cannot end this paragraph without discussing the complex Monge–Ampère Equation.

Definition. A function φ with value in $[-\infty, +\infty[$, $(\varphi \not\equiv -\infty)$ is plurisubharmonic if it is lower semi-continuous and if the restriction of φ to any complex line is either a subharmonic function or else equal to $-\infty$. In case φ is C^2, φ is plurisubharmonic if the Hermitian form $\partial_{\lambda\bar{\mu}}\varphi\, dz^\lambda\, dz^{\bar{\mu}}$ is non-negative.

Henceforth Ω will be a strictly *pseudoconvex* bounded open set in $\mathbb{C}^m$ defined by a strictly plurisubharmonic function $h \in C^\infty(\bar{\Omega})$: $h/\partial\Omega = 0$ and let $u \in C^k(\partial\Omega)$ $(k \geq 0)$.

We consider the *Dirichlet Problem*

$$\det((\partial_{\lambda\bar{\mu}}\varphi)) = f(x, \varphi), \qquad \varphi/\partial\Omega = u \tag{42}$$

where $f(x, \varphi)$ is a non-negative function such that $f^{1/m}(x, \varphi) \in C^r(\bar{\Omega} \times \mathbb{R})$, $r \geq 0$. $\partial_{\lambda\bar{\mu}}\varphi$ denotes the second derivative of φ with respect to z^λ and $z^{\bar{\mu}}$ $(1 \leq \lambda, \mu \leq m)$.

This problem was studied by Bedford and Taylor [29] and [30], who use a very special method. They consider the upper envelope of the set of plurisubharmonic functions which are lower solutions of (42). They first prove that this upper envelope is a solution of (42) in a generalized sense and then try to prove that this solution is regular.

Remark 8.25. As boundary condition we can use $\varphi(x) \to +\infty$ when $x \to \partial\Omega$. Cheng and Yau [90] solve a problem of this kind:

They set $\varphi = u - \log(-h)$ and write the problem using the Kähler metric g defined by $g_{\lambda\bar{\mu}} = -\partial_{\lambda\bar{\mu}} \log(-h)$.
They now solve a Monge–Ampère equation on a complete Kähler manifold, using the continuity method. The estimates are obtained by using the methods for compact Kähler manifolds (Chapter 7) thanks to their generalized maximum principle (Theorem 3.76).

6.1. Bedford and Taylor's Results

8.26 Bedford and Taylor [29] consider the case where the function f does not depend on φ and in Bedford and Taylor [30] the following is proved:

Theorem. *If $f^{1/m}(x, \varphi)$ is convex and nondecreasing in φ, then there exists a unique plurisubharmonic function $\varphi \in C(\bar{\Omega})$ which solves the Dirichlet problem (42) in a generalized sense. If $k \geq 2$ and $f^{1/m}(x, \varphi)$ is Lipschitz on $\bar{\Omega} \times \mathbb{R}$, then the solution φ is Lipschitz on $\bar{\Omega}$.*
In case Ω is the unit ball B, if in addition $r = k = 2$, then the function φ has second partial derivatives almost everywhere which are locally bounded.

6.2. The Measure $\mathfrak{M}(\varphi)$

8.27 Recall that Ω is a strictly pseudoconvex bounded open set in $\mathbb{C}^m$. For a continuous plurisubharmonic function φ on Ω, it is possible to define a measure $\mathfrak{M}(\varphi)$ which is equal to

$$\frac{i^m}{m!}\Lambda^m\, dd''\varphi = \det((\partial_{\lambda\bar{\mu}}\varphi))\, dz_1 \wedge d\bar{z}_1 \wedge \cdots \wedge dz_m \wedge d\bar{z}_m$$

in case φ is C^2. Recall Λ^m is m-times the exterior product and $d''\varphi = \partial_{\bar{\lambda}}\varphi\, dz^{\bar{\lambda}}$.

The method used in an earlier article by Chern, Levine, and Nirenberg [93] is similar to that of 8.13.
The main point is: for all compact $K \subset \Omega$ there is a constant $C(K)$ such that $\int_K \Lambda^m\, dd''\varphi \leq C(K)\sup_\Omega|\varphi|^m$ for all plurisubharmonic function $\varphi \in C^2(\Omega)$.

6.3. The Functional $\mathfrak{I}(\varphi)$

8.28 For a plurisubharmonic function $\varphi \in C(\overline{\Omega}) \cap C^2(\Omega)$ which is zero on the boundary $\partial\Omega$ we set:

$$\mathfrak{I}(\varphi) = -\frac{i^m}{(m-1)!}\int_\Omega \varphi\Lambda^m\, dd''\varphi. \tag{43}$$

If the plurisubharmonic function belongs only to $C(\overline{\Omega})$ we can extend this definition by the same procedure as in 8.15. For $a < 0$ set $\Omega_a = \{x \in \Omega/\varphi(x) \leq a\}$. Since $\mathfrak{M}(\varphi)$ is a non-negative Radon measure $-\int_{\Omega_a}\varphi\mathfrak{M}(\varphi) = \tau(a)$ makes sense and is an increasing function of a. We define

$$\mathfrak{I}(\varphi) = -m\lim_{a\to 0}\int_{\Omega_a}\varphi\mathfrak{M}(\varphi).$$

The set of the continuous plurisubharmonic functions on $\overline{\Omega}$ vanishing on $\partial\Omega$ for which $\mathfrak{I}(\varphi)$ is finite will play an important role.

6.4. Some Properties of $\mathfrak{I}(\varphi)$

8.29 Suppose $\varphi \in C^2(\overline{\Omega})$ is a strictly plurisubharmonic function vanishing on $\partial\Omega$. Set $g_{\lambda\bar{\mu}} = \partial_{\lambda\bar{\mu}}\varphi$ and let $g^{\lambda\bar{\mu}}$ be the components of the inverse matrix of $((g_{\lambda\bar{\mu}}))$. Integrating (43) by parts leads to:

$$\mathfrak{I}(\varphi) = i^m\int_\Omega g^{\lambda\bar{\mu}}\,\partial_\lambda\varphi\,\partial_{\bar{\mu}}\varphi\,\det((\partial_{\lambda\bar{\mu}}\varphi))\, dz^1 \wedge dz^{\bar{1}} \wedge \cdots \wedge dz^{\bar{m}}. \tag{44}$$

Thus $\mathfrak{I}(\varphi)$ is the integral on Ω of the square of the gradient of φ in the Kähler metric $g_{\lambda\bar{\mu}}$.
To carry out the integration by parts we need, in fact, $\varphi \in C^3$; however we obtain the result for $\varphi \in C^2$ by a density argument.

Proposition. *$\mathfrak{I}(\varphi)$ is convex on the set of the strictly plurisubharmonic functions $\varphi \in C^2(\bar{\Omega})$ vanishing on $\partial\Omega$.*

Proof. Let $\psi \in C^2(\Omega)$ be a function which is zero on the boundary. $\varphi + t\psi$ is strictly plurisubharmonic for $|t|$ small enough. Thus we have to verify that the second derivative with respect to t at $t = 0$ of $\mathfrak{I}(\varphi + t\psi)$ is non-negative. Integrating by parts enough times yields the following expression for this second derivative:

$$\left[\frac{\partial^2}{\partial t^2}\mathfrak{I}(\varphi + t\psi)\right]_{t=0}$$
$$= m(m-1)i^m \int_\Omega g^{\lambda\bar{\mu}}\,\partial_\lambda\psi\,\partial_{\bar{\mu}}\psi\,\det((\partial_{\lambda\bar{\mu}}\varphi))\,dz^1 \wedge dz^{\bar{1}} \wedge dz^2 \wedge \cdots \wedge dz^{\bar{m}}.$$

This is obviously non-negative. ∎

Theorem. *$\mathfrak{I}(\varphi)$ is lower semi-continuous: if $\{\varphi_p\}_{p\in\mathbb{N}}$ is a sequence of plurisubharmonic functions continuous on $\bar{\Omega}$ and vanishing on $\partial\Omega$ which converges uniformly to φ on $\bar{\Omega}$, then*

$$\mathfrak{I}(\varphi) \le \liminf_{p\to\infty} \mathfrak{I}(\varphi_p).$$

Proof. It is similar to that in 8.18. φ, being the uniform limit of the φ_p, is continuous on $\bar{\Omega}$, vanishes on $\partial\Omega$, and also is plurisubharmonic according to the definition. Thus $\mathfrak{I}(\varphi)$ makes sense. ∎

§7. The Case of Radially Symmetric Functions

8.30 If Ω is a ball of radius τ in $\mathbb{C}^m$ and φ a C^2 radially symmetric function plurisubharmonic on $\bar{\Omega}$ vanishing on the boundary, we can write $\varphi(z) = g(\|z\|)$ with $g(\tau) = 0$. In this paragraph we suppose that φ has these properties. In this case, by (44)

$$\mathfrak{I}(\varphi) = \omega_{2m-1}\int_0^\tau \frac{1}{4}g'^2\left(\frac{g'}{2r}\right)^{m-1} 2^m r^{2m-1}\,dr$$
$$= \frac{1}{2}\omega_{2m-1}\int_0^\tau g'^{m+1}r^m\,dr$$

and we can apply Proposition 2.48 with $q = m + 1$.

Similarly we can associate to g the function ψ defined on the ball B_τ of $\mathbb{R}^{m+1}$ by $\psi(x) = g(\|x\|)$. Then $\mathfrak{I}(\varphi) = \frac{1}{2}\omega_{2m-1}\omega_m^{-1}\|\nabla\psi\|_{m+1}^{m+1}$ and we can apply all the results of 2.46–2.50 by noting that $r^{2m-1} \leq \tau^{m-1}r^m$. In particular, from Theorem 2.47 we get

Theorem 8.30. *If φ satisfies $\mathfrak{I}(\varphi) \leq 1$, then*

$$\int_\Omega \exp[\nu_m|\varphi|^{(m+1)/m}]\,dV \leq C\int_\Omega dV$$

where the constant C depends only on m and where

$$\nu_m = (m+1)2^{-1/m}\omega_{2m-1}^{1/m}.$$

From Corollary 2.49 we get

Corollary 8.30. *Set* $\xi_m = 2m^m(m+1)^{-2m-1}\omega_{2m-1}^{-1}$. *Then all φ satisfy*

$$\int_\Omega e^\varphi\,dV \leq C\exp[\xi_m\mathfrak{I}(\varphi)]\int_\Omega dV$$

where C depends only on n.

Proposition 8.30. *Let $\mathscr{A}$ be a set of functions φ for which $\mathfrak{I}(\varphi) \leq$ Const. Then the set $\{e^\varphi\}_{\varphi\in\mathscr{A}}$ is precompact in L_1.*

The proof is similar to that of (2.46δ).

Using these results, under certain assumptions we can solve the Dirichlet Problem (42) on a ball of $\mathbb{C}^m$ with $u = 0$ on the boundary, assuming the dependence of $f(x, \varphi)$ on x only involves $\|x\|$.

By seeking radially symmetric solutions we actually obtain all solutions, provided f is decreasing in $r = \|x\|$, by a result of Gidas, Ni, and Nirenberg [124].

7.1. Variational Problem

8.31 Let Ω be a ball of radius τ in $\mathbb{C}^m$. We can consider the following problem: Find the Inf of $\mathfrak{I}(\varphi)$ for all plurisubharmonic function $\varphi \in C(\overline{\Omega})$ vanishing on the boundary such that $\|\varphi\|_q = 1$ for some $q \geq 1$ (or $\int_\Omega e^{-\varphi}\,dV = 1 + \int_\Omega dV$, for instance). We know that $\mathfrak{I}(\varphi)$ is convex. So for $q = 1$ we can consider only radially symmetric functions and it is possible to solve the problem in one dimension. The same result holds for $q \leq m$. For $q > m$ (or if the constraint is $\int_\Omega e^{-\varphi}\,dV = 1 + \int_\Omega dV$) it is conjectured that the inf of $\mathfrak{I}(\varphi)$ is unchanged if we consider only radially symmetric functions.

7.2. An Open Problem

8.32 The regularity of the solution of (42) obtained by Bedford and Taylor is an open problem.

If f is allowed to vanish, there is a counterexample of Bedford and Fornaess [27] showing that the solution may not be C^2.
Therefore, let us simply assume that $k = r = \infty$, $\Omega = B$, and $f < 0$ everywhere on $\bar{B}$. The problem of regularity is nevertheless open, not only on $\bar{B}$, of course (this case is not even solved for the real analog), but also on B.

One of the reasons we can get the interior regularity for the real Monge–Ampère equation, is that for each compact $\mathfrak{K} \subset \Omega$ it is possible to obtain a sequence of C^∞ convex functions φ_j converging uniformly to the generalized solution φ on $\mathfrak{K}$ while $M(\varphi_j) \to M(\varphi)$ on $\mathfrak{K}$ in C^r for large r. This result is obtained by geometrical considerations. At the present time a similar result has not been established in the complex case.[3]

§8. A New Method

8.33 In [190b] P. L. Lions presents a very interesting method for solving the Dirichlet problem for the real Monge–Ampère equation on a bounded strictly convex set Ω of $\mathbb{R}^n$:

$$(45) \qquad \log M(\varphi) = f(x), \qquad \varphi/\partial\Omega = 0$$

where f belongs to $C^\infty(\bar{\Omega})$. $\partial\Omega$ is supposed to be C^∞.

The method consists to exhibit a sequence of functions $\varphi_k \in C^\infty(\bar{\Omega})$ which are solutions of equation (45): $\log M(\varphi_k) = f(x)$, but with approximated boundary data: $\varphi_k/\partial\Omega = u_k$, u_k being an increasing sequence of functions converging uniformly to zero when $k \to \infty$.

Let $\tilde{f} \in \mathscr{D}(\mathbb{R}^n)$ be such that $\tilde{f}/\Omega = f$ and let $p(x) \in C^\infty(\mathbb{R}^n)$ be a function which are equal to zero on $\bar{\Omega}$ and to 1 outside a compact set. Moreover p satisfies: for all $\eta > 0$ there exists $\gamma > 0$ such that $p(x) > \gamma$ when $\operatorname{dist}(x, \bar{\Omega}) > \eta$.

The idea is to consider instead of (45) the following equation on $\mathbb{R}^n$:

$$(46) \qquad \log \det\left(\left(\partial_{ij}\varphi - \frac{p}{\varepsilon}\varphi\mathscr{E}_{ij}\right)\right) = \tilde{f}$$

with $\varepsilon > 0$ and $\mathscr{E}_{ij}$ the euclidean tensor.

8.34 We can prove that equation (46) has a unique solution belonging to $C_B^\infty(\mathbb{R}^n)$ for which the tensor $g_{ij} = \partial_{ij}\varphi - (p/\varepsilon)\varphi\mathscr{E}_{ij}$ defines a Riemannian metric.

[3] See 8.35 and 8.36 for new results.

The sketch of P. L. Lions' proof is the following. First he solves in $C_B^\infty(\mathbb{R}^n)$ an approximated equation of equation (46):

$$\log \det\left(\left(\partial_{ij}\varphi - \left(\frac{p}{\varepsilon} + \lambda\right)\varphi\mathscr{E}_{ij}\right)\right) = \tilde{f}$$

with $\lambda > 0$. For that he proves the existence of a solution in $H_2^\infty(\mathbb{R}^n)$ of an associated stochastic control problem; and for the regularity he uses the results of Evans (112b). These results yield uniform bounds with respect to λ for the second derivatives. So he obtains a solution $\varphi_\varepsilon \in C_B^\infty(\mathbb{R}^n)$ of equation (46), the regularity being given by Evans' results.

Instead of using the diffusion processes, it is possible by using the continuity method to solve directly equation (46).
In any case we must construct a subsolution $\varphi_0 \in C_B^\infty(\mathbb{R}^n)$ of (46) such that $(g_0)_{ij} = \partial_{ij}\varphi_0 - (p/\varepsilon)\varphi_0\mathscr{E}_{ij}$ defines a Riemannian metric. Then the maximum principle implies that any solution of (46) satisfies $\varphi_0 \leq \varphi < 0$.

For $t \geq 0$ a parameter we consider the equation:

$$\text{(47)}\quad \log \det\left(\left(\partial_{ij}\varphi - \frac{p}{\varepsilon}\varphi\mathscr{E}_{ij}\right)\right) = t\tilde{f} + (1-t)\log\det\left(\left(\partial_{ij}\varphi_0 - \frac{p}{\varepsilon}\varphi_0\mathscr{E}_{ij}\right)\right)$$

and for $\alpha \in]0, 1[$ the operator Γ:

$$C_B^{2,\alpha}(\mathbb{R}^n) \supset \Theta \ni \varphi \to \log\det\left(\left(\partial_{ij}\varphi - \frac{p}{\varepsilon}\varphi\mathscr{E}_{ij}\right)\right) \in C_B^\alpha(\mathbb{R}^n)$$

where Θ is the subset of functions φ for which $((g_{ij}))$ is a positive definite bilinear form. Γ is continuously differentiable and its differential $d\Gamma_\varphi$ at φ is invertible. Indeed the equation

$$\text{(48)}\qquad g^{ij}\,\partial_{ij}\psi - \frac{p}{\varepsilon}\psi g^{ij}\mathscr{E}_{ij} = F \in C_B^\alpha(\mathbb{R}^n)$$

has a unique solution belonging to $C_B^{2,\alpha}(\mathbb{R}^n)$.
To establish this result we consider for instance the solution $\psi_k \in C^{2,\alpha}(\bar{B}_k)$ of (48) on $\bar{B}_k$ for Dirichlet's data equal to zero on the boundary. It is easy to prove that the set of functions $\{\psi_k\}_{k\in\mathbb{N}}$ is uniformly bounded, then we use the Schauder Interior Estimates 3.61. At $x \in \mathbb{R}^n$ we have for $k > \|x\| + 2$:

$$\|\psi_k\|_{C^{2,\alpha}(K)} \leq C[\|\psi_k\|_{C^0(\mathbb{R}^n)} + \|F\|_{C^\alpha(\mathbb{R}^n)}] \leq \text{Constant}$$

with $K = \bar{B}_x(1)$ and the constant C independant on x.
It follows that a subsequence of ψ_k converges to a solution of (48) which belongs to $C_B^{2,\alpha}(\mathbb{R}^n)$. The generalized maximum principle 3.76 implies the uniqueness assertion. By Theorem 3.56 the solution belongs to $C_B^\infty(\mathbb{R}^n)$.

The inverse function theorem 3.10 establishes that the set $\mathscr{C}$ of the $t \in [0, 1]$, for which (47) has a solution, is open in $[0, 1]$. To prove that $\mathscr{C}$ is closed we need uniform estimates of the solutions of (47) in $C^3(\mathbb{R}^n)$. To get them we do similar computations to those done at the beginning of this chapter, but here we use the generalized maximum principle 3.76.

8.35 Pogorelov, Cheng and Yau get C^∞ approximated solutions of (45) by geometrical considerations. Here we get the functions $\varphi_k = \psi_k/\overline{\Omega}$ by analysis. It is the distinction between both proofs, because we proceed as Pogorelov [235] to have uniform estimates in $C^3(K)$ with the compact $K \subset \Omega$. That is why Lions' result is not an improvement for equation (45).

But we can apply the method to the Dirichlet problem for the complex Monge–Ampère equation 8.25. By a similar approach, we get C^∞ approximated solutions of the complex equation.
Then we need estimates. The C^0 and C^1 estimates are not difficult to obtain. But if there exists Aubin's estimate 7.22 for the third derivatives of mixed type (and so for the gradient of the laplacian) when we assume the estimate for the laplacian, there is no complex equivalent of Pogorelov's C^2-estimate [235 pp. 73–75]. This is still an open problem.

8.36 Note added in proofs. In June 1982, Cafarelli L., Nirenberg L. and Spruck J. announced that in the general case they have obtained the estimate of the third derivatives of the functions φ_σ on the boundary (see 8.11). They claim that there exists a modulus of continuity for the second derivatives of the function φ_σ:

$$|\partial_{ij}\varphi_\sigma(x) - \partial_{ij}\varphi_\sigma(y)| \leq C[1 + |\log \|x - y\| |]^{-1}$$

with C a constant. They say that it is also possible to obtain this modulus of continuity for the complex equation 8.25.

Bibliography

[1] Adams R.— *Sobolev Spaces*. Academic Press, New York, 1975.

[2] Agmon S.— The L_p approach to the Dirichlet Problem. *Ann. Scuola Norm. Sup. Pisa* **13**, (1959) 405–448.

[3] Agmon S.— *Lectures on elliptic boundary value problems*. Van Nostrand, New York, 1965.

[4] Agmon S., Douglis A., and Nirenberg L.— Estimates near the boundary for solutions of elliptic partial differential equations satisfying general boundary conditions I. *Comm. Pure Appl. Math.* **12**, (1959) 623–727.

[5] Alexandrov A.D.— Dirichlet's problem for the equation

$$\text{Det } \|z_{ij}\| = \varphi(z_1, \ldots, z_n, z, x_1, \ldots, x_n).$$

Vestinik Leningrad *Univ. Ser. Math. Meh. Astr.* **13**, (1958) 5–24. MR 20 ♯ 3385.

[6] Amann H. and Crandall M.— On some existence theorems for semi-linear elliptic equations. Math. Research Center, Uni. Wisconsin, Madison 1977.

[7] Atiyah M. and Bott R.— On the Yang–Mills Equations over Riemannian Surfaces. (to appear).

[8] Atiyah M. F. and F. Hirzebruch— Riemann–Roch theorems for differentiable manifolds. *Bull. Amer. Math. Soc.* **65**, (1959) 276–281.

[9] Atiyah M., Hitchin N., and Singer I.— Self-duality in four-dimensional Riemannian geometry. *Proc. Roy. Soc. Lond.* **362**, (1978) 425–461.

[10] Aubin T.— Sur la fonction exponentielle. *C.R. Acad. Sci. Paris* **270A**, (1970) 1514.

[11] Aubin T.— Métrique riemanniennes et courbure. *J. Diff. Geo.* **4**, (1970) 383–424.

[12] Aubin T.— Fonction de Green et valeurs propres du Laplacien. *J. Math. pures et appl.* **53**, (1974) 347–371.

[13] Aubin T.— Problèmes isopérimétriques et espaces de Sobolev. *J. Diff. Géo.* **11**, (1976) 573–598. *Abstract in C.R. Acad. Sc. Paris* **280A**, (1975) 279.

[14] Aubin T.— Equations différentielles non linéaires et Problème de Yamabe concernant la courbure scalaire. *J. Math. Pures et appl.* **55**, (1976) 269–296.

[15] Aubin T.— Inégalités concernant la lère valeur propre non nulle du laplacien *C.R.A.S. Paris* **281A**, (1975) 979.

[16] Aubin T.— Equations différentielles non linéaires. *Bull. Sc. Math.* **99**, (1975) 201–210.

[17] Aubin T.— Espaces de Sobolev sur les variétés Riemanniennes. *Bull. Sc. Math.* **100**, (1976) 149–173.

[18] Aubin T.— Equations du type Monge–Ampère sur les variétés kählériennes compactes. *C.R.A.S. Paris* **283A**, (1976) 119.

[19] Aubin T.— The Scalar Curvature. *Differential Geometry and Relativity*, edited by M. Cahen and M. Flato, Reidel Publ. Co. Dordrecht, Holland (1976).

[20] Aubin T.— Equations du type Monge–Ampère sur les variétés kählériennes compactes. *Bull. Sc. Math.* **102**, (1978) 63–95.

[21] Aubin T.— Meilleures constantes dans le théorème d'inclusion de Sobolev et un théorème de Fredholm non linéaire pour la transformation conforme de la courbure scalaire. *J. Funct. Ana.* **32**, (1979) 148–174.

[22] Aubin T.— Inégalités d'Interpolation. *Bull. Sc. Math.* **105**, (1981) 229–234.

[23] Aubin T.— Equations de Monge–Ampère réelles. *J. Funct. Ana.* **41**, (1981) 354–377.

[24] Avez A.— Characteristic Classes and Weyl Tensor: Applications to general Relativity. *Proc. Nat. Acad. Sci. U.S.A.* **66**, (1970) 265–268.

[25] Avez A.— Le laplacien de Lichnerowicz sur les tenseurs *C.R. Acad. Sc. Paris* **284A**, (1977) 1219.
[26] Avez A.— Le laplacien de Lichnerowicz. *Rend. Sem. Mat. Univ. Poli. Torino* **33**, (1976) 123–127.

[27] Bedford E. and Fornaess J.— Counterexamples to règularity for the complex Monge–Ampère Equation. *Invent. Math.* **50**, (1979) 129–134.
[28] Bedford E. and Gaveau B.— Hypersurfaces with bounded Levi Form. *Indiana Uni. Math. J.* **27**, (1978) 867–873.
[29] Bedford E. and Taylor B.— The Dirichlet Problem for a complex Monge–Ampère equation. *Invent. Math.* **37**, (1976) 1–44.
[30] Bedford E. and Taylor B.— The Dirichlet Problem for an equation of complex Monge–Ampère Type. Partial differential equations and Geometry. *Proc. Conf. Park City Utah* (1977), 39–50. *Lectures Notes in Pure and Appl. Math.* **48**, Dekker, New-York (1979).
[31] Bedford E. and Taylor B.— Variational Properties of the complex Monge–Ampère equation. I Dirichlet Principle. *Duke Math. J.* **45**, (1978) 375–403.
[32] Bedford E. and Taylor B.— Variational Properties of the complex Monge–Ampère equation. II Intrinsic Norms. *Am. J. Math.* **101**, (1979) 1131–1166.
[33] Berard P. et Meyer D.— Inégalités isopérimétriques et applications. *Acta Math.* (to appear).
[34] Berard Bergery L.— La courbure scalaire des variétés riemanniennes. Séminaire N. Bourbaki (1980) **556**.
[35] Berger Marcel—Blaschke's Conjecture for Spheres. Appendix D in the book below by A. Besse.
[36] Berger M.— Lectures on geodesics in Riemannian geometry. *Lecture Notes in Math.* **33**, Tata Inst. Bombay (1965).
[37] Berger M., Gauduchon P. et Mazet E.— Le spectre d'une variété riemannienne. *Lecture Notes in Math.* **194** Springer-Verlag, Berlin, Heidelberg, New York, 1971.
[38] Berger M. et B. Gostiaux— *Géométrie Différentielle.* Armand Colin, Paris, 1972.
[39] Berger Melvyn S.— An eigenvalue problem for nonlinear elliptic partial differential equations. *Trans. Amer. Math. Soc.* **120**, (1965) 145–184.
[40] Berger M. S.— On Riemannian structures of prescribed Gaussian curvature for compact 2-manifolds. *J. Diff. Geo.* **5**, (1971) 325–332.
[41] Berger M. S.— Constant scalar curvature metrics for complex manifold. *Proc. Amer. Math. Soc. Inst. Diff. Geo.* **27**, (1975) 153–170.
[42] Berger M. S.— *Nonlinearity and Functional Analysis.* Academic Press, New York, 1977.
[43] Berger M. S.— Nonlinear Problems with exactly three Solutions. *Ind. Univ. Math. J.* **28**, (1979) 689–698.
[44] Berger M. S.— Two minimax problems in the Calculus of variations. In Partial differential equations. *Proc. Sympos. Pure Math.* **23**, Berkeley (1971) 261–267.
[45] Berger M. S. and Bombieri E.— On Poincaré's isoperimetric problem for simple closed geodesic. *J. Funct. Ana.* **42**, (1981) 274–298.
[46] Berger M. S. and Church P.— On the complete integrability of nonlinear partial differential equations. *Bull. Am. Math. Soc.* **1**, (1979) 698–701.
[47] Berger M. S. and Fraenkel L.— Nonlinear desingularization in certain free-boundary Problems. *Com. Math. Phys.* **77**, (1980) 149–172.
[48] Berger M. S. and Plastock R.— On the singularities of nonlinear Fredholm Operators of positive index. *Proc. Am. Math. Soc.* **79**, (1980) 217–221.
[49] Berger M. S. and Podolak E.—On the solutions of a nonlinear Dirichlet Problem. *Indiana J. Math.* **24**, (1975) 837–846.
[50] Bers L., John F., and Schechter M.— *Partial Differential Equations.* Wiley, New York (1964).
[51] Besse A.—Manifolds all of whose Geodesics are closed. *Ergebnisse der Mathematik* **93**, Springer-Verlag, Berlin 1978.
[52] Bidal P. et De Rham G.— Les formes différentielles harmoniques. *Comm. Math. Helv.* **19**, (1946) 1–49.
[53] Bishop R. and Crittenden R.— *Geometry of Manifolds.* Academic Press, New York (1964).

[54] Blaschke W.— *Vorlesungen über Differential Geometrie*, Vol. 1. Springer-Verlag, 1924.
[55] Bliss G.— An integral inequality. *J. London. Math. Soc.* **5**, (1930) 40–46.
[56] Bombieri E.— Variational problems and elliptic equations. *Proceedings of the international Congress of Math., Vancouver* **1**, (1974) 53–63.
[57] Bombieri E. and Giusti E.— Harnack's inequality for elliptic differential equations on minimal surfaces. *Invent. Math.* **15**, (1972) 24–46.
[58] Bombieri E. and Giusti E.— Local estimates for the gradient of nonparametric surfaces of prescribed mean curvature. *Comm. Pure Appl. Math.* **26**, (1973) 381–394.
[59] Bourguignon J. P.— Séminaire Palaiseau. Première classe de Chern et courbure de Ricci: preuve de la conjecture de Calabi. *Astérisque* **58**.
[60] Bourguignon J. P.— Premières formes de Chern des variétés kählériennes compactes. Séminaire N. Bourbaki (1977) **507**.
[61] Bourguignon J. P.— Deformations des métriques d'Einstein. *Asterisque* **80**, (1980) 21–32.
[62] Bourguignon J. P., Lawson H. B., and Simons J.— Stability and gapphenomena in Yang–Mills fields. *Proc. Nat. Acad. Sci. USA* **76**, (1979) 1550–1553.
[63] Brezis H. et Browder F.— Existence theorems for nonlinear integral equations of Hammerstein type. *Bull. Am. Math. Soc.* **81**, (1975) 73–78.
[64] Brezis H. and Lions P. L.— Boundary regularity for some nonlinear elliptic degenerate equation. *Comm. Math. Phy.* **70**, (1979) 181–185.
[65] Brezis H. and Nirenberg L.— Characterizations of the ranges of some nonlinear operators and applications to boundary value problems. *Ann. Scuola Norm. Sup. Pisa Cl. Sci.* **5**, (1978) 225–326.
[66] Brezis H. and Veron L.— Removable singularities of some nonlinear elliptic equations. *Arch. Rat. Mech. and Analysis* **75**, (1980) 1–6.
[67] Brezis H. and Waigner S.— A note on limiting cases of Sobolev Imbeddings and Convolution Inequalities. *Comm. Part. Diff. Eq.* **5**, (1980) 773–789.
[68] Browder F.— Nonlinear operators in Banach spaces. *Proc. Symp. P.M. 18 pt. 2—Amer. Math. Soc.* Providence, Rhode Island.
[69] Browder F.— Problèmes non-linéaires. *Presses de l'Université de Montreal*, Montreal, 1966.
[70] Browder F.— Functional Analysis and related fields. *Proceedings of a Conference in honor of Professor Marshall Stone, University of Chicago 1968*. Springer-Verlag, New York, 1970.
[71] Buseman H.— Convex surfaces. *Interscience Tracts in Pure and Applied Mathematics* **6** (1958).
[72] Buser P.— On Cheeger inequality $\lambda_1 \geq h^2/4$. *Proc. Symp. Pure Math. A.M.S.* **36**, (1980) 29–77.

[73] Calabi E.— The space of Kähler metrics, *Proc. Internat. Congress Math. Amsterdam* **2**, (1954) 206–207.
[74] Calabi E.— On Kähler manifolds with vanishing canonical class. Algebraic geometry and topology. A symposium in Honor of S. Lefschetz, Princeton University Press, Princeton (1955) 78–79.
[75] Calabi E.— Improper affine hyperspheres and a generalization of a theorem of K. Jörgens. *Mich. Math. J.* **5**, (1958) 105–126.
[76] Calabi E.— On Ricci curvature and geodesics. *Duke Math. J.* **34**, (1967) 667–676.
[77] Calabi E.— Examples of Bernstein Problems for some nonlinear equations. *Proceedings of symposia in pure Mathematics. Amer. Math. Soc.* **15**, (1970) 223–230.
[78] Calabi E.— Complete hyperspheres I, Ist. *Naz. Alta Mat. Symp. Math.* **10**, (1972), 19–38.
[79] Calabi E.— On manifolds with non-negative Ricci curvature II. *Notices Amer. Math. Soc.* **22**, (1975) A 205.
[80] Calabi E.— Métriques Kählériennes et fibrés holomorphes. *Ann. Sc. Ec. Norm. Sup. Paris* **12**, (1979) 269–294.
[81] Chaljub-Simon A. et Choquet-Bruhat Y.— Solutions asymptotiquement euclidiennes de l'équation de Lichnerowicz. *C.R. Acad. Sc. Paris* **286**, (1978) 917.
[82] Cheeger J.— A lower bound for the smallest eigenvalue of the Laplacian. Problems in Analysis. *Symposium in Honor of S. Bochner*. Princeton University Press, Princeton (1970) 195–199.
[83] Cheeger J. and Ebin D.— *Comparison Theorems in Riemannian Geometry*. North Holland, Amsterdam 1975.

[84] Cheeger J. and Gromoll D.— The splitting theorem for manifolds of non-negative Ricci curvature. *J. Diff. Geo.* **6**, (1971) 119–128.
[85] Cheng S. Y.— Eigenfunctions and eigenvalues of Laplacian. *Proc. Symp. Pure Math. Am. Math. Soc.* **27**, (1975) 185–193.
[86] Cheng S. Y.— Eigenvalue comparison theorems and its geometric applications. *Math. Z.* **143**, (1975) 289–297.
[87] Cheng S. Y.— On the Hayman–Osserman–Taylor Inequality. (to appear).
[88] Cheng S. Y. and Yau S. T.— On the regularity of the solution of the n-dimensional Minkowski problem. *Comm. Pure Appl. Math.* **29**, (1976) 495–515.
[89] Cheng S. Y. and Yau S. T.— On the regularity of the Monge–Ampère equation $\det((\partial^2 u/\partial x^i \partial x^j)) = F(x, u)$. *Comm. Pure Appl. Math.* **30**, (1977) 41–68.
[90] Cheng S. Y. and Yau S. T.— On the Existence of a complete Kähler metric on non-compact complex Manifolds and the regularity of Fefferman's equation. *Comm. Pure Appl. Math.* **33**, (1980) 507–544.
[91] Chern S. S.— Characteristic classes of Hermitian Manifold. *Ann. of Math.* **47**, (1946) 85–121.
[92] Chen S. S.— On curvature and characteristic classes of a Riemannian manifold. *Abh. Math. Sun. Univ. Hamburg* **20**, (1956) 117–126.
[93] Chern S. S., Levine and Nirenberg L.— Intrinsic norms on a complex manifold. In "Global Analysis" Papers in Honor of Kodaira K. ed. Spencer D. and Iyanaga S. University of Tokyo Press and Princeton University Press 1969.
[94] Chern S. S. and Moser J.— Real hypersurfaces in complex manifolds. *Acta. Math.* **133**, (1974) 219–271.
[95] Cherrier P.— Une inégalité de Sobolev sur les variétés Riemanniennes. *Bull. Sc. Math.* **103**, (1979) 353–374.
[96] Cherrier P.— Cas d'exception du théorème d'inclusion de Sobolev sur les variétés Riemanniennes et applications. *Bull. Sc. Math.* **105**, (1981) 235–288.
[97] Cherrier P.— Problèmes de Neumann non linéaires sur les Variétés Riemanniennes. *J. Diff. Geo.* (to appear).
[98] Cherrier P.—Etude d'une équation du type Monge–Ampère sur les variétés kählériennes compactes. *J. Funct. Ana.* (to appear).
[99] Choquet-Bruhat Y.— *Geométrie Différentielle et systèmes extérieurs.* Dunod, Paris (1968).
[100] Choquet-Bruhat Y. et J. Leray— Sur le problème de Dirichlet quasi-linéaire d'ordre 2. *Compt. Rendu Acad. Sc. Paris* **274**, (1972) 81–85.
[101] Choquet-Bruhat Y. et Marsden J.— Sur la positivité de la masse. *Compt. Rendu Acad. Sc. Paris* **282**, (1976) 609.
[102] Choquet-Bruhat Y. and Marsden J.—Solution of the local mass problem in general relativity. *Comm. Physics* **51**, (1976) 283–296.
[103] Croke C.— Some isoperimetric inequalities and eigenvalue estimates. *Ann. Sc. Ec. Norm. Sup.* **13**, (1980) 419–435.

[104] Delanoë P.— Equations du type de Monge–Ampère sur les variétés Riemanniennes compactes. *J. Funct. Anal.* Part I, **40**, (1980) 358–386. Part II, **41**, (1981) 341–353. Part III, **45**, (1982) 403–430.
[105] De Rham G.— Sur la théorie des formes différentielles harmoniques. *Ann. Univ. de Grenoble* **22**, (1946) 135–152.
[106] De Rham G.— *Variétés différentielles.* Hermann, Paris (1955).
[107] Deturck D. and Kazdan J.— Some regularity theorems in Riemannian Geometry. *Ann. Sc. Ec. Norm. Sup.* **14**, (1981).
[108] Dieudonné J.— *Panorama des Mathématiques Pures* Gauthier Villars, Paris (1979).
[109] Dieudonne J.— *Treatise on Analysis.* Academic Press, New York.
[110] Dinghas A.— Einfacher Beweis der isoperimetrischen Eigenschaft der Kugel in Riemannschen Räumen konstanter Krümmung. *Math. Nachr.* **2**, (1949) 148–162.
[111] Dunford N. and Schwartz J.— *Linear Operators.* Wiley, New York. Part I (1958), Part II (1963), Part III (1971).

[112a] Eells J. Jr. and Sampson J.— Harmonic mappings of Riemannian manifolds. *Amer. J. Math.* **86**, (1964) 109–160.
[112b] Evans L. C.— Classical solutions of fully non linear, convex second order, elliptic equations (to appear).

[113] Federer H.— *Geometric Measure Theory*. Springer-Verlag, New York (1969).
[114] Fefferman C.— Monge–Ampère equations, the Bergman kernel, and geometry of pseudoconvex domains. *Ann. of Math.* **103**, (1976) 395–416.
[115] Fischer A. and Marsden J.— Deformations of the scalar curvature. *Duke Math. J.* **42**, (1975) 519–547.
[116] Friedman A.— *Partial Differential Equations*. Holt, New York (1969).
[117] Friedman A.— *Partial differential equations of parabolic type*. Prentice-Hall, Englewood Cliffs, New Jersey (1964).

[118] Gagliardo E.— Properietà di alcune classi di funzioni in piu variabili. *Ric. Mat.* **7**, (1958) 102–137.
[119] Gallot S.— Un theorème de pincement et une estimation sur la première valeur propre du laplacien d'une variété Riemannienne. *C.R. Acad. Sc. Paris* **289A**, (1979) 441.
[120] Gallot S.— Estimées de Sobolev quantitatives sur les variétés Riemanniennes et applications. *C.R. Acad. Sc. Paris* **292A**, (1981) 375.
[121] Gallot S.— Minorations sur le λ_1 des variétés Riemanniennes. Séminaire N. Bourbaki (1981) **569**.
[122] Gaveau B.— Méthode de contrôle optimal en analyse complexe, I: Résolution d'équations de Monge–Ampère, *J. Funct. Anal.* **25**, (1977) 391–411, abstract in *C.R. Acad. Sc. Paris* **284A**, (1977) 593–596.
II: Equation de Monge–Ampère dans certains domaines faiblement pseudoconvexes. *Bull. Sc. Math.* **102**, (1978) 101–128.
III: Diffusion des fonctions convexes, plurisousharmoniques et des courants positifs fermés. *J. Funct. Anal.* **32**, (1979) 228–253. Abstract in *C.R. Acad. Sc. Paris* **287A**, (1978) 413.
[123] Giaquinta M. and Giusti E.— Nonlinear elliptic systems with quadratic growth. *Manuscripta Math.* **24**, (1978) 323–349.
[124] Gidas B., Ni W-M. and Nirenberg L.— Symmetry and related properties via the Maximum Principle. *Comm. Math. Phys.* **68**, (1979), 209–243.
[125] Gilbarg D., Trüdinger N. S.— *Elliptic Partial Differential Equations of Second Order*. Springer-Verlag, New York (1977).
[126] Giraud G.— Sur la problème de Dirichlet généralisé. *Ann. Sc. Ecole Norm. Sup.* **46**, (1929) 131–245.
[127] Giraud G.— Sur différentes questions relatives aux équations du type elliptique. *Ann. Sc. Ecole Norm. Sup.* **47**, (1930) 197–266.
[128] Glimm J. and Jaffe A.— Multiple meron solutions of the Classical Yang–Mills equation. *Physics Letters* **73B**, (1978) 167–170.
[129] Glimm J. and Jaffe A.— Droplet model for quark confinement. *Phys. Rev. D* **18**, (1978) 463–467.
[130] Glimm J. and Jaffe A.— Meron Paris and quark confinement. *Phy. Rev. Letters* **40**, (1978) 277–278.
[131] Golberg S.— Mappings of nonpositively curved manifolds. *Nagoya Math. J.* **61**, (1976) 73–84.
[132] Gray A.— Chern numbers and curvature. *Am. J. Math.* **100**, (1978) 463–476.
[133] Gray A. and Vanhecke L.— Riemannian geometry as determined by the volumes of small geodesic balls. *Acta Math.* **142**, (1979) 157–198.
[134] Gromov M.— Homotopical effects of dilatation. *J. Diff. Geo.* **13**, (1978) 303–310.
[135] Gromov M.— Paul Levy's isoperimetric inequality, *Ann. Sc. Ec. Norm. Sup. Paris.* (to appear).
[136] Gromov M., Lawson H. B. Jr.— Spin and scalar curvature in the presence of a fundamental group I. *Ann. of Math.* **111**, (1980) 209–230.
[137] Gromov M., Lawson H. B. Jr.— The classification of a simply connected manifolds of positive scalar curvature. *Ann. of Math.* **111**, (1980) 423–434.

[138] Hardy G., Littlewood J. et Polya G.— *Inequalities*. Cambridge Univ. Press, London and New York 1943.
[139] Helgason S.— *Differential Geometry and Symmetric Spaces*. Academic Press, New York, 1962.
[140] Hilbert D.— *Grundzüge einer allgemeinen Theorie der linearen Integralgleichungen*. Teubner (1912) Chapter XVIII.
[141] Hilbert Problems. Edited by F. Browder. *Proc. Symp. Pure Math. A.M.S. V.28* Providence, Rhode Island 1976.

[142] Hirzebruch F.— *Topological Methods in Algebraic Geometry*, Springer-Verlag, 1962.

[143] Hirzebruch F. and Kodaira K.— On the complex projective spaces. *J. Math. Pures Appl.* **36**, (1957) 201–216.

[144] Hitchin N.— Compact four-dimensional Einstein manifolds. *J. Diff. Geom.* **9**, (1974) 435–441.

[145] Hitchin N.— The space of harmonic spinors. *Advances in Math.* **14**, (1974) 1–55.

[146] Hitchin N.— On the curvature of rational surfaces. In Differential Geometry (*Proc. Sympos. Pure Math.* **27**, Stanford 1973) 65–80.

[147] Hopf E.— Uber den functionalen insbesondere den analytischen Charakter der Lösung elliptischer Differentialgleichungen zweiter Ordnung. *Math. Zeit.* **34**, (1931) 194–233.

[148] Hörmander L.— *An introduction to several complex variables.* Van Nostrand-Reinhold, Princeton.

[149] Hsiung C-C.— Some integral formulas for closed hypersurfaces. *Math. Scand.* **2**, (1954) 286–294.

[150] Inoue A.— On Yamabe's problem—by a modified direct method (unpublished).

[151] Itô S.— The Fundamental solution of the Parabolic Equation in a differentiable Manifold II. *Osaka Math. J.* **6**, (1954) 167–185.

[152] Itô S.— Fundamental solutions of parabolic differential equations and boundary value problems. *Japan J. Math.* **27**, (1957) 55–102.

[153] Jaffe A.— Introduction to gauge theories. *Proceedings of the International Congress of Mathematicians*, Helsinki (1978) 905–916.

[154] Jonsson T., McBryan O., Zirilli F., and Hubbard J.— An existence theorem for multimeron solution to classical Yang–Mills field equations. *Comm. Math. Phys.* **68**, (1979) 259–273.

[155] Kazdan J.— An isoperimetric inequality and wiedersehen manifolds. Seminar on Differential Geometry, ed. S. T. Yau *Annals of Math. Studies* **102**.

[156] Kazdan J. and Kramer R.— Invariant criteria for existence of solutions to second order quasilinear elliptic equations. *Comm. Pure Appl. Math.* **31**, (1978) 619–645.

[157] Kazdan J. and Warner F.— Curvature functions for compact 2-manifolds. *Ann. of Math.* **99**, (1974) 14–47.

[158] Kazdan J. and Warner F.— Curvature functions for open 2-manifolds. *Ann. of Math.* **99**, (1974) 203–219.

[159] Kazdan J. and Warner F.— Existence and conformal deformation of metrics with prescribed Gaussian and scalar curvatures. *Ann. of Math.* **101**, (1975) 317–331.

[160] Kazdan J. and Warner F.— Scalar Curvature and conformal deformation of Riemannian structure. *J. Diff. Geom.* **10**, (1975) 113–134.

[161] Kazdan J. and Warner F.— Prescribing curvatures. *Proc. Symposia Pure Math.* **27**. *Am. Math. Soc.* (1975).

[162] Kazdan J. and Warner F.— A direct approach to the determination of Gaussian and scalar curvature functions. *Inv. Math.* **28**, (1975) 227–230.

[163] Kazdan J. and Warner F.— Remarks on some quasilinear elliptic equations. *Comm. Pure Appl. Math.* **38**, (1975) 567–579.

[164] Kinderlehrer D., Nirenberg L., and Spruck J.— Regularity in elliptic free boundary Problems. I, *J. d'Analyse Math.* **34**, (1978) 86–119. II, *Ann. Sc. Norm. Sup. Pisa* **6**, (1979) 637–683.

[165] Kobayashi S.— The First Chern class and holomorphic symmetric tensor fields. *J. Math. Soc. Japan* **32**, (1980) 325–329.

[166] Kobayashi S.— First Chern class and holomorphic tensor fields. *Nagoya Math. J.* **77**, (1980) 5–11.

[167] Kobayashi S. and Nomizu K.— *Foundations of Differential Geometry I and II.* Interscience Publishers, New York, 1963.

[168] Kodaira K.— *Collected Works I–III.* Princeton University Press 1975.

[169] Kodaira K. and Morrow J.— *Complex manifolds.* Holt, New York (1971).

[170] Kramer R.— Sub- and Super-Solutions of quasilinear elliptic boundary value problems. *J. Diff. Equat.* **28**, (1978) 278–283.

[171] Krasnoselki M.— Topological Methods in the theory of nonlinear integral equations. *Pergamon*, Oxford.

[172] Kuiper N. H.— On conformally flat spaces in the large. *Ann. of Math.* **50**, (1949) 916–924.

[173] Ladyzenskaya O. and Uralsteva N.— *Linear and Quasilinear Elliptic Partial Differential Equations*. Academic Press, New York, 1968.

[174] Lelong P.— *Fonctions plurisousharmoniques et formes différentielles positives*. Gordon et Breach, Paris.

[175] Lelong-Ferrand J.— Transformations conformes et quasi conformes des variétés riemanniennes compactes (démonstration de la conjecture de Lichnerowicz), *Mem. Acad. Roy. Belg. Cl. Sc. Mem. Coll.*, in *8°*, **39** (1971).

[176] Lelong-Ferrand J.— Construction de modules de continuité dans le cas limite de Sobolev et applications à la Géométrie Différentielle. *Arch. Rat. Mech. Ana.* **52**, (1973) 297–311.

[177] Lemaire L.— Applications harmoniques de surfaces Riemanniennes. *J. Diff. Geo.* **13**, (1978) 51–78.

[178] Leray J.— Etude de diverses équations intégrales non linéaires. *J. Math.* **12**, (1933) 1–82.

[179] Leray J.— La théorie des points fixes et ses applications en analyse, (1952) 202–208. *Amer. Math. Soc.*, Providence, Rhode Island.

[180] Leray J. and Schauder J.— Topologie et équations fonctionnelles. *Ann. Sci. Ecole Norm. Sup.* **51**, (1934) 45–78.

[181] Li P.— A lower bound for the first eigenvalue of the Laplacian on a compact Riemannian manifold. *Ind. J. Math.* **28**, (1979) 1013–1019.

[182] Li P.— On the Sobolev constant and the p-spectrum of a compact Riemannian manifold. *Ann. Ec. Norm. Sup.* **13A**, (1980) 451–469.

[183] Li P. and Yau S-T.— Estimates of eigenvalues of a compact Riemannian manifold. *Proc. Symp. Pure Math. A.M.S.* **36**, (1980) 205–239.

[184] Lichnerowicz A.— *Theorie globale des connexions et des groupes d'holonomie*, Edizioni Cremonese, Roma (1962).

[185] Lichnerowicz A.— *Geométrie des groupes de transformations*. Dunod, Paris (1958).

[186] Lichnerowicz A.— Spineurs harmoniques. *C.R. Acad. Sc. Paris* **257A**, (1963) 7–9.

[187] Lichnerowicz A.— Propagateurs et commutateurs en relativité générale. *Publications Math. Inst. H.E.S.* **10**, (1961).

[188] Lions J. L.— *Equations différentielles opérationnelles et problèmes aux limites*. Springer-Verlag, Berlin (1961).

[189] Lions J. L.— *Quelques methodes de résolution des problèmes aux limites non-linéaires*. Dunod, Paris (1969).

[190a] Lions P. L.—Résolution analytique des problèmes de Bellman–Dirichlet. *Acta Math.* **146**, (1981) 151–166.

[190b] Lions P. L.— Une méthode nouvelle pour l'existence de solutions régulières de l'équation de Monge–Ampère. *C.R. Acad. Sc. Paris* **293**, (1981) 589.[4]

[191a] Lions J. L. and Magenes F.— *Problèmes aux limites non homogènes*, Vols. 1, 2. Dunod, Paris 1968.

[191b] Lions J. L. and Magenes F.— On a mixed problem for $\square$ with a discontinuous boundary condition II. An example of moving boundary. *J. Math. Soc. Japan* **30**, (1978) 633–651.

[192] Ljusternik L.— Topology of the Calculus of Variations in the Large. *Amer. Math. Soc. Transl.* **16**. *Amer. Math. Soc.*, Providence, Rhode Island (1966).

[193] Ljusternik L. and Schnirelmann L.— Methode topologique dans les Problèmes Variationnelles. *Actualités Sci. Indust.* **188**. Hermann, Paris (1930).

[194] Loewner C. and Nirenberg L.—Partial differential equations invariant under conformal or projective transformations. *Contributions to Analysis* 245–272, Academic Press, New York, 1974.

[195] Malliavin P.— *Géométrie différentielle intrinsèque*. Hermann, Paris (1972).

[196] Malliavin P.— Formules de la moyenne, calcul des perturbations et theorèmes d'annulation pour les formes harmoniques. *J. Funct. Anal.* **17**, (1974) 274–291.

[197] Malliavin P.— Annulation de cohomologie et calcul des perturbations dans L_2. *Bull. Sc. Math.* **100**, (1976) 331–336.

[198] Meyer D.— Courbure des sommes de métriques. *Bull. Soc. Math. France* **109**, (1981) 123–142.

[199] Michael J. and Simon L.— Sobolev and mean-value inequalities on generalized submanifolds of $\mathbb{R}^n$. *Comm. Pure Appl. Math.* **26**, (1973) 361–379.

[4] The details will appear *Manuscripta Math.* and *Arch. Rat. Mech. Anal.*

[200] Milnor J.— *Morse Theory*. Princeton Univ. Press, Princeton, New Jersey (1963).
[201] Milnor J.— *Topology from the differentiable viewpoint*. Univ. of Virginia Press, Charlottesville, Virginia (1965).
[202] Minakshisundaram S.— Eigenfunctions on Riemannian manifolds. *J. Indian Math. Soc.* **17**, (1953) 158–165.
[203] Minakshisundaram S. and Pleijel A.— Some properties of the eigenfunctions of the Laplace operator on Riemannian manifolds. *Canad. J. Math.* **1**, (1949) 242–256.
[204] Morrey C. Jr.— *Multiple Integrals in the Calculus of Variations*. Springer-Verlag, New York (1966).
[205] Morse M.— The Calculus of Variations in the Large (Colloq. Publ. 18), *Amer. Math. Soc.*, Providence, Rhode Island (1934).
[206] Morse M.— Global variational Analysis: Weierstrass Integrals on a Riemannian manifold. *Princeton Math. Notes*, Princeton University Press, Princeton (1976).
[207] Morse M. and Cairns S.— *Critical Point Theory in Global Analysis*. Academic Press, New York (1969).
[208] Moser J.— A rapidly convergent iteration method and nonlinear partial differential equations, I, *Ann. Scuola Norm. Sup. Pisa* **20**, (1966) 226–315, II, 449–535.
[209] Moser J.— A sharp form of an inequality of N. Trudinger. *Indiana Univ. Math. J.* **20**, (1971) 1077–1092.
[210] Moser J.— On a nonlinear problem in differential geometry. In *Dynamical Systems* edited by M. Peixoto. Academic Press, New York (1973).
[211] Myers S.— Riemannian manifolds with positive mean curvature. *Duke Math. J.* **8**, (1941) 401–404.

[212] Narasimhan R.—*Analysis on real and complex manifolds*. Masson, New York (1971).
[213] Nash J.— The imbedding problem for Riemannian manifolds. *Ann. of Math.* **63**, (1956) 20–63.
[214] Ni W-M.— Conformal metrics with zero scalar curvature and a symmetrization process via Maximum Principle. Seminar on Differential Geometry, ed. S-T. Yau. *Annals of Math. Studies* **102**.
[215] Ni W-M.— Some Minimax principles and their applications in nonlinear elliptic equations. *J. d'Analyse Math.* **37**, (1980) 248–275.
[216] Nirenberg L.— On nonlinear elliptic differential equations and Hölder continuity. *Comm. Pure Appl. Math.*, **6**, (1953) 103–156.
[217] Nirenberg L.— Contributions to the theory of partial differential equations. *Ann. Math. Studies* **33**, Princeton (1954) 95–100.
[218] Nirenberg L.— Remarks on strongly elliptic partial differential equations. *Comm. Pure Appl. Math.* **8**, (1955) 649 675.
[219] Nirenberg L.— Estimates and existence of solutions of elliptic Equations. *Comm. Pure Appl. Math.* **9**, (1956) 509–530.
[220] Nirenberg L.—On elliptic partial differential equations. *Ann. Scuola Norm. Sup. Pisa Sci. Fis. Mat.* **13**, (1959) 116–162.
[221] Nirenberg L.— Topics in nonlinear functional analysis (Lecture Notes), New York Univ. (1974).
[222] Nirenberg L.— Monge–Ampère equations and some associated problems in geometry. *Proceedings of the International Congress of Mathematicians*, Vancouver (1974) 275–279.
[233] Nirenberg L.— Variational and topological methods in nonlinear problems. *Bull. Amer. Math. Soc.* **4**, (1981) 267–302.

[224] Obata M.— Certain conditions for a Riemannian manifold to be isometric with a sphere. *J. Math. Soc. Japan* **14**, (1962) 333–340.
[225] Obata M.— The Conjectures on conformal transformations of Riemannian Manifolds. *J. Diff. Geom.* **6**, (1971) 247–258.
[226] Osserman R.— A note on Hayman's theorem on the bass note of a drum. *Comm. Math. Helvetici* **52**, (1977) 545–555.
[227] Osserman R.— The isoperimetric inequality. *Bull. Am. Math. Soc.* **84**, (1978) 1182–1238.
[228] Osserman R.— Bonnesen-style isoperimetric inequalities. *Am. Math. Month.* **86**, (1979) 1–29.

[229] Palais R.— *Foundations of Global nonlinear analysis*. Benjamin, Reading, Mass. (1967).

[230] Palais R.— The principle of symmetric criticality. *Comm. Math. Phy.* **69**, (1979) 19–30.
[231] Pham Mau Quan— *Introduction a la géométrie des variétés différentiables*—Dunod, Paris, 1969.
[232] Pinsky M.— The spectrum of the laplacian on a manifold of negative curvature I. *J. Diff. Geo.* **13**, (1978) 87–91.
[233] Pogorelov A. V.— *Monge–Ampère Equations of Elliptic Type*. Noordhoff, Groningen (1964).
[234] Pogorelov A. V.—Extrinsic geometry of convex surfaces. *Translations of Mathematical Monographs V.* **35**, A.M.S. Providence (1973).
[235] Pogorelov A. V.— *The Minkowski Multidimensional Problem*, translated by V. Oliker, Winston et Sons (1978).
[236] Pohozaev S.— Eigenfunctions of the equation $\Delta u + \lambda f(u) = 0$. *Doklady* **165**, (1965) 1408–1410.
[237] Pohozaev S.— The set of critical values of a functional. *Math. USSR* **4**, (1968) 93–98.
[238] Polyakov A.— Compact Gauge fields and the infrared catastrophe. *Phys. Lett.* **59B**, (1975) 82–84.
[239] Protter M. and Weinberger H.— *Maximum Principles in Partial Differential Equations*. Prentice-Hall, Englewood Cliffs, New Jersey (1967).

[240] Rabinowitz P. H.— Some global results for nonlinear eigenvalue problems. *J. Funct. Ana.* **7**, (1971) 487–513.
[241] Rabinowitz P. H.—Some aspects of nonlinear eigenvalue problems. *Rocky Mountain J. Math.* **3**, (1973) 161–202.
[242] Rauch J. and Taylor B. A.— The Dirichlet Problem for the multidimensional Monge–Ampère Equation. *Rocky Mountain J. of Math.* **7**, (1977) 345–363.
[243] Rudin W.— *Functional analysis*. McGraw-Hill, New York (1973).

[244] Sacks J. and Uhlenbeck K.— The existence of minimal immersions of 2-spheres. *Ann. of Math.* **113**, (1981) 1–24.
[245] Scheffer V.— The Navier–Stokes equations on a bounded domain. *Comm. Math. Phy.* **73**, (1980) 1–42.
[246] Schoen R. and Yau S-T.— Existence of incompressible minimal surfaces and the topology of manifolds with non-negative scalar curvature. *Ann. of Math.* **110**, (1979) 127–142.
[247] Schoen R. and Yau S-T.— On the structure of manifolds with positive scalar curvature. *Man. Math.* **28**, (1979) 159–183.
[248] Schoen R. and Yau S-T.— Proof of the positive mass theorem II. *Comm. Math. Phy.* **79**, (1981) 231–260.
[249] Schwartz J.— *Nonlinear functional analysis*. Gordon and Breach, New York (1969).
[250] Schwartz L.— *Théorie des distributions*. Hermann, Paris (1966).
[251] Schwarz A.— On regular solutions of Euclidean Yang–Mills equations. *Phys. Lett.* **67B**, (1977) 172–174.
[252] Seminar on Differential Geometry, ed. S-T. Yau. *Annals of Math. Studies* ♯ **102**.
[253] Sin Y-T.— Pseudo-convexity and the Problem of Levi. *Bull. Amer. Math. Soc.* **84**, (1978) 481–512.
[254] Sin Y-T. and Yau S-T.— Compact Kähler manifolds of positive bisectional curvature. *Invent. Math.* **59**, (1980) 189–204.
[255] Sobolev S. L.— Sur un théorème d'analyse fonctionnelle. *Math. Sb.* (*N.S*) **46**, (1938) 471–496.
[256] Sternberg S.— *Lectures on Differential Geometry*. Prentice-Hall, Englewood Cliffs, N.J. (1964).

[257] Talenti G.— Best constant in Sobolev inequality. *Ann. Mat. Pura Appl.* **110**, (1976) 353–372.
[258] Talenti G.— Elliptic equations and rearrangements. *Ann. Scuola Norm. Sup. Pisa* **3**, (1976) 697–718.
[259] Talenti G.— Nonlinear elliptic equations, rearrangement of functions and Orlitz spaces. *Ann. Mat. Pura Appl.* **120**, (1979) 159–184.
[260] Talenti G.— Some estimates of solutions to Monge–Ampère type equations in dimension two. *Ann. Scuola Norm. Sup. Pisa* **7**, (1981) 183–230.
[261] Trüdinger N.— On imbedding into Orlitz spaces and some applications. *J. Math. Phys.* **17**, (1967) 473–484.

[262] Trüdinger N.— Remarks concerning the conformal deformation of Riemannian structures on compact manifolds. *Ann. Scuola Norm. Sup. Pisa* **22**, (1968) 265–274.

[263] Trüdinger N.— Maximum principles for linear, non-uniformly elliptic operators with measurable coefficients. *Math. Z.* **156**, (1977) 299–301.

[264] Uhlenbeck K.— Removable singularities in Yang–Mills fields. *Bull. Amer. Math. Soc.* **1**, (1979) 579–581.

[265] Vaugon M.— Equations différentielles non linéaires sur les variétés Riemanniennes compactes. *Bull. Sc. Math.* (to appear).

[266] Ward R.— On self-dual gauge fields. *Phys. Lett.* **61A**, (1977) 81–82.

[267] Warner F.— *Foundations of differentiable manifolds and Lie Groups.* Academic Press, New York (1971).

[268] Wells R. O.— *Differential Analysis on Complex Manifolds.* Springer-Verlag, New York (1973).

[269] Yamabe H.— On the deformation of Riemannian structures on compact manifolds. *Osaka Math. J.* **12**, (1960) 21–37.

[270] Yang P. and S-T. Yau— Eigenvalues of the Laplacian of compact Riemannian surfaces and minimal submanifolds. *Ann. Scuola Norm. Sup. Pisa* **7**, (1980) 55–63.

[271] Yano K.— *Differential geometry on complex and almost complex spaces.* Pergamon Press, Oxford (1965).

[272] Yano K. and Bochner S.— Curvature and Betti numbers. *Annals of Math. Studies* **32**. Princeton University Press, 1953.

[273] Yano K. and Ishihara S.— *Tangent and Cotangent Bundles.* Marcel Dekker, New-York 1973.

[274] Yau S-T.— Isoperimetric inequalities and the first eigenvalue of a compact Riemannian manifold. *Ann. Sci. Ecole Norm. Sup. Paris* **8**, (1975) 487–507.

[275] Yau S-T.— On the curvature of compact hermitian manifolds. *Inv. Math.* **25**, (1974) 213–239.

[276] Yau S-T.— Calabi's conjecture and some new results in algebraic geometry. *Proc. Nat. Acad. Sc. U.S.A.* **74**, (1977) 1798–1799.

[277] Yau S-T.— On the Ricci curvature of a compact Kähler manifold and the complex Monge–Ampère equation I. *Comm. Pure Appl. Math.* **31**, (1978) 339–411.

[278] Yosida K.— *Functional Analysis*, Springer-Verlag, Berlin (1965).

Subject Index

Notation

Basic Notation

We use the Einstein summation convention.
Compact manifold means compact manifold without boundary unless we say otherwise.
$\mathbb{N}$ is the set of positive integers. $n \in \mathbb{N}$
$\mathbb{R}^n$: Euclidean n-space $n \geq 2$ with points $x = (x^1, x^2, \ldots, x^n)$ $x^i \in \mathbb{R}$ real numbers.
$\mathbb{C}^m$: Complex space with real dimension $2m$. z^λ $(\lambda = 1, 2, \ldots, m)$ are the complex coordinates $\bar{z}^\lambda = z^{\bar{\lambda}}$.
We often write ∂_i for $\partial/\partial x^i$, ∂_λ for $\partial/\partial z^\lambda$.
$\mathbb{H}_n$: hyperbolic space

Notation Index

Other symbols